VLSI MODULATION CIRCUITS

Signal Processing, Data Conversion and Power Management

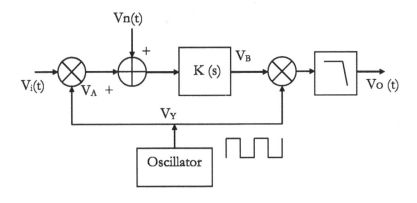

Hongjiang Song

Copyright © 2014 by Hongjiang Song.

ISBN: 978-1-312-21861-1

All rights reserved. No part of this book may be reproduced or transmitted in any form or by any means, electronic or mechanical, including photocopying, recording, or by any information storage and retrieval system, without permission in writing from the copyright owner.

This book was printed in the United States of America.

To order additional copies of this book, contact the publisher or copyright owner.

Foreword

This text is based on the class notes of a VLSI circuit design course (EEE598) professor Hongjiang Song developed for his graduate students in the engineering school at Arizona State University. The contents cover a special VLSI signal processing circuit family that is based on the VLSI modulation circuit techniques. These circuit techniques have been widely used in VLSI signal processing, data conversions, and power managements for dc offset cancellation, 1/f noise minimization, device mismatch compensation, and quality and efficiency enhancements in data conversion and amplification. These VLSI circuits traditionally belong to relatively isolated VLSI circuit areas. Fortunately, as we will see in this text, most of them share same theoretical basis and highly similar circuit design implementations. Therefore, as one of the goals, this text is to provide a unified framework for better understanding the underline circuit operation theories and circuit implementations of these VLSI modulation circuit techniques. It is important to note that such a goal is also closely aligned with the trend of modern SOC VLSI circuit applications for integrating various VLSI circuit structures on the same chip using the same process technology for a specific electronic system application. This unified theory is expected to provide an effective path for future VLSI circuit design and development of the targeted electronic systems.

Editor

2014

Preface

This is one of the book modules within my VLSI signal process circuit technique series. This material covers a special VLSI signal processing circuit family that employs the modulation/demodulation circuit techniques. Such circuit operations are featured by the time varying circuit topologies using the controlling clock or using the nonlinear behaviors of the VLSI circuit elements (such as VLSI diodes).

In recent years, there have been increasing interests in the industrial and academic communities for using VLSI modulation circuit techniques in highly integrated electronic systems for signal processing, data conversion and power management. This material was originally developed for my EEE598 course series in the engineering school at Arizona State University. This course module is dedicated to the VLSI signal processing operation and circuit techniques that are based on the modulation principles. This course module should extend my VLSI signal processing circuit approaches from the frequency domain circuit techniques (covered in book module: VLSI Analog Signal Processing Circuits), to the time domain circuit techniques (covered in the book module: VLSI High-Speed I/O Circuits), and the space domain circuit technique (covered in the book module: The Arts of VLSI Circuit Design) to the nonlinear and time varying circuit techniques.

From my personal experience, it is interesting to mention that the first practical circuit I have built in my junior high year was actually related to such modulation (and demodulation) circuits. It was a simple AM radio receiver that was built using a single semiconductor diode, an earphone and a pair of wire of about half a meter length. This simple "zero power" radio was with me almost daily during my high school years for listening to the news and the music. I should mention that such story happened long before other electronic devices such as the TV, the tape recorder, and the iPod became available many years later. During my school years at Yunnan University as student and late as faculty member, I have had chances to design and build high stability power supply circuits, where the dc/dc and charge pump circuits were used to generate stable high voltage (>100kV) supplies for the electron microscopes. During these circuit design practices, the modulation circuit theory and the circuit implementation were used to achieve these design goals.

I would like to thank my family for their supports during this text development. I also would like to thank many of my professors especially these in Yunnan University and in Arizona State University for the course works and labs that helped my learning and practice in this unique research field. I also would like to thank many friends and students for much informative discussion in this text development.

Hongjiang Song, Ph.D.

Mesa, Arizona, July 2013

CONTENTS

Preface   5

Chapter 1   Introduction   19
- 1.1   Characteristics of VLSI Modulation   20
  - 1.1.1   Linear VLSI Signal Processing Circuits   20
  - 1.1.2   Linear VLSI Circuit Model and Design   23
  - 1.1.3   Modulation Based Linear VLSI Circuits   25
- 1.2   Basic Modulation Operations   26
  - 1.2.1   Limiting Operations   27
  - 1.2.2   Multiplication Operations   30
  - 1.2.3   Switched-Network   31
- 1.3   Modeling of Modulation Circuits   32
- 1.4   Key Benefits of VLSI Modulations   36
  - 1.4.1   High Precision Data Conversions   36
  - 1.4.2   Low Noise Siganl Amplifications   37
  - 1.4.3   High Efficiency Power Amplifications   37
  - 1.4.4   High Efficiency Power Conversions   38
  - 1.4.5   Other Special Circuit Functions   38
- 1.5   Applications of VLSI Modulation Circuits   38
  - 1.5.1   Frequency Translation   38
  - 1.5.2   Pulse Density Modulation   39
  - 1.5.3   Pulse Width Modulation   40
  - 1.5.4   Pulse Code Modulation   40
  - 1.5.5   Phase and Delay Modulation   40
  - 1.5.6   Chopping   41
- 1.6   Organization of Book Chapters   41

Chapter 2		Basic VLSI Modulation Theory	47
	2.1	Pulse Density Modulation (PDM)	48
		2.1.1 PDM Representation of Analog Signals	48
		2.1.2 Demodulation of PDM Signals	49
	2.2	Pulse Coded Modulation (PCM)	49
	2.3	Pulse Width Modulation (PWM)	53
		2.3.1 Two-Level PWM Modulation	53
		2.3.2 Multi-Level PWM Modulation	54
	2.4	Frequency Translation	55
		2.4.1 Analog Multiplication	57
		2.4.2 Switching Frequency Translation	58
	2.5	Phase Translation	58
Chapter 3		VLSI Sigma-Delta Converter Circuits	61
	3.1	Sigma-Delta Modulation Principle	62
		3.1.1 Quantization Model	62
		3.1.2 Dynamic Range of Ideal A/D Converter	64
		3.1.3 Oversampling	65
		3.1.4 Noise-Shaping	66
		3.1.5 Performance Metrics	68
		3.1.6 Performance of Ideal Sigma-Delta Modulator	70
	3.2	Sigma-Delta Modulator Architectures	71
		3.2.1 First-Order Modulator	71
		3.2.2 Second-Order Modulator	75
		3.2.3 High-Order Modulator	79
		3.2.4 Cascade Modulation Topologies	83
		3.2.5 Multi-Bit Modulation	85
		3.2.6 Other Circuit Techniques	85
	3.3	Sigma-Delta D/A Conversion	85

		3.3.1	First-Order Modulation	86
		3.3.2	High-Order Modulation	87
		3.3.3	Modulator Implementation	88
Chapter 4		VLSI Class-D Amplifier Circuits		97
	4.1	Class-D amplifier Principle		98
	4.2	Analog Class-D Amplifier Architectures		101
		4.2.1	Open-Loop Analog Class-D Amplifier	101
		4.2.2	Closed-Loop Analog Class-D Amplifier	106
		4.2.3	Self-Oscillation Class-D Amplifiers	112
		4.2.4	Alternative class-D Amplifier Stricture	114
	4.3	Digital Class-D Amplifier Architectures		114
		4.3.1	Hybrid Class-D Amplifier	116
		4.3.2	Direct PCM-PWM Class-D Amplifier	117
		4.3.3	VLSI Class-D Amplifier Topologies	119
	4.4	VLSI PWM Circuit Implementations		120
		4.4.1	Triangular Ramp Signal Generation	120
		4.4.2	PWM Signal Generation Circuits	122
		4.4.3	Ramp-Less PWM Circuit	125
		4.4.4	Direct Input PWM Signal Generation	127
	4.5	PWM Signal Power Spectra		128
Chapter 5		VLSI Phase (Delay) Interpolation Circuits		137
	5.1	Phase Interpolation Principle		138
		5.1.1	Sinusoidal Phase Interpolation	138
		5.1.2	Triangular Phase Interpolation	140
	5.2	VLSI Phase Interpolator Implementations		143
	5.3	VLSI Phase Interpolator Noise Model		149

Chapter 6		VLSI Auto-Zero and Chopper Stabilization Circuits	155
	6.1	VLSI Noise Effects and Compensation	156
	6.2	VLSI Auto-Zero Circuits	158
		6.2.1 Auto-Zero Circuit Principle	159
		6.2.2 Auto-Zero Circuit Implementation	162
		6.2.3 Aliasing Effect of Auto-Zero Circuits	165
	6.3	VLSI Chopper Stabilization Circuits	167
		6.3.1 Chopper Circuit Principle	168
		6.3.2 VLSI Chopper Circuit Structures	171
	6.4	VLSI Circuit Implementations	174
		6.4.1 Inverted Gain Chopper Stage	174
		6.4.2 Open-Loop CDS Circuits	175
		6.4.3 Closed-Loop CDS Circuits	177
		6.4.4 Inverted CDS Comparator Circuits	179
		6.4.5 CDS Sample/Hold Circuit	181
		6.4.6 CDS Amplifier Circuits	182
	6.5	VLSI Compound Zero-Drift Amplifier	186
		6.5.1 Compound Circuit Structures	187
		6.5.2 Compound Amplifier Characteristics	196
		6.5.3 Design Considerations	197
	6.6	Comparison of Circuit Techniques	198
	6.7	Basic VLSI Chopper Circuit Elements	200
		6.71 Basic Chopper Modulator Circuit	200
		6.72 VLSI Chopper Opamp Circuits	200
Chapter 7		Introduction to VLSI Lock-In Amplifier Circuits	205
	7.1	Lock-In Amplification Principle	206
		7.1.1 Weak Signal Detection Concept	206
		7.1.2 Phase Sensitive Signal Detection	208

		7.1.3	Basic Lock-In Amplifier Model	210
	7.2		VLSI Lock-In Amplifier Implementations	211
		7.2.1	Low –Noise Amplifier (LNA)	212
		7.2.2	Analog PSD	213
		7.2.3	Switching PSD	214
		7.2.4	Digital PSD	215
		7.2.5	Lowpass Filter	216
		7.2.6	Reference Signal	216
	7.3		Noise Effects and Dynamic Reserve	217
		7.3.1	Intrinsic Noise Sources	217
		7.3.2	External Noise Sources	218
		7.3.3	Dynamic Reserve	220
Chapter 8			VLSI Switched-Capacitor DC/DC Converter Circuits	225
	8.1		VLSI Charge Pump Circuit Principle	226
		8.1.1	Basic Charge Pump Circuit Concept	226
		8.1.2	Basic Charge Pump Circuit Components	227
		8.1.3	Charge Pump Circuit Classifications	228
	8.2		Charge Pump Circuit Structures	230
		8.2.1	H-Bridge Charge Pump Circuits	232
		8.2.2	Cockcroft-Walton Charge Pumps	242
		8.2.3	Villard Charge Pump Circuit	248
		8.2.4	Greinacher Charge Pump Circuit	249
		8.2.5	Delon Charge Pump Circuit	250
		8.2.6	Marx Charge Pump Circuit	251
		8.2.7	Dickson Charge Pump Circuits	252
		8.2.8	Cross-Coupled Charge Pump Circuits	261
	8.3		Charge Pump Circuit Configurations	263
		8.3.1	Regulated VLSI Charge Pump Circuits	263

		8.3.2	Buck-Boost Regulator	265
		8.3.3	Inverter Regulator	266
		8.3.4	Configurable Charge Pumps	267
		8.3.5	Hybrid Charge Pump Circuit	269
	8.4		Charge Pump Circuit Modeling	271
		8.4.1	Circuit Performance Parameters	271
		8.4.2	Micro-, Macro- and Simulation Modeling	272
		8.4.3	Charge Pump Circuit Operation Modes	278
	8.5		Voltage Source and Impedance Models	279
		8.5.1	H-Bridge Charge Pumps	279
		8.5.2	Dickson Charge Pumps	284
		8.5.3	Cockcroft-Walton Charge Pumps	286
		8.5.4	Cross-Coupled Charge Pumps	287
		8.5.5	Parasitic Capacitance Effects	290
	8.6		Transformer Based Charge Pump Circuit Models	296
		8.6.1	Ideal Transformer Characteristics	296
		8.6.2	H-Bridge Charge Pumps	299
		8.6.3	Dickson Charge Pumps	304
		8.6.4	Cross-Coupled Charge Pumps	305
	8.7		Gain Hopping Techniques	306
Chapter 9			VLSI Switched-Capacitor Filter Circuits	309
	9.1		VLSI Switched-Capacitor Circuit Principle	310
		9.1.1	Charge Redistribution Analysis Method	313
		9.1.2	Non-Inverted Switched-Capacitor Integrator	314
		9.1.3	Inverted Switched-Capacitor Integrator	317
		9.1.4	Doubly Pumped Switched-Capacitor Integrator	319
	9.2		Basic VLSI Switched-Capacitor Circuit Elements	321
		9.2.1	Analog Switch Circuit Structures	321

	9.2.2	Non-Overlap Clock Generation Circuits	324
	9.2.3	VLSI OTA Circuit Structures	325
9.3	VLSI Sample/Hold Circuit Structures		326
	9.3.1	Buffered Sample/Hold Circuit	328
	9.3.2	The Correlated Double Sampling Techniques	329
	9.3.3	Non-Unity Gain Sample/Hold Circuit	332
	9.3.4	Charge Injection Compensation Techniques	333
	9.3.5	Fully Differential Sample/Hold Circuit	334
Chapter 10	VLSI Switched-Current Circuits		347
10.1	VLSI Switched-Current Circuit Principle		348
10.2	Basic Switched-Current Signal Processing Elements		349
	10.2.1	Unit Delay Circuit	349
	10.2.2	Adder Circuit	350
	10.2.3	Scaler Circuit	351
10.3	VLSI Dynamic Switched-Current Circuits		352
10.4	Charge Injection Compensation Techniques		354
	10.4.1	Charge Injection Attenuation Circuits	354
	10.4.2	Charge Injection Cancellation Circuits	356
	10.4.3	Algorithmic Charge Injection Cancellation	357
	10.4.4	Full Differential Switched-current Circuits	359
	10.4.5	Controlled Clocking and Input Techniques	359
Chapter 11	VLSI Switched-Inductor DC/DC Converter Circuits		367
11.1	Switched-Inductor DC/DC Converter Principle		368
	11.1.1	Voltage Divider Based Model	368
	11.1.2	LC Filter Based Model	372
	11.1.3	Sync- and Asynchronous Modes	374
11.2	Basic DC/DC Converter Architectures		375

	11.2.1	Step Down Converter	375
	11.2.2	Step Up Converter	379
	11.2.3	Step Up/Down Converter	381
	11.2.4	Cuk DC/DC Converter	383
	11.2.5	Sepic DC/DC Converter	386
	11.2.6	Flyback Converter	388
	11.2.7	Forward Converter	389
	11.28	Resonant Converter	390
11.3		VLSI Switched-Inductor Regulators	391
	11.3.1	Hysteretic Regulator	392
	11.3.2	Voltage-Mode PWM Control Loop	394
	11.3.3	Current-Mode PWM Control Loop	397
11.4		Special Operation Modes	402
	11.4.1	Discontinuous Mode	402
	11.4.2	Skip Mode	406
	11.4.3	PFM Mode	407
	11.4.4	LDO Mode	408
	11.4.5	Comparison of Operation Modes	408
11.5		Behavioral DC/DC Converter Model	409
	11.5.1	2-S Switched –Inductor Model	409
	11.5.2	4-S Switched-Inductor Model	412
	11.5.3	Conversion between 2-S and 4-S Models	413
	11.5.4	Voltage Domain Conversion	416
	11.5.5	Steady-State DC/DC Converter Model	417
11.6		Closed-Loop Modeling of Regulators	417
	11.6.1	Voltage Mode Regulators	419
	11.6.2	Current Mode Regulators	421
11.7		Converter Performance Models	426
	11.7.1	Energy Factor	426

	11.7.2	Power Efficiency	427
	11.7.3	Time Constant	428
	11.7.4	Dumping Time Constant	428
	11.7.5	Time Constant Ratio	429
11.8		Converter Parameter Selections	429
	11.8.1	Switch Selection	430
	11.8.2	Inductor Selection	431
	11.8.3	Capacitor Selection	432
	11.8.4	Loss Control in DC/DC Converters	433
	11.8.5	Zero Voltage Switching (ZVS)	435
Chapter 12		Introduction to VLSI Mixer Circuits	439
12.1		VLSI Mixer Circuit Principle	440
12.2		VLSI Mixer Performance Parameters	441
	12.2.1	Conversion Gain	441
	12.2.2	Noise figure	442
	12.2.3	Signal Isolation	442
	12.2.4	Linearity	443
	12.2.5	Spurs	446
	12.6.5	Dynamic Range	446
	12.2.7	Image	446
	12.2.8	DC-Offset	447
	12.2.9	LO Drive Level	447
	12.2.10	Voltage Standing Wave Ratio	447
12.3		VLSI Mixer Examples	448
12.4		VLSI Mixer Circuit Implementations	449
	12.4.1	Single Device Mixer	450
	12.4.2	Balanced Mixer	452
	12.4.3	Image Rejection Mixer	454

	12.4.4 Sub-Harmonic Mixer	454
	12.4.5 Phase Selection Mixer	455
Chapter 13	**VLSI Spread Spectrum Clocking Circuits**	**461**
13.1	Spread Spectrum Clocking Principle	462
	13.1.1 Electromagnetic Interference (EMI)	462
	13.1.2 Spread Spectrum Clocking Basis	463
13.2	Spread Spectrum Clocking Modeling	465
	13.2.1 Modulation Frequency	466
	13.2.2 Modulation Index	466
	13.2.3 Rate of Modulation	467
	13.2.4 Modulation Profile	467
	13.2.5 Spread of Spectrum under SSC	468
13.3	VLSI SSC Circuit Implementations	469
	13.3.1 Direct VCO Modulation SSC Circuit	469
	13.3.2 Feedback Modulation SSC Circuit	472
Chapter 14	**VLSI Fractional-N PLL Circuits**	**483**
14.1	VLSI Phase-Locked Loop Principle	484
	14.1.1 Integer Phase-Locked Loop	485
	14.1.2 Fractional-N Phase-Locked Loop	487
14.2	VLSI Fractional-N PLL Implementations	488
	14.2.1 Multi-Modulus Divider	488
	14.2.2 Modulator	496
Appendix A	**VLSI Amplifier Families**	**503**
A.1	VLSI Amplifier Basis	504
	A.1.1 Amplifier Gain	504
	A.1.2 Power Efficiency	506

		A.1.3	Amplifier Linearity	506
		A.1.4	Intermodulation	509
A.2		VLSI Amplifier Classes		511
		A.2.1	Class-A Amplifier	512
		A.2.2	Class-B Amplifier	514
		A.2.3	Class-AB Amplifier	515
		A.2.4	Class-C Amplifier	516
		A.2.5	Class-D Amplifier	517
		A.2.6	Class-E and Class-F Amplifiers	519
		A.2.7	Class-G and Class-H Amplifiers	522

CHAPTER 1
INTRODUCTION

VLSI modulation circuits (and their inversion form: demodulation circuits) belong to a special VLSI circuit family that employ nonlinear or time varying (or switched) circuit structures and operations to provide linear signal processing operations. Such VLSI circuit family offer attractive circuit features, such as noise cancellation, noise shaping, data conversion resolution enhancement, design robustness and power efficiency improvements, that conventional linear signal process circuits cannot offer.

VLSI modulation circuit techniques rely on special VLSI circuit structures and operations, such as VLSI switched-capacitor circuits, switched-inductor circuits, chopper circuits, lock-in amplifier, and class-D amplifier and mixer circuits.

VLSI modulation circuit operations are usually featured by the frequency domain signal manipulations where VLSI modulations purposely translate the signal (and noise) in the circuits to new frequency bands that are more suitable for VLSI signal processing circuit operations, such as signal amplification, noise minimization, and high power efficiency signal amplification and processing. In the signal aspect, VLSI modulation circuits offer highly linearity mapping between the inputs and the output signals even though the circuits are based on the nonlinear or time varying circuit operations.

1.1 CHARACTERISTICS OF VLSI MODULATION

As a special family of VLSI signal processing circuits, VLSI modulation circuits are constructed based on nonlinear and time varying VLSI signal processing circuit operations, such as clamping, multiplication, and switched-network operations. These basic VLSI circuit operations can be combined with other basic linear VLSI signal processing circuit operations, including the addition, scaling and integration to realize various specific VLSI circuit operations in signal processing, data conversions and power managements.

VLSI modulation circuits have some special properties in common:

- In time domain, these circuits usually employ the nonlinear or time varying circuit networks. The circuit topology changes can be realized using either the nonlinear or the switched circuit networks. Examples of such circuits are the switched-capacitor filter circuits, the dc/dc (switched-capacitor or switched-inductor) converter circuits, the chopper circuits, the lock-in amplifier circuits, the sigma-delta modulator circuits, and the VLSI passive mixer circuits.

- In frequency domain, these circuits usually employ frequency manipulation circuit operations. VLSI modulation circuits purposely translate the input signal (and noise) frequencies to frequency bands that are more suitable for VLSI signal processing circuit operations, such as the signal amplification, the noise minimization, and the power transformations, etc.

- In signal aspect, these circuits offer high linearity in the signal path between the signal inputs and the modulated signal outputs even though the circuits are inherently time varying or non-linear.

1.1.1 LINEAR VLSI SIGNAL PROCESSING CIRCUITS

For a given VLSI linear signal process circuit and any two signals $x_1(t)$ and $x_2(t)$ shown in figure 1.1, the two signal process circuit operations with respect to two constant a and b are equivalent. Both circuit provide the same output $y(t)$ (i.e. $y_1(t) = y_2(t)$).

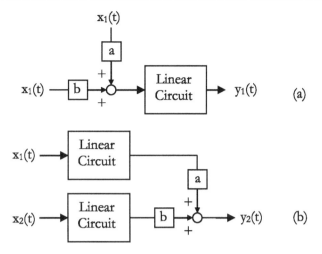

Fig. 1.1 Basic linear VLSI circuit operation

Some of the important VLSI linear time invariant (LTI) signal processing circuit properties include:

- The circuit operations are linear: For linear circuit operation F with signal x_1, x_2 and constant a and b we have that:

$$F(ax_1(t) + bx_2(t)) = aF(x_1(t)) + bF(x_2(t)) \tag{1.1}$$

- The steady state output signal frequency band is the same as the input signal frequency band.

$$F(e^{j\omega t}) = |H(j\omega)|e^{j\omega t + \phi(j\omega)} \tag{1.2}$$

- The circuit is time invariant: circuit structure and parameters independent of time.

$$\text{IF } y(t) = F(x(t)) \text{ then } y(t - t_0) = F(x(t - t_0)) \tag{1.3}$$

- Such circuit can be modeled and designed in s-domain employing Laplace Transform.

Example 1.1 The following RC circuit is a linear time invariant circuit.

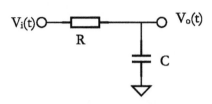

Fig.1.2 Linear RC filter circuit

This circuit can be modeled in time-domain using linear differential equation:

$$[RC\frac{d}{dt}+1]V_o(t)=V_i(t) \tag{1.4}$$

This circuit is linear since if

$$[RC\frac{d}{dt}+1]V_{o1}(t)=V_{i1}(t) \tag{1.5}$$

$$[RC\frac{d}{dt}+1]V_{o2}(t)=V_{i2}(t) \tag{1.6}$$

and if a and b are constant, then

$$[RC\frac{d}{dt}+1][aV_{o1}(t)+bV_{o2}(t)]=aV_{i1}(t)+bV_{i2}(t) \tag{1.7}$$

This circuit is also time invariant since:

$$[RC\frac{d}{dt}+1]V_o(t-t_0)=V_i(t-t_0) \tag{1.8}$$

In s-domain, this circuit can be modeled employing a transfer function as:

$$H(s)=\frac{1}{RCs+1} \tag{1.9}$$

1.1.2 LINEAR VLSI CIRCUIT MODEL AND DESIGN

For LTI circuit, it's convenient to model and design the circuit in s-domain or z-domain employing Laplace transform and z-transform, where the inductor is mapped to Ls, capacitor to $1/(Cs)$ and unity delay to $1/z$.

Example 1.2 The s-domain representation of the RC circuit is given as figure 1.3.

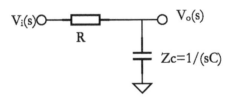

Fig.1.3 Linear RC filter circuit

Consequently continuous-time VLSI linear (or LTI) circuit can be represented based on three basic linear signal processing operations of Addition, Scaling, and Integration as shown in figure 1.4.

In similar method, the discreet-time VLSI linear (or LTV) circuit can be represented based on the basic linear signal processing operations of Addition, Scaling, and Unit delay as shown in figure 1.5.

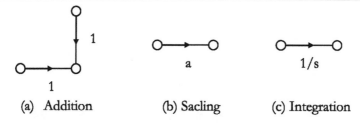

Fig.1.4 Basic continuous-time linear signal processing operations

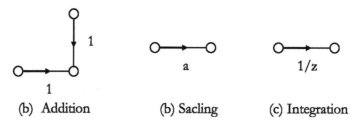

Fig.1.5 Basic discreet-time linear signal processing operations

These basic linear signal processing operations can be mapped to VLSI circuit directly for the VLSI circuit implementation.

Example 1.3 Shown in figure 1.6 are the prototype VLSI active-RC circuit representation of the basic linear signal processing operations.

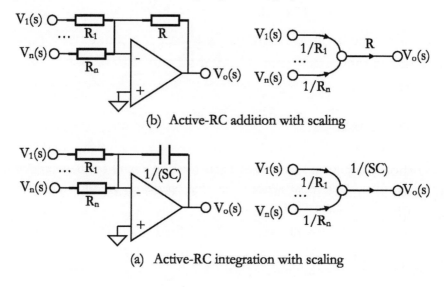

Fig.1.6 VLSI active-RC linear signal processing elements

VLSI Modulation Circuits

Example 1.4 Shown in figure 1.7 are the prototype VLSI Gm-C circuit representation of the basic linear signal processing operations.

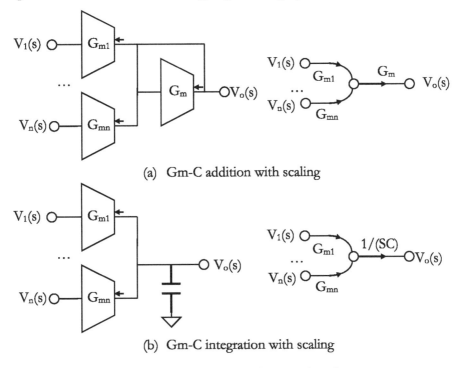

Fig.1.7 VLSI Gm-C linear signal processing elements

1.1.3 MODULATION BASED LINEAR VLSI CIRCUITS

In VLSI circuit implementation, nonlinear (NL) or linear time variant (LTV) modulation circuit structures are used to realize linear time invariant VLSI signal processing operations as shown in figure 1.8.

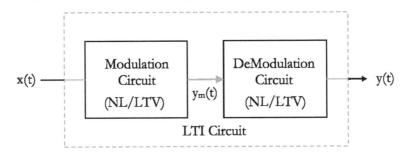

Fig.1.8 Modulation based linear signal processing circuit

In such VLSI signal processing approach, even though both modulation and demodulation subcircuit circuits are inherently nonlinear or linear time variant (LTV), the entire circuit structure is still linear time invariant (LTI). In many circuit analysis the demodulation portion of the circuit many design are not clearly specified and signal $y_m(t)$ is used for the circuit analysis and modeling.

Example 1.5 The given passive switched-capacitor circuit implements a linear lowpass filter. This circuit inherently time varying since the circuit has different topology for the two clock phases. In this circuit, the capacitor at the output serves as demodulation circuit to filter out the sampling noise for the circuit, such that the output is smooth signal even though the input signal is sampled in the first part of the circuit.

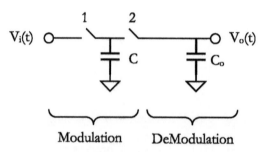

Fig.1.9 Passive VLSI switched-capacitor lowpass filter

1.2 BASIC MODULATION OPERATIONS

In additional to the all basic linear signal processing operations of Addition, Scaling, and Integration (or Unit-Delay), the nonlinear and time-variant linear operations are added to the basic modulation signal processing operations.

- **Linear Operations (LTI):** Addition, Scaling and Integration (or Unit Delay).
- **Nonlinear Operations (NL):** Nonlinear operations include unidirectional conduction or signal limiting operations.

- **Time-Variant Operations (LTV):** This is typically realized by synchronous switched-network where circuits are operated in phases and each phase is represented by a dedicated linear circuit.

1.2.1 LIMITING OPERATIONS

Limiting is basic nonlinear signal processing operation. There are three types of limiting operations:

- **High-Threshold Limiting** is used to limit the high value of a given signal to a specific high-side threshold as shown in figure 1.10. The mathematic model of a single-threshold clamping operation is given as:

$$y = \begin{cases} x & x < X_{ThH} \\ X_{ThH} & x \geq X_{THH} \end{cases} \quad (1.10)$$

X_{ThH} is the high-side threshold of the limiting circuit.

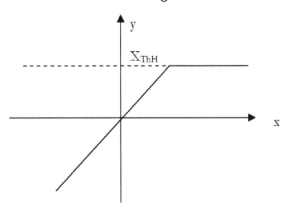

Fig.1.10 High-side limiting operation

Example 1.6 A commonly used VLSI limiting circuit is the diode clamper circuit as shown in figure 1.11, where the signal is limited when its value is higher than the limiting threshold (ignoring the forward diode drop in this conceptual description for simplicity).

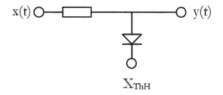

Fig.1.11 Diode-based VLSI high-side limiting circuit

- **Low-Threshold Limiting** is used to limit the low value of a given signal to a specific low-side threshold as shown in figure 1.12. The mathematic model of a single-threshold clamping operation is given as:

$$y = \begin{cases} x & x < X_{ThH} \\ X_{ThH} & x \geq X_{THH} \end{cases} \quad y = \begin{cases} x & x > X_{THL} \\ X_{THL} & x \leq X_{THL} \end{cases} \quad (1.11)$$

X_{ThH} is the high-side threshold of the limiting circuit.

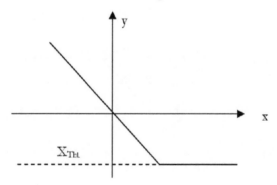

Fig.1.12 Low-side limiting operation

- **Dual-Threshold Limiting** is used to limit both the high and the low value of a given signal to within the specific thresholds as shown in figure 1.12. The mathematic model of a Dual threshold limiting operation is given as:

VLSI Modulation Circuits

Note that many practical VLSI nonlinear circuits inherently offer two clamping thresholds as shown in figure 1.12. The mathematic model of such two-level clamping circuit operation can be expressed as:

$$y = \begin{cases} X_{ThH} & x \geq X_{ThH} \\ x & X_{ThH} > x > X_{THL} \\ X_{ThL} & x \leq X_{THL} \end{cases} \quad (1.12)$$

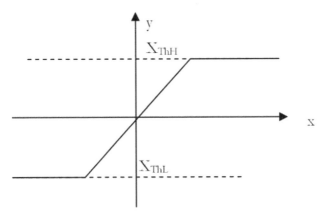

Fig.1.12 Dual threshold limiting operation

Example 1.7 It is interesting to see that a CMOS inverter shown in figure 1.4 also offers a two-level clamping operation where the supply rails serve as the clamping thresholds.

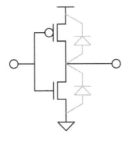

Fig.1.13 VLSI circuit structure of dual-threshold limiting

VLSI limiting operations offer several attractive signal processing functions, such as noise elimination, slew rate enhancement, and data pattern dependent timing variations minimization.

It is important to point out that when a single-tone signal is limited, the harmonics tones are usually generated.

1.2.2 MULTIPLICATION OPERATIONS

Signal multiplication is the basic modulation nonlinear signal operation. Signal multiplication involves multiplication of two or more signals as shown in figure 1.14.

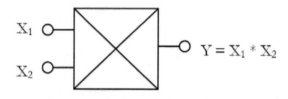

Fig. 1.14 Signal multiplication operation

The multiplication circuit operation results in single-tone signal energy to be split into energy bands with frequencies given as

$$f = nf_1 \pm mf_2 \qquad (1.13)$$

Where f_1 and f_2 are the signal and the control clock frequency respectively. Parameter n and m are integers.

One example of the VLSI multiplication circuit is the VLSI mixer circuit that is widely used in the communication circuits and systems.

1.2.3 SWITCHED-NETWORK

A switched-network offers the basic linear time-variant (LTV) signal processing operation for modulation where the control clock is used to switch the circuit connectivity among multiple linear topologies.

The switched-network can also be used to enable energy transfer among VLSI circuit components.

Example 1.8 Shown in figure 11.15 is VLSI switched-capacitor integrate circuit where the switched-network is used to activate the charge redistribution among the capacitor network.

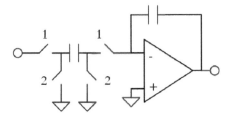

Fig. 11.15 Switched-network in switched-capacitor integrator

Example 1.8 Shown in figure 11.16 is switched-inductor DC/DC converter circuit.

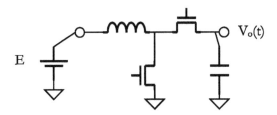

Fig. 11.16 Switched-inductor boost DC/DC converter

1.3 MODELING OF MODULATION CIRCUITS

Laplace, Z- and Fourier transforms have been used very effectively for modeling the linear time-invariant VLSI circuits and systems.

For nonlinear and linear time-variant signal processing circuits, these well-developed LTI modeling techniques and methodologies must be modified for modeling the LTV circuits and systems.

There are major types of models for the LTV circuits and systems including:

- **Micro-Modeling**: This modeling approach zooms into the details of the nonlinear circuit operation of the modulation circuits at the carrier frequency. It provides the detail information of the circuit operations at the cost of the modeling complexity, time, and resources.

- **Macro-Modeling**: This modeling approach ignores the detail operation of the carrier signal band and focus on the signal frequency band of the interest. It provides highly simplified linear model for the behavioral of high level linear time invariant circuit and system of the modulation circuit.

Example 1.9. For the switched-capacitor filter shown in figure 1.17, a micro-modeling is realized by partitioning the circuit into two equivalent linear circuit for two operation carrier clock phases.

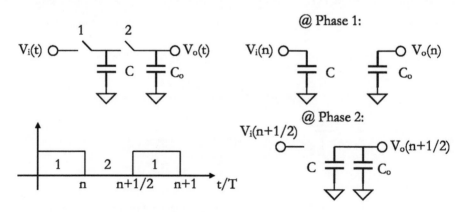

Fig.1.17 Micro-modeling of VLSI switched-capacitor lowpass filter

- In phase 1, the input capacitor is charged to the input signal and the output capacitor the charge of the last clock cycle. The charge on the capacitors are given as:

$$Q_C = CV_i(n) \quad (1.14)$$

$$Q_{Co} = C_o V_o(n) \quad (1.15)$$

- In phase 2, the charge on the two capacitors are shared. The output voltage can be determined as

$$V_o(n+\tfrac{1}{2}) = \frac{1}{C_o + C}[CV_i(n) + C_o V_o(n)] \quad (1.16)$$

By the above analysis, the output voltage at t= (n+1)T can be derived based on the value at the nT and the input voltage as:

$$V_o(n+1) = \frac{1}{C_o + C}[CV_i(n) + C_o V_o(n)] \equiv \alpha V_i(n) + \beta V_o(n) \quad (1.17)$$

Where

$$\alpha \equiv \frac{C}{C_o + C} \quad (1.18)$$

$$\beta \equiv \frac{C_o}{C_o + C} \quad (1.19)$$

The above difference equation can be solved by the following procedures:

- For n=0, he initial condition is given as:

$$V_o(0) = V_o(0) \qquad (1.20)$$

- For n=1, we have that

$$V_o(1) = \alpha V_i(0) + \beta V_o(0) \qquad (1.21)$$

- For n=2, we have that

$$V_o(2) = \alpha V_i(1) + \beta V_o(1) = \alpha V_i(1) + \beta[\alpha V_i(0) + \beta V_o(0)]] \qquad (1.22)$$

Or

$$V_o(2) = \alpha \sum_{k=0}^{1} \beta^{1-k} V_i(k) + \beta^2 V_o(0) \qquad (1.23)$$

- For n=3, we have that

$$V_o(3) = \alpha V_i(2) + \beta V_o(2) = \alpha V_i(2) + \beta[\alpha \sum_{k=0}^{1} \beta^{1-k} V_i(k) + \beta^2 V_o(0)] \qquad (1.24)$$

Or

$$V_o(3) = \alpha \sum_{k=0}^{2} \beta^{2-k} V_i(k) + \beta^3 V_o(0) \qquad (1.25)$$

- For n=n, we have that

$$V_o(n) = \alpha \sum_{k=0}^{n-1} \beta^{n-1-k} V_i(k) + \beta^n V_o(0) \qquad (1.26)$$

Or

For the special case of $V_i(n)=1$ and $V_o(0)=0$, we have that:

$$V_o(n) = \alpha\sum_{k=0}^{n-1}\beta^{n-1-k}V_i(k) + \beta^n V_o(0) = \alpha\sum_{k=0}^{n-1}\beta^{n-1-k} = \alpha\frac{1-\beta^n}{1-\beta} \quad (1.27)$$

- For n→infinite, we have that:

$$\lim_{n\to\infty} V_o(n) = \alpha\frac{1}{1-\beta} = 1 \quad (1.28)$$

Such micro-model solution can also be done through the computer aided simulation, such as the SPICE simulation.

Example 1.10. For the above switched-capacitor filter, a macro-modeling can be realized by employing the macro-model as shown in figure 11.18.

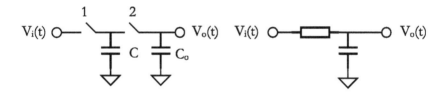

(a) Switched-Capacitor circuit   (b) Active-RC equivalent

Fig.1.18 Macro-modeling of VLSI switched-capacitor lowpass filter

For the input and the initial condition of $V_i(t)=1$ and $V_o(0)=0$, the output signal can be solved by the following differential equations as:

$$[RC\frac{d}{dt}+1]V_o(t) = V_i(t) \quad (1.29)$$

$$V_o(t) = (1 - e^{-\frac{t}{RC}})$$ (1.30)

1.4 KEY BENEFITS OF VLSI MODULATIONS

VLSI modulation circuits serve as alternatives to the linear VLSI signal processing circuit solution. However from the VLSI circuit technology aspects, VLSI linear signal processing circuit offer circuit solution that conventional linear circuits cannot be offer. Or in the other worlds, some of the specific VLSI signal processing functions can only be realized using VLSI modulation circuit techniques.

There are a few classic VLSI modulation circuit applications to the linear VLSI signal processing problems:

- High precision data conversions.
- Low noise and low dc offset signal amplification.
- High efficiency high performance signal power amplification.
- High efficiency power conversion and power management.
- Other special circuit functions.

1.4.1 HIGH PRECISION DATA CONVERSIONS

All conventional VLSI data conversion circuits, such as the flash, the pipeline, the SAR and single or dual-slope converters are typically limited by the achievable dc offset caused by the random mismatch of the VLSI devices. Typical values of VLSI comparator circuits are in the order of one tenth of mV to tens of mV. This limits the typical data conversion precision of within 14bits even with calibration circuit techniques. On the other hand high-resolution (>20 bits) data conversion (ADC and DAC) can be realized employing the noise shaping operation of sigma-delta modulation circuit techniques. Such circuit techniques can be used to minimize the non-linearity and the mismatch effects in VLSI circuit elements.

1.4.2 LOW NOISE SIGNAL AMPLIFICATIONS

VLSI modulation circuit techniques can be used to minimize the VLSI device non-idealities, such as the device mismatch induced the dc offset, the $1/f$ noise, the low frequency cross-coupling and supply noise effects. VLSI modulation can be used to significantly improve the signal-to-noise ratio of the specific VLSI signal processing operations.

In the CDS, chopper, and lock-in amplifier circuits, VLSI modulation can be used to translate the circuit noise and non-idealities and then eliminated using signal processing circuit techniques. Typical CDS and chopper circuit technique can be used to reduce the random mismatch and $1/f$ noise effects to uV range and provide high precision signal amplification and processing. Employing the phase sensitive lock-in amplifier circuit techniques, noise level of sub-uV can be realized for high dynamic range signal detections.

1.4.3 HIGH EFFICIENCY POWER AMPLIFICATIONS

Conventional high linearity VLSI signal power amplifiers require devices to bias at class-A mode, leads to low power efficiency of lower than 30%. Such circuit may significantly reduce the battery time of the mobile products and the circuit reliability. It will also increase the product cost for adding effective heat sinker to dissipate the produced heat from the power amplifier.

VLSI Class-D and other switching amplifier circuit techniques based on modulation circuit techniques offers solutions of high power efficiency beyond 90% with better than 0.01% THD that can be used for mobile and high power amplification product applications.

1.4.4 HIGH EFFICIENCY POWER COVERSIONS

VLSI modulation offers capability of power source conversion among the dc and the ac sources, at the different voltage levels and for the voltage and the current regulation. The frequency translations convert the power source into the frequency bands that matches the characteristics of the VLSI capacitive or inductive circuit elements and therefore suitable of the VLSI integration.

VLSI charge pump and switched-inductor DC/DC converters offer power management solutions of better than 90% for various power supply applications of electronic devices and equipment.

1.4.5 OTHER SPECIAL CIRCUIT FUNCTIONS

There are a class of circuit functions that conventional VLSI linear circuit cannot be realized and that require specific VLSI modulation circuit techniques, such as frequency translation for RF signals, phase modulation for VLSI high-speed I/O circuits, switched-capacitor and switched-current for ratio-based and high precision analog filters.

1.5 APPLICATIONS OF VLSI MODULATION CIRCUITS

VLSI modulations can be implemented in various forms targeted at the specific applications.

1.5.1 FREQUENCY TRANSLATION

Frequency translation is a basic operation of the VLSI modulation that can be used to translate the original signal frequency band into a specific frequency band that is more suitable for VLSI circuit implementation. For example, in the VLSI chopper circuit techniques, a low frequency signal input is translated into a higher frequency band where the signal can be effectively amplified with minimized impacts on the dc and 1/f noises of the VLSI amplifier circuit elements. The amplified noise-free ac signal is then frequency translated back (i.e. demodulated) to the original signal frequency band using a second chopper. Such circuit approach can be used to significantly improve the performance of the VLSI amplifier circuit.

Frequency translations are the basis of VLSI RF mixers where the baseband signals are frequency translated into the RF band that can be transmitted effectively in air. The received RF signal can then be translated back to the baseband using an inverse frequency translation employing a down conversion mixer in the receiver circuit.

Frequency translations have been used in many other VLSI signal processing circuit applications, such as the lock-in amplifier where frequency translation is used to detect a weak signal in strong noise background. In the switched-capacitor circuits, the baseband signal is frequency translated to a frequency band near the carrier clock frequency such that the charge redistribution circuit operation can be realized. In the dc/dc converter circuit applications, the frequency translations are used to convert the dc voltage into the ac form such that the voltage level can be converted effectively using the energy storage circuit elements such as the capacitors and the inductors.

1.5.2 PULSE DENSITY MODULATION

The pulse density modulation (PDM) provides a method to represent the analog signal in the digital or digital-like forms, where the average of the discrete level signal over a given time period represents the input analog signal value.

The PDM serves as the basis of the high efficiency signal power amplification and the power conversion such as the class-D amplifiers and the dc/dc conversion that allows the analog input to be amplified in the digital domain for better power efficiency and SDNR. In a VLSI class-D amplifiers, the analog signal is first converted into the pulse density signal form using either the pulse width modulation (PWM) or the pulse code modulation (PCM) technique such that the signal amplification operation can be implemented in the full digital form for better fidelity and efficiency.

1.5.3 PULSE WIDTH MODULATION

The pulse width modulation (PWM) is a special type of the pulse density modulation. Under the pulse width modulation, an analog signal is represented by the duty-cycle of a digital (or digital-like) signal with discrete signal magnitude values, where the analog signal is the local average of the PWM signal across a given time period.

The PWM signal is commonly used in the VLSI class-D amplification and the dc/dc supply voltage conversion circuits.

1.5.4 PULSE CODE MODULATION

The pulse code modulation (PCM) is another form of pulse density modulation (PDM). A major difference of a signal in the PCM form versus the PWM form is that the transition of the PCM is synchronous to the carrier clock. In the VLSI circuit implementations, the PCM signal can be generated from an analog signal using a sigma-delta modulation circuit.

Due to their high VLSI implementation simplicities and superior performances, the PCM techniques are also commonly used in the VLSI class-D amplifications and the data conversions (A/D or D/A conversion) operations.

1.5.5 PHASE AND DELAY MODULATION

The VLSI phase and delay modulation are special VLSI modulation circuit techniques that are commonly used in VLSI high-speed I/O circuit applications where the phase interpolation (PI) and the delay manipulation are critical circuit operations in VLSI high-speed I/O data recovery circuits. A phase modulation is also commonly used for spread spectrum clocking (SSC) in the PLL circuits.

1.5.6 CHOPPING

The VLSI chopper circuit techniques employ the time varying (switched) networks that swap the two signal paths (usually the two signal paths in a differential signal) in the circuit operation.

1.6 ORGANIZATION OF BOOK CHAPTERS

The materials in this text are organized into 15 specific chapters covering the basic modulation theory, the circuit and the system modeling methods, the VLSI circuit implementations and their applications.

Chapter 1 provides an overview to the VLSI modulation techniques. The VLSI modulations are special VLSI signal processing operations that are related to signal frequency spectrum manipulations. VLSI modulations are widely used for VLSI circuit non-ideality compensation, for ease of the VLSI signal processing and for high effective data conversion and power conversion.

The basic VLSI modulation theories are discussed in detail in Chapter 2, where the mathematic models of the VLSI modulations are provided. Various VLSI modulation techniques, such as the frequency translation, the PDM, the PWM and the PCM are analyzed and the applications are also discussed.

The VLSI sigma-delta modulation circuits are presented in Chapter 3. The sigma-delta modulations have been widely used in applications such as the data conversion, the PCM coding and the random pattern generation. A sigma-delta modulation offers the attractive noise shaping operation for VLSI data conversion that pushes the quantization noise away from the signal band into higher frequency band. Such a noise shaping technique serves as the basis of high accuracy data conversions. In the sigma-delta modulator an analog signal is presented in the digital domain synchronous to the control clock where the relative density of the pulses that corresponds to the analog amplitude of the signal (also called pulse-code modulation-PCM).

VLSI class-D amplifier circuit techniques are discussed in Chapter 4. The class-D amplifier employs either PWM or PCM technique to provide high efficiency analog power amplification with high linearity that is insensitive to VLSI device non-idealities. Such circuit techniques are widely used in VLSI audio and RF band signal amplifications. The class-D amplifiers offer very high power efficiency ($\geq$ 90%) with compact size/light weight for high power amplification where high efficiency is achieved by operating the output stages in switching modes that are either fully on or fully off, thereby minimizing the power dissipation. The switching (or digital) operation of the class-D output stage also avoids the requirement of linear characteristic of power device to achieve high amplification linearity.

VLSI phase interpolation (PI) circuit is introduced in Chapter 5. VLSI phase interpolation circuits employ VLSI phase modulation techniques to manipulate the phase (or delay) of a clock. VLSI PI circuit is widely used in VLSI systems such as VLSI high speed I/O circuit and VLSI clocking circuits.

VLSI zero-drift amplifier circuit techniques are discussed in Chapter 6. The dc offset and 1/f noise as the major performance limitation factors in CMOS analog circuits and systems can be minimized using special VLSI circuit techniques such as the auto-zero circuits, the synchronous choppers and the chopper stabilize circuits. An auto-zero circuit uses the correlative double sampling (CDS) circuit techniques to cancel the dc offset and low frequency noise such as 1/f noise. A chopper amplifier relies on shifting the dc and low frequency noise to higher frequency band such that they can be eliminated by lowpass filtering. A chopper stabilized amplifier uses the compound amplifier of auto-zero and chopper circuit techniques to separate the normal high bandwidth signal amplification path from the low bandwidth offset and drift compensation path therefore offers excellent performance in dc and low frequency noise compensation.

VLSI lock-in amplifier circuit techniques are introduced in Chapter 7. Lock-in amplifiers can be used to measure very weak ac signals down to few nano-volts under extremely high noise background of a few orders magnitude higher than the signal. Lock-in amplifier separates the noise from signal in the frequency domain based on the phase-sensitive detection, where noise signals, at frequencies other than the reference frequency, can be rejected and therefore will not affect the measurement.

VLSI switched-capacitor dc/dc converter (also known as the charge pump) circuits are discussed in Chapter 8. VLSI charge pumps offer the best choice for power management applications that require both low power and low cost, where the available supply rails are not directly usable, nor are the direct use of battery voltage. VLSI charge pumps are very suitable for dc/dc voltage conversion applications that require some combination of low power, simplicity, and low cost. VLSI charge pump circuits use capacitors (instead of inductor) as the main energy storage elements to create either a higher, lower or inverted power supply sources. Switched-capacitor charge pump circuits are capable of high efficiencies, sometimes as high as 90-95% while being electrically simple.

VLSI switched-capacitor signal processing circuit techniques are discussed in Chapter 9. VLSI switched-capacitor circuits based on the discrete-time (DT) signal processing operation offer high analog signal processing accuracy at voice-band frequencies in fully integrated form. VLSI switched-capacitor signal processing circuits also offer attractive features such as low power, high accuracy, compact design and fully compactable to VLSI digital process technologies.

VLSI switched-current circuit techniques are discussed in Chapter 10 where current mode signals are used as functional circuit elements for signal processing. VLSI switched-current circuits employing MOS current mirror and switches offer circuit operations that are fully compatible to the VLSI digital process technologies.

VLSI switched-inductor dc/dc converter circuit is introduced in Chapter 11. Such dc/dc converters employ switched-inductor circuit networks to realize the circuit topology change by turning on and off specific switches according to appropriate PWM feedback control. VLSI voltage regulators based on switched-inductor dc/dc converters offer several advantages such as better switching efficiency, smaller components and less thermal management requirement and flexible dc/dc conversion, such as boost, buck and inverter conversions and isolations.

VLSI fractional-N phase locked-loop circuit is introduced in Chapter 12. VLSI phase-locked loop (PLL) circuits offer control clock sources for various integrated circuits such as microprocessors, high-speed I/Os, memory interfaces, audio, and video ports. The fractional-N PLLs can be used to generate multiple frequencies from one clock source with fast settling time, low frequency error, better spurious performance, and low phase noise that are suitable for various VLSI communication circuit applications.

VLSI spread spectrum clocking (SSC) circuits are introduced in Chapter 13. SSC circuit techniques allow spreading signal power spectrum in the frequency domain for the purposes of establishing secure communications, increasing resistance to natural interference and jamming, preventing detection, and limiting power flux density. SSC techniques are commonly used to reduce the electronic emission interference (EMI) or electronic emission compliance (EMC) effects of the digital signals and associated harmonics. SSC circuits make a narrowband signal broadband through the frequency or phase modulation of the control clock of the system. SSC circuit spreads the radiation energy in the frequency domain such that peak energy of the system is minimized. As the result of the frequency and phase modulation, the amplitudes of the harmonics of all the digital signals resulted from this clock can be reduced. These circuit techniques can be used to reduce the EMI effects by typically 2-22dB, depending of the circuit implementation and measurement methods.

VLSI mixer circuit is introduced in Chapter 14. Mixer circuit can be used to translate signal frequency band through the up-conversion or the down-conversion. Such frequency conversion is the basis for both receiver and transmitter in radio frequency (RF) systems.

An appendix chapter is included in this text. The basis of VLSI switched amplifier circuits is discussed. Compared with conventional class-A, class-B and class-AB amplification types, VLSI switched amplifiers offer the ability to amplify a relatively small input signal into a larger output signal very effectively.

Reference:

[1] J. Holtz, "Pulse width modulation-a survey," *IEEE* Transactions on Industrial Electronics, Volume: 39, Issue: 5, 1992, Page(s): 410-420.

[2] B. Putzeys, "Digital audio's final frontier," IEEE Spectrum, Volume: 40, Issue: 3, 2003, Page(s): 34-41.

[3] F. Raab, "Radio Frequency Pulse width Modulation," *IEEE* Transactions on Communications, Volume: 21, Issue: 8, 1973, Page(s): 958-966.

[4] C. C Enz, G. C.Temes, "Circuit techniques for reducing the effects of op-amp imperfections: autozeroing, correlated double sampling, and chopper stabilization," Proceedings of the IEEE, Volume: 84, Issue: 11, 1996, Page(s): 1584-1614.

[5] Miguel Gabal, Nicolas Medrano, Belen Calvo, Santiago Celma, "A Low-Voltage Single-Supply Analog Lock-in Amplifier for Wireless Embedded Applications," 2010 European Workshop on Smart Objects: Systems, Technologies and Applications (RFID Sys Tech), Page(s): 1-6.

[6] R. T. Baird, T. S.Fiez, "Linearity enhancement of multibit Delta Sigma; A/D and D/A converters using data weighted averaging," IEEE Transactions on Circuits and Systems II: Analog and Digital Signal Processing, Volume: 42, Issue: 12, 1995, Page(s): 753-762.

[7] On-Cheong Mak, Yue-Chung Wong, A. Ioinovici, "Step-up dc power supply based on a switched-capacitor circuit," IEEE Transactions on Industrial Electronics, Volume: 42, Issue: 1, 1995, Page(s): 90-97.

[8] L. J. Bloom, "Past, present and future dynamics within the power supply industry," Conference Proceedings 1998. Thirteenth Annual Applied Power Electronics Conference and Exposition. APEC '98. Volume: 1, 1998, Page(s): 278-283.

[9] M. Klapfish, "Trends in AC/DC switching power supplies and dc/dc converters," Conference Proceedings 1993. Eighth Annual Applied Power Electronics Conference and Exposition, 1993. APEC '93. 1993, Page(s): 87-91.

[10] Ju-Ming Chou, Yu-Tang Hsieh, Jieh-Tsorng Wu, "Phase averaging and interpolation using resistor strings or resistor rings for multi-phase clock generation," IEEE Transactions on Circuits and Systems I: Regular Papers, Volume: 53, Issue: 5, 2006, Page(s): 984-991.

[11] J. Crols, M. Steyaert, "Switched-opamp: an approach to realize full CMOS switched-capacitor circuits at very low power supply voltages," IEEE Journal of Solid-State Circuits, Volume: 29, Issue: 8, 1994, Page(s): 936-942.

[12] J. B.Hughes, K. W. Moulding, "Switched-current signal processing for video frequencies and beyond," IEEE Journal of Solid-State Circuits, Volume: 28, Issue: 3, 1993, Page(s): 314 – 322.

CHAPTER 2
BASIC VLSI MODULATION THEORY

Modulation provides a special VLSI signal processing approach where the frequency contents of the signal under processing is manipulated in specific way for optimizing the VLSI circuit operation, mitigating the VLSI circuit component non-ideality, and for the manufacture and application compatibilities.

Several modulation operations are very suitable for VLSI circuit implementation, such as the frequency translation, phase modulation (PM), pulse density modulation (PDM), pulse width modulation (PWM) and pulse code modulation (PCM). These modulation techniques have been widely adopted in VLSI circuits and systems to support wide range of applications.

2.1 PULSE DENSITY MODULATION (PDM)

A pulse-density modulation can be used to represent the analog signal in digital-like (i.e. discrete in values) form. For a PDM signal, the relative density of the pulses represents the analog signal value.

Pulse-width modulation (PWM) and pulse code modulation (PCM) are two special types of PDM where a PWM signal represents continuous time PDM analog signal and a PCM signal represents a discrete time PDM analog signal.

2.1.1 PDM PRESENTATION OF ANALOG SIGNAL

Shown in figure 2.1 is a 2-level PDM bit stream X(nT), where a 1 represents a pulse of positive polarity (V_{ref+}) and a 0 represents a pulse of negative polarity (V_{ref-}) where V_{ref+} and V_{ref-} are predefined voltage references for the PDM. The analog signal represented by such a PDM signal stream is given as:

$$V(nT) = [V_{ref+} - V_{ref-}]X(nT) + V_{ref-} \qquad (2.1)$$

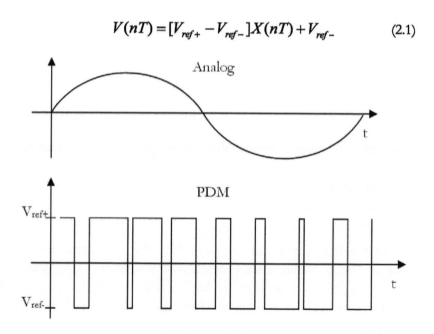

Fig.2.1 2-level PDM modulation

VLSI Modulation Circuits

In the PDM stream all 1s represent the maximum (V_{ref+}) amplitude value, all 0s represent the minimum (V_{ref-}) amplitude value, and alternating 1s and 0s represent an average of the two reference values. A data stream of the combinations of 1s and 0s can be used to represent the analog values between the two specified references.

The PDM modulation can also be realized using multi-level coding methods where multiple digital values correspond to specific multiple analog references. A 3-level PDM signal example is shown in figure 2.2.

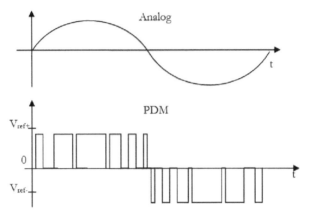

Fig.2.2 3-level PDM modulation

2.1.2 DEMODULATION OF PDM SIGNALS

Since the pulse density of a PDM signal corresponds to the magnitude of the value of an analog signal, a PDM data stream can be demodulated (i.e. recovered) back to the analog signal using a low-pass filter.

2.2 PULSE CODE MODULATION (PCM)

The pulse-code modulation (PCM) represents the sampled analog signals where the value of the analog signal is sampled regularly at uniform intervals, with each sample being quantized to the nearest value within a range of digital steps.

PCM streams have two basic properties that determine their fidelity to the original analog signal including the sampling rate and the quantization bit. Sampling rate is the number of times per second that samples are taken. On the other hand, the quantization bit specifies the number of possible digital values that each sample can take.

The pulse-code modulation (PCM) can be realized using the sigma-delta modulator. Shown in figure 2.3 is a simple sigma-delta modulator employing a first-order continuous-time loop filter.

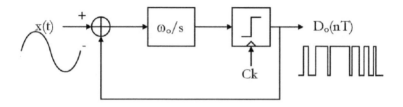

Fig.2.3 1st order continuous-time Sigma-Delta modulator

Such a modulator can be modeled mathematically using a linear system by introducing a quantization noise term as shown in figure 2.4.

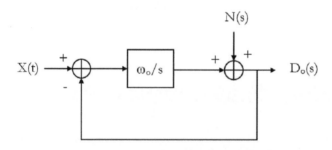

Fig.2.4 Linear 1st order continuous-time Sigma-Delta modulator model

The signal and noise transfer functions of such modulator can be derived respectively as:

$$\frac{D_o}{X}\bigg|_{N=0} = \frac{1}{1+\left(\dfrac{s}{\omega_o}\right)} \qquad (2.2)$$

$$\frac{D_o}{N}\bigg|_{X=0} = \frac{\left(\dfrac{s}{\omega_o}\right)}{1+\left(\dfrac{s}{\omega_o}\right)} \qquad (2.3)$$

The frequency responses of the signal and noise transfer functions are plotted in figure 2.5. It is important to see that the passband of the signal and noise transfer functions is separated in frequency. This interesting behavior (known as noise sharping) is the basis of all sigma-delta modulation circuit techniques.

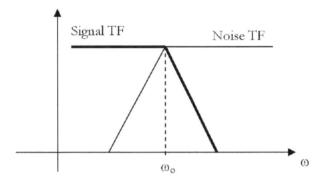

Fig.2.5 Typical signal and noise transfer functions of sigma-delta modulator

In general case, the PCM signal can be generated using single-loop sigma-delta modulation shown as figure 2.6.

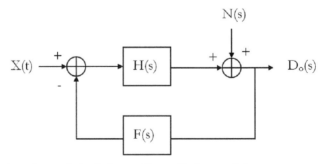

Fig.2.6 General single-loop Sigma-Delta modulator structure

The signal and noise transfer functions of the circuit can be expressed respectively as:

$$\frac{D_o}{X}\bigg|_{N=0} = \frac{H(s)}{1+H(s)F(s)} \tag{2.4}$$

$$\frac{D_o}{N}\bigg|_{X=0} = \frac{1}{1+H(s)F(s)} \tag{2.5}$$

We can see that if H(s) has a lowpass response with infinite dc gain the signal transfer function will have lowpass frequency response as:

$$\begin{cases} |H(s)|_{s\to 0} \to \infty \\ |H(s)|_{s\to\infty} \to 0 \end{cases} \tag{2.6}$$

On the other hand, the noise transfer function will have the highpass frequency response.

A PCM signal can also be generated using discrete-time VLSI circuit such as the switched-capacitor circuit. Shown in figure 2.7 is a typical sigma-delta modulator employing the discrete-time circuit.

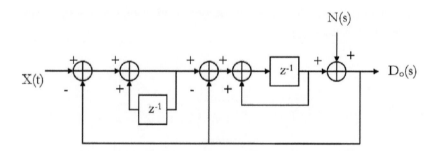

Fig.2.7 2nd order discrete-time Sigma-Delta modulator

The signal and noise transfer functions of such modulator can be derived respectively as:

$$\left.\frac{D_o}{X}\right|_{N=0} = z^{-1} \tag{2.7}$$

$$\left.\frac{D_o}{N}\right|_{X=0} = (1-z^{-1})^2 \tag{2.8}$$

The frequency response of the modulator has an allpass frequency response for the signal path:

$$\left|\frac{D_o}{X}\right|_{N=0} = 1 \tag{2.9}$$

On the other hand, the noise will be highpass shaped by the modulator as:

$$\left|\frac{D_o}{N}\right|_{X=0}\bigg|_{z=e^{j\omega T}} = (2\sin(\frac{\omega T}{2}))^2 \tag{2.10}$$

2.3 PULSE WIDTH MODULATION (PWM)

A pulse width modulation represents an analog signal using the local duty-cycle of the signal between two or more digital reference levels. Such duty-cycle can be continuously selected (analog PWM) or has very high resolution (digital PWM) compared with the modulation control clocks.

2.3.1 TWO-LEVEL PWM MODULATIONS

A two-level PWM signal can be generated by subtracting the signal from a triangular (or sawtooth) carrier and clamp it to power rails after a high gain amplification as shown in figure 2.8.

Such PWM signal is commonly used in the power amplification circuits due to the fact that the power loss in the switching devices is very low, since when a switch is off there is practically no current, and when it is on, there is almost no voltage drop across the switch. Power loss, being the product of voltage and current, is thus in both cases close to zero. PWM also works well with digital controls, which, because of their on/off nature, can easily set the needed duty-cycle.

The PWM has also been used in the communication systems where its duty-cycle is used to convey the information over a communication channel.

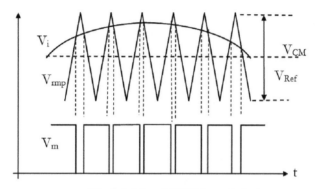

Fig.2.8 2-level PWM modulation

2.3.2 MULTI-LEVEL PWM MODULATION

Shown in figure 2.9 is a multi-level pseudo-differential PDM scheme where the effective output is expressed by the difference of the positive and the negative outputs.

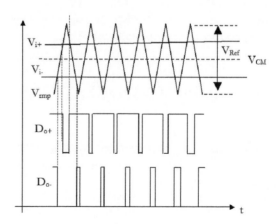

Fig.2.9 Multi-level pseudo-differential analog PWM modulation

This modulation method is equivalent to a 3-level modulation of [V+-V-] = [-1, 0, 1] with the combination of the output polarity [D_{o+}, D_{o-}] = [0, 1], [1, 1] or [0, 0], [1, 0]. An attractive feature of such modulation is that the differential output has a very small equivalent duty-cycle when the input is close to zero even though each output has a duty-cycle close to 50%. This effect is that the switching frequency at such condition is shaped significantly higher than the single-end modulation that has the 50% duty-cycle. Consequently, the switching noise of the PWM can be significantly attenuated by the parasitic lowpass response of the load without using an additional filter (i.e. the filter-less class-D solution). Shown in figure 2.10 is an alternative multi-level PWM that uses slightly different sawtooth waveform.

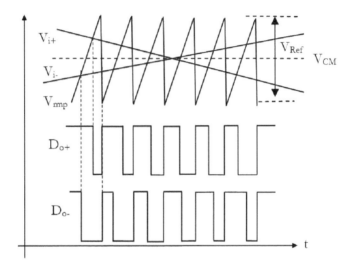

Fig.2.10 Alternative multi-level pseudo-differential analog PWM modulation

2.4 FREQUENCY TRANSLATION

A frequency translation can be used to shift the signal power spectrum by a given frequency offset f_m as shown in figure 2.11.

The frequency translation operations serve as basis of VLSI frequency mixer circuits where two signals at frequencies f and f_m are used to produce new signals with frequency at the sum $f + f_m$ or difference $f - f_m$ of the original frequencies. Frequency translations are also commonly used in VLSI circuits such as the chopper stabilize circuit, the lock-in amplifier circuit, the switched-capacitor and switched-current circuits where the signals are frequency translated for the offset and 1/f noise compensation, for the signal detection and for ease of VLSI signal processing.

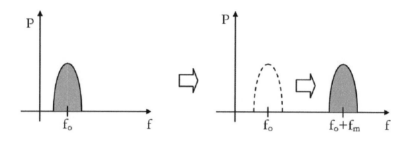

Fig.2.11 Frequency translation

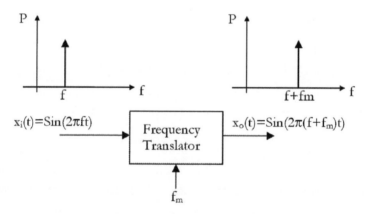

Fig.2.12 Basic frequency translation circuit model

In the RF transmitter mixer applications, frequency translation is used to shift the baseband signal up to RF frequencies (passbands of air) such that it can be transmitted effectively in the air. On the other hand, the received RF signal in the RF receiver is frequency translated down to baseband to recover the original signal contents.

In the chopper stabilized amplifier applications, the low frequency input signals are translated to higher frequencies such that the signals can be amplified effectively with minimized impacts of the amplifier dc offset and 1/f noise.

In the lock-in amplifier application a weak low frequency signal within a high noise floor can be modulated to a higher frequency band and therefore can be amplified effectively. The amplified signal is then detected using the demodulation circuit technique employing a phase sensitive detection circuit.

In the switched-capacitor and switched-current circuit, the low frequency (baseband) signal is frequency translated to control frequency such that the switched-capacitor resistance or switched-current circuit can be used for effective VLIS signal processing. The modulated signal will be modulated back to baseband after the processing.

2.4.1 ANALOG MULTIPLICATION

A simple way for realizing the frequency translation is through the analog multiplication that generates the product of two input signals.

In the simplest form, a mixer is a multiplier. The mixer multiplies the input signal to produce output signal at new frequencies. Such mixing operation is used as modulator and demodulator in the transmitter and the receiver path of a communication system, where the baseband signal is converted into RF signal and vice versa.

A mixer based on the analog multiplication is shown in figure 2.13. The output of the mixer contains two frequency terms including the sum and difference of the frequencies given as:

$$V_o = A_1 \cos(\omega_1 t) A_2 \cos(\omega_2 t) = \frac{A_1 A_2}{2}[\cos(\omega_1 + \omega_2)t + \cos(\omega_1 - \omega_2)t] \quad (2.11)$$

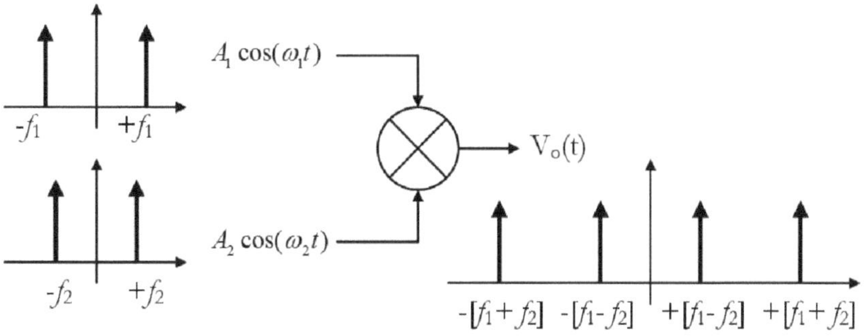

Fig.2.13 Analog multiplication frequency translation

2.4.2 SWITCHING FREQUENCY TRANSLATION

Since analog multiplier suffers from noise effect as it depends on modulation control signal magnitude, a switching frequency translation can be used to improve the performance of the frequency translation operation.

Shown in figure 2.14 is a switching frequency translation circuit.

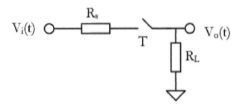

Fig.2.14 Sampling frequency translation

The switch in this mixer is controlled by a LO (local oscillator) circuit. For a switch control clock with period T and 50% duty-cycle, the transfer function of this circuit can be expressed as:

$$S(t) = \frac{1}{2} + \frac{2}{\pi} \sum_{k=1}^{\infty} \sin[(2k+1) \cdot 2\pi \frac{t}{T})] \qquad (2.12)$$

It can be seen that a single tone input will be multiplied by multiple modulation frequency tones.

2.5 PHASE TRANSLATION

Shown in figure 2.15 is a phase translation circuit that translates the signal to a new value with a modulation phase offset. Such signal processing operation is the basis of the VLSI phase interpolation (PI) and the delay locked-loop circuits.

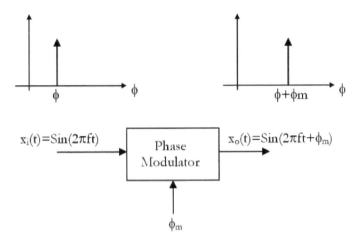

Fig.2.15 Basic phase translation circuit model

Reference:

[1] J. Holtz, "Pulse width modulation-a survey," IEEE Transactions on Industrial Electronics, Volume: 39, Issue: 5, 1992, Page(s): 410-420

[2] B. Putzeys, "Digital audio's final frontier," IEEE Spectrum, Volume: 40, Issue: 3, 2003, Page(s): 34-41.

[3] C. C. Benz, G. C. Temes, "Circuit techniques for reducing the effects of op-amp imperfections: auto zeroing, correlated double sampling, and chopper stabilization," Proceedings of the IEEE Volume: 84, Issue: 11, Page(s): 1584 - 1614, 1996.

[4] Hongjiang Song, Yan Song, Tai-hua Chen, "VLSI passive switched-capacitor signal processing circuits: Circuit architecture, closed form modeling and applications," 2008 IEEE International SOC Conference, Page(s): 297 - 300, 2008.

CHAPTER 3
VLSI SIGMA-DELTA DATA CONVERTER CIRCUITS

The VLSI sigma-delta (ΣΔ) modulators offer the VLSI circuit realization fully compatible to the digital VLSI technologies and therefore they are very suitable for SOC implementations. The oversampling and noise shaping techniques employed in the sigma-delta converters allow trading circuit speed for system accuracy to resolve the issues of the traditional data converters for higher overall conversion accuracy. In this way, a device imperfection insensitive signal processing operation can be obtained at the cost of increasing complexity and speed in the associated digital circuit forms.

In the sigma-delta modulator an analog signal is presented in the digital form where the relative density of the pulses corresponds to the analog signal's amplitude (also known as pulse code modulation - PCM). The sigma-delta modulation is similar to the pulse-width modulation (PWM) where the pulse density of the output represents the input analog signal amplitude. However in the sigma-delta modulator, the output can only change discretely as determined by the control clock.

3.1 SIGMA-DELTA MODULATION PRINCIPLE

The sigma-delta modulator is based on the basic techniques of the oversampling, the error processing and the feedback to improve the effective resolution from a coarse quantizer.

3.1.1 QUANTIZATION MODEL

Shown in figure 3.1 is the quantization error model of an ideal conventional 2-bit A/D converter.

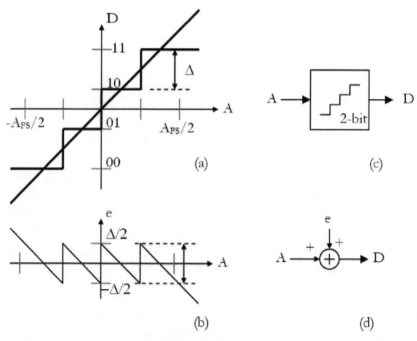

Fig.3.1 Quantization error model of ideal A/D converter

Such converter maps an analog signal into the digital form based on the mapping rule shown in figure 3.1a. Such a mapping will introduce the signal amplitude dependent error (called quantization error) as shown in figure 3.1b where the highest quantization error is determined by resolution Δ of the A/D converter. This quantization error can be modeled as shown in figure 3.1d where the quantization error in time domain is given as:

$$e(t) \equiv D(t) - A(t) \qquad (3.1)$$

Assuming the input signal is fully random between the full scale inputs of +/- $A_{FS}/2$, the quantization error will be fully uncorrelated from sample to sample. We will assume the error is uniformly distributed in the range of $+/-\Delta/2$ as shown in figure 3.2a. The power associated with such quantization error can be expressed as:

$$\overline{e^2} \equiv \sigma^2(e) = \int_{-\infty}^{+\infty} e^2 PDF(e)de = \frac{1}{\Delta}\int_{-\frac{\Delta}{2}}^{+\frac{\Delta}{2}} e^2 de = \frac{\Delta^2}{12} \qquad (3.2)$$

The quantization error will be distributed in the frequency band $[-f_s/2, +f_s/2]$. The assumption of uniform PDF distribution of random quantization error also implies that its power spectral density PSF is uniform as shown in figure 3.2b.

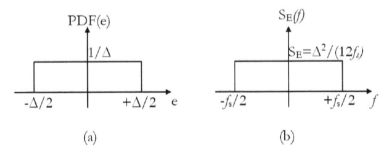

Fig.3.2 Quantization error distribution functions

The error power can also be derived as:

$$\overline{e^2} = \int_{-\infty}^{+\infty} S_E(f)df = S_E \int_{-\frac{f_s}{2}}^{+\frac{f_s}{2}} df = S_E f_s = \frac{\Delta^2}{12} \qquad (3.3)$$

The power spectral density of the additive white noise from the quantization can then be approximately expressed as:

$$S_E(f) = \frac{\Delta^2}{12 f_s} \qquad (3.4)$$

3.1.2 DYNAMIC RANGE OF IDEAL A/D CONVERTER

The dynamic range (DR) of an ideal A/D converter can be expressed by the ratio of the output power at the input signal frequency with the maximum amplitude ($A_{FS}/2$) to the inband quantization error power.

For full-scale sinusoidal signal, the signal power is given as:

$$P_s = \frac{1}{2}(A_{FS}/2)^2 = A_{FS}^2/2^3 \qquad (3.5)$$

For an M-bit A/D converter, the full scale signal magnitude can be expressed as:

$$A_{FS} = 2^M \cdot \Delta \qquad (3.6)$$

Since the inband quantization error power is given as:

$$P_Q = \frac{\Delta^2}{12}, \qquad (3.7)$$

The dynamic range of an ideal A/D converter can be expressed as:

$$DR \equiv \frac{P_s}{P_Q} = [2^{2M} \cdot \Delta^2 / 2^3] / \frac{\Delta^2}{12} = \frac{3}{2} 2^{2M} \quad (3.8)$$

$$DR|_{dB} \equiv 10\log(\frac{P_s}{P_Q}) = 6.02 \cdot M + 1.76 \quad (3.9)$$

It can be seen that each additional bit in the ideal Nyquist A/D converter results in an increase of approximately 6dB in the A/D conversion DR.

3.1.3 OVERSAMPLING

The oversampling data converters use sampling frequencies (f_s) much higher than the Nyquist frequency (f_N) that is the minimum sampling rate required avoiding aliasing effect. An oversampling ratio (OSR) can be defined to specify the level of oversampling as:

$$OSR \equiv \frac{f_s}{f_N} = \frac{f_s}{2f_B} \quad (3.10)$$

Where f_B is the bandwidth of the signal.

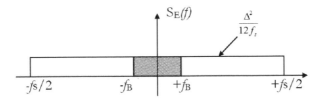

Fig.3.3 Quantization error distribution in oversampling converter

Based on above discussion, the inband power of the quantization error is given as:

$$P_Q = \int_{-f_B}^{+f_B} S_E(f)df = \frac{\Delta^2}{12 \cdot OSR} \qquad (3.11)$$

The DR of the oversampling data converter in term of the OSR can be expressed as:

$$DR = \frac{3}{2} 2^{2M} \cdot OSR \qquad (3.12)$$

$$DR|_{dB} = 10\log(\frac{P_s}{P_Q}) = 6.02 \cdot M + 1.76 + 10\log(OSR) \qquad (3.13)$$

It can be seen that the DR can be improved using higher OSR.

3.1.4 NOISE-SHAPING

For the oversampling data converter shown in figure 3.4, a feedback loop is used to modulate the error transfer function of the data conversion.

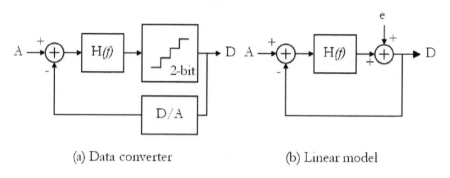

(a) Data converter (b) Linear model

Fig.3.4 Oversampling data converter model

The signal and noise transfer functions of the converter can be derived respectively as:

$$H_s(f) \equiv \frac{D(f)}{A(f)} = \frac{H(f)}{1+H(f)} \qquad (3.14)$$

$$H_N(f) \equiv \frac{D(f)}{N(f)} = \frac{1}{1+H(f)} \quad (3.15)$$

Where H(f) usually has a lowpass frequency response with high dc gains such that Hs(f) and H_N(f) have the lowpass and highpass frequency responses respectively.

For example, for the first order continuous-time modulator employing a lossless integrator as:

$$H = \frac{\omega_o}{s} \quad (3.16)$$

The signal and the noise transfer functions of the converter are given respectively as:

$$H_s(s) = \frac{1}{s/\omega_o + 1} \quad (3.17)$$

$$H_N(s) = \frac{s/\omega_o}{s/\omega_o + 1} \quad (3.18)$$

It has a first-order highpass response for frequency band higher than the characteristic frequency ω_o as shown in figure 3.5.

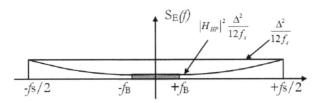

Fig.3.5 Quantization error distribution of noise shaping modulator

The dynamic range of the oversampling data conversion can be further improved by noise shaping. For an M-th order discrete-time sigma-delta modulator with noise transfer function given as:

$$H_N(s) = (1-z^{-1})^N \tag{3.19}$$

The quantization noise power is given as:

$$P_Q = \int_{-f_B}^{+f_B} \frac{\Delta^2}{12 f_s} |H_N(j2\pi f/f_s)|^2 \, df \approx \frac{\Delta^2}{12} \frac{\pi^{2N}}{(2N+1)OSR^{2N+1}} \tag{3.20}$$

The dynamic range of the ideal sigma-delta modulator can be derived as:

$$DR \equiv \frac{(A_{SF}/2)^2/2}{P_Q} \approx \frac{3}{2} 2^{2M} \frac{(2N+1)}{\pi^{2N}} OSR^{2N+1} \tag{3.21}$$

$$DR \approx 6.02M + 1.76 + 10\log(\frac{2N+1}{\pi^{2N}}) + (2N+1)10\log(OSR) \tag{3.22}$$

3.1.5 PERFORMANCE METRICS

A few performance metrics are commonly used to specify a sigma-delta modulator, including the signal-to-noise ratio (SNR), signal-to-noise plus distortion ratio (SNDR), the dynamic range (DR), the effective number of bits (ENOB) and the overload level (X_{OL}).

- Signal-to-Noise Ratio (SNR) is the ratio of output power at the frequency of the input sinusoidal signal to the total in-band noise power at the output. The SNR of an ideal sigma-delta modulator with signal amplitude A_m and the only the quantization error is given approximately as

$$SNR(dB) = 10\log(A_m/(2P_Q)) \tag{3.23}$$

- Signal-to-Noise+Distortion Ratio (SNDR) is the ratio of output power at the frequency of the input sinusoidal signal to the total in-band noise power including both noise and harmonics at the output

- Dynamic Range (DR) is the ratio of output power at the frequency of the input sinusoidal signal with maximum amplitude (i.e. full-scale range) to the output power for a small input for which SNR =0dB.

  For an ideal sigma-delta modulator, the dynamic range can be approximately expressed as:

$$DR(dB) = 10\log((A_{FS}/2)^2/(2P_Q)) \quad (3.24)$$

- Effective Number of Bits (ENOB) is the number of bits needed for an ideal Nyquist-rate converter to achieve the same DR of the sigma-delta converter. The ENOB is related to the DR of the converter as:

$$ENOB = \frac{DR(dB) - 1.76}{6.02} \quad (3.25)$$

- Overload Level (X_{OL}) is used to specify the maximum input signal amplitude (that is lower than full-scale input) that offers the maximum converter SNR as described in figure 3.6.

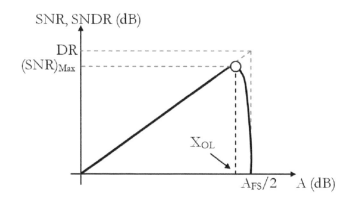

Fig.3.6 Sigma-Delta converter overload level

3.1.6 PERFORMANCE OF IDEAL ΣΔ MODULATOR

For an N-th order discrete-time sigma-delta modulator the output can be expressed as the signal and the noise inputs as:

$$D(z) = z^{-N} A(z) + (1-z^{-1})^N E(z) \qquad (3.26)$$

The dynamic range of such sigma-delta modulator employing oversampling and noise shaping techniques is given as:

$$DR \approx \frac{3}{2}(2^M - 1)^2 \cdot \frac{(2N+1)OSR^{(2N+1)}}{\pi^{2N}} \qquad (3.27)$$

$$DR \approx 20\log(2^M - 1) + 1.76 + 10\log(\frac{2N+1}{\pi^{2N}}) + (2N+1)\cdot 10\log(OSR)$$

$$(3.28)$$

Where M is the bits of the quantizer of the modulator. It can be seen that the order of modulation N, the oversampling ratio OSR and the number of quantization bits determine the dynamic range of the sigma-delta modulation.

The increase in modulation order N leads to high dynamic range. This is achieved by the increased order of noise shaping that push the quantization noise to high frequency band. However such design approach is subject to the modulation stability limitations. We can see the increase of the dynamic range in dB as the result of the modulation order as:

$$\frac{\Delta(DR)}{\Delta N} \approx 20[\frac{1}{\ln 10 \cdot [2N+1]} + \log(\frac{OSR}{\pi})] > 0 \qquad (3.29)$$

Alternatively a higher oversampling ratio can also be used to improve the dynamic range of the converter. The dynamic range improvement as result of higher OSR can be derived as:

$$\frac{\Delta(DR)}{\Delta OSR} \approx (2N+1) \cdot \frac{10}{\ln 10 \cdot OSR} > 0 \qquad (3.30)$$

The major limitations of higher OSR for a given signal band are higher sampling frequency, faster circuit operation speed, and increased power dissipation.

The increase in the quantization resolution using multi-bit quantization also leads to higher modulation dynamic range as the result of the reduction in the quantization error. However such approach requires a multi-bit D/A converter feedback that usually suffers from the linearity limitation of the circuit.

3.2 SIGMA-DELTA MODULATOR ARCHITECTURES

The sigma-delta modulation can be realized in various ways, either in continuous- or discrete-time, employing single- or multiple-feedback loops.

3.2.1 FIRST ORDER MODULATOR

Among the variety of existing sigma-delta modulators, the first order architecture shown in figure 3.7 is the simplest one. It consists of an integrator, an one bit A/D converter and a D/A converter in the feedback path.

The difference between the input and the analog representation of the quantized feedback is integrated in the integrator.

Such a modulator belongs to a nonlinear circuit due to a high baseband gain block (i.e. A/D converter) in the feedforward path.

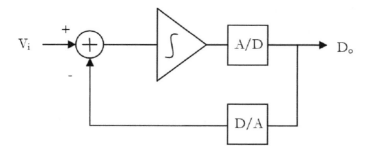

Fig.3.7 1st order continuous-time Sigma-Delta modulator

Shown in figure 3.8 is the linear s-domain model of the first order continuous-time sigma-delta modulator, where parameter k, k_I, F_1 are the input path gain, the integration gain and the feedback gains respectively.

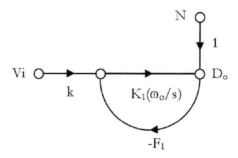

Fig.3.8 S-domain model of 1st order continuous-time Sigma-Delta modulator

The signal and noise transfer functions of this first order sigma-delta modulator under the unity loop gain constrain (i.e. $K_1F_1=1$) are given respectively as:

$$H_s(s) \equiv \frac{D_o(s)}{V_i(s)}\bigg|_{N=0} = \frac{kk_1}{1+\dfrac{s}{\omega_o}} \qquad (3.31)$$

$$H_N(s) \equiv \frac{D_o(s)}{N(s)}\bigg|_{A_i=0} = \frac{\dfrac{s}{\omega_o}}{1+\dfrac{s}{\omega_o}} \qquad (3.32)$$

The frequency responses of the signal and noise transfer functions of the first order continuous-time sigma-delta modulator are given in figure 3.9. It can be seen that the signal transfer function has a lowpass frequency response that allows the inband signal to pass without attenuation. On the other hand, the noise transfer function has a high pass frequency response. Therefore the quantization error is pushed out to the high frequency band by so called noise-shaping, separated from the inband signal.

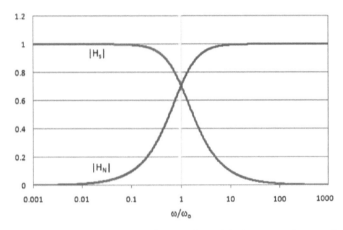

Fig. 3.9 Frequency responses of 1st order CT Sigma-Delta modulator

The first order sigma-delta modulator can also be realized in discrete-time form as shown in figure 3.10. Such a modulator can be modeled in z-domain linear model as shown in figure 3.11.

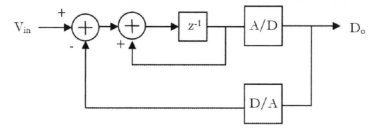

Fig.3.10 1st order discrete-time Sigma-Delta modulator

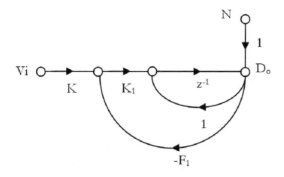

Fig.3.11 Z-domain model of 1st order discreet-time Sigma-Delta modulator

The signal and noise transfer functions of the first order discrete-time sigma-delta modulator under the unity loop gain constraint (i.e. $K_1F_1 = 1$) are given respectively as:

$$H_s(z) \equiv \frac{D_o(z)}{V_i(z)}\bigg|_{N=0} = KK_1 z^{-1} \qquad (3.33)$$

$$H_N(z) \equiv \frac{D_o(z)}{N(z)}\bigg|_{V_i=0} = 1 - z^{-1} \qquad (3.34)$$

The frequency response of the noise transfer function at low frequency band is given as

$$\left|H_N(z)\right|_{z=e^{-j\omega T}} = 2\left|\sin(\frac{\omega T}{2})\right| \qquad (3.35)$$

The frequency responses of signal and noise transfer functions are plotted in figure 3.12. It can be seen that such circuit offers an allpass frequency response to signal and a highpass frequency response to quantization noise that shapes the quantization noise away from the signal band.

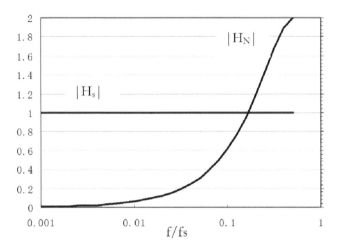

Fig.3.12 Frequency responses of 1st order DT Sigma-Delta modulator

3.2.2 SECOND-ORDER MODULATOR

Shown in figure 3.13 is a second-order continuous-time sigma-delta modulation circuit employing two integrators in the loop.

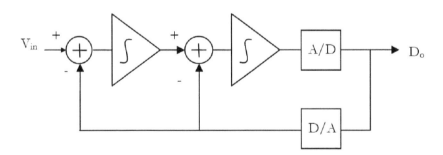

Fig.3.13 2nd order continuous-time Sigma-Delta modulator

A linear s-domain model of such modulator is shown in figure 3.14, where K, K_{I1}, K_{I2}, F_1, F_2 are the gain factors.

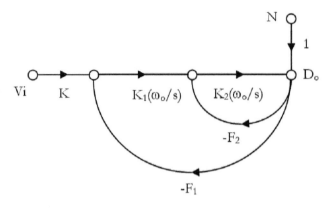

Fig.3.14 S-domain 2nd order CT Sigma-Delta modulator model

The signal and noise transfer functions of the second-order continuous-time sigma-delta modulator are given respectively as:

$$H_s(s) \equiv \frac{D_o(s)}{V_i(s)}\Big|_{N=0} = \frac{K \cdot K_1 K_2}{K_1 K_2 F_1 + K_2 F_2(\frac{s}{\omega_o}) + (\frac{s}{\omega_o})^2} \quad (3.36)$$

$$H_N(s) \equiv \frac{D_o(s)}{N(s)}\Big|_{V_i=0} = \frac{(\frac{s}{\omega_o})^2}{K_1 K_2 F_1 + K_2 F_2(\frac{s}{\omega_o}) + (\frac{s}{\omega_o})^2} \quad (3.37)$$

It can be seen that such a modulator is always stable. By further applying the constraint to the gain factors such that the signal and noise transfer function can be further expressed respectively as:

$$\begin{cases} K_1 K_2 F_1 = 1 \\ K_2 F_2 = 1/Q \end{cases} \quad (3.38)$$

$$H_s(s) \equiv \frac{D_o(s)}{V_i(s)}\Big|_{N=0} = \frac{K \cdot K_1 K_2}{1 + \frac{1}{Q}(\frac{s}{\omega_o}) + (\frac{s}{\omega_o})^2} \quad (3.39)$$

$$H_N(s) \equiv \frac{D_o(s)}{N(s)}\Big|_{V_i=0} = \frac{(\frac{s}{\omega_o})^2}{1 + \frac{1}{Q}(\frac{s}{\omega_o}) + (\frac{s}{\omega_o})^2} \quad (3.40)$$

Shown in figure 3.15 are the frequency responses of the signal and noise transfer functions of above second-order continuous-time sigma-delta modulator. It can be seen that such circuit also offers a lowpass frequency response for signal and a highpass frequency response for the quantization noise.

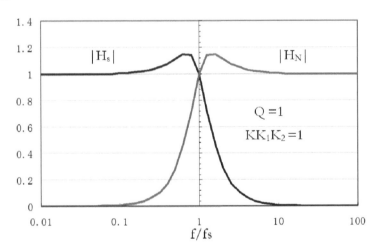

Fig.3.15 Frequency responses of 2nd order DT Sigma-Delta modulator

Note that the Q factor of such modulator can be used to control the roll off and the peaking responses of the signal and noise transfer functions.

A second-order discrete-time sigma-delta modulator is shown in figure 3.16. Such circuit uses two discrete-time integrators with K, K_1, K_2, F_1, and F_2 as the gain factors.

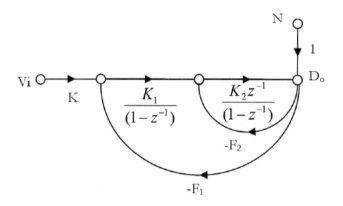

Fig.3.16 Z-domain 2nd order discrete-time Sigma-Delta modulator model

The signal and noise transfer functions of such circuit can be derived respectively as:

$$H_s(s) \equiv \frac{D_o(s)}{V_i(s)}\bigg|_{N=0} = \frac{K \cdot K_1 K_2 z^{-1}}{1+(F_2 K_2 + F_1 K_2 K_1 - 2)z^{-1} + (1 - F_2 K_2)z^{-2}}$$
(3.41)

$$H_N(s) \equiv \frac{D_o(s)}{N(s)}\bigg|_{V_i=0} = \frac{(1-z^{-1})^2}{1+(F_2 K_2 + F_1 K_2 K_1 - 2)z^{-1} + (1 - F_2 K_2)z^{-2}}$$
(3.42)

We may apply the following gain constraints for the second-order noise shaping response:

$$\begin{cases} F_2 K_2 = 1 \\ F_1 K_1 K_2 = 1 \end{cases}$$
(3.43)

The signal and noise transfer function under such design constraints are given as

$$H_s(s) = K \cdot K_1 K_2 z^{-1} \qquad (3.44)$$

$$H_N(s) = (1-z^{-1})^2 \qquad (3.45)$$

This noise transfer function has a second-order highpass frequency response to the quantization noise that will push the noise away from the inband signal.

3.2.3 HIGH ORDER MODULATION TOPOLOGIES

Shown in figure 3.17 is a general high order continuous-time sigma-delta modulator with distributed feedbacks.

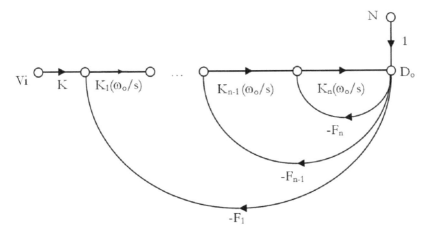

Fig.3.17 S-domain high order CT Sigma-Delta modulator model

Such a circuit can be converted into an equivalent circuit as shown in figure 3.18 where

$$\begin{cases} G \equiv K \cdot K_1 \cdot K_2 ... K_n \\ G_1 \equiv F_1 \cdot K_1 \cdot K_2 ... K_n \\ G_2 \equiv F_2 \cdot K_2 ... K_n \\ ... \\ G_n \equiv F_n \cdot K_n \end{cases} \quad (3.46)$$

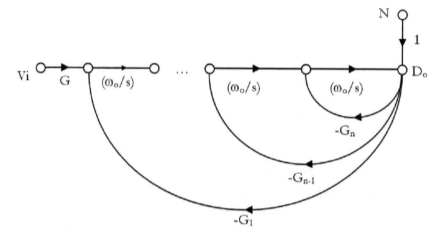

Fig.3.18 S-domain high-order CT Sigma-Delta modulator model

The signal and noise transfer functions of such modulator can be expressed respectively as:

$$H_s(s) \equiv \frac{D_o(s)}{V_i(s)}\bigg|_{N=0} = \frac{G}{G_1 + G_2(\frac{s}{\omega_o}) + ... + G_n(\frac{s}{\omega_o})^{n-1} + (\frac{s}{\omega_o})^n} \quad (3.47)$$

$$H_N(s) \equiv \frac{D_o(s)}{N(s)}\bigg|_{V_i=0} = \frac{(\frac{s}{\omega_o})^n}{G_1 + G_2(\frac{s}{\omega_o}) + ... + G_n(\frac{s}{\omega_o})^{n-1} + (\frac{s}{\omega_o})^n} \quad (3.48)$$

It can be seen that the signal transfer function has an n-th order lowpass frequency response. On the other hand, the noise transfer function has an n-th order highpass frequency response.

Similarly the above high order sigma-delta modulator can also be implemented in the discrete-time form as shown in figure 3.19. This circuit can be transformed into the equivalent circuit shown in figure 3.20.

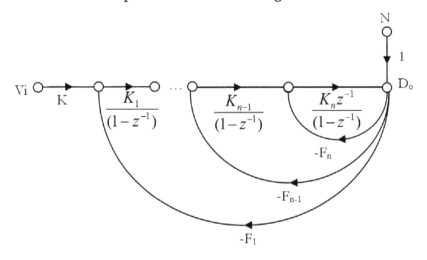

Fig. 3.19 Z-domain high-order discrete-time Sigma-Delta modulator model

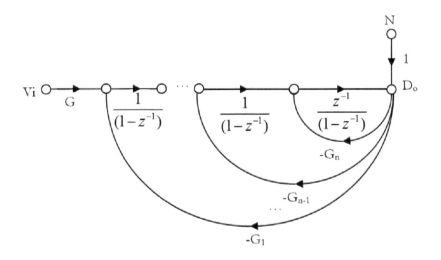

Fig. 3.20 Equivalent high-order discrete-time Sigma-Delta modulator model

It can be proved that such circuit offers an M-th order noise shaping response given as:

$$H_N(s) \equiv \frac{D_o(s)}{N(s)}\Big|_{V_i=0} \propto (1-z^{-1})^N \qquad (3.49)$$

The signal transfer function of such circuit can be simplified by applying the gain constraints as:

$$\begin{cases} G \equiv K \cdot K_1 \cdot K_2 ... K_n = 1 \\ G_1 \equiv F_1 \cdot K_1 \cdot K_2 ... K_n = 1 \\ G_2 \equiv F_2 \cdot K_2 ... K_n = 1 \\ ... \\ G_n \equiv F_n \cdot K_n = 1 \end{cases} \qquad (3.50)$$

$$H_s(s) \equiv \frac{D_o(s)}{V_i(s)}\Big|_{N=0} = z^{-1} \qquad (3.51)$$

An alternative single-path M-th order sigma-delta modulator structure is show in figure 3.21, where G can be continuous- or discrete-time integrators for the continuous- or discrete-time modulators respectively.

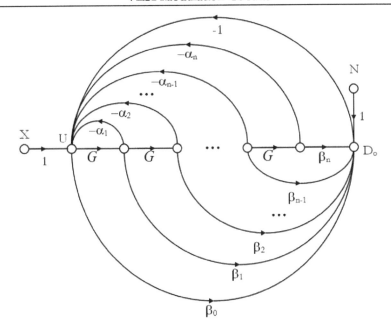

Fig.3.21 Alternative high-order Sigma-Delta modulator structure

3.2.4 CASCADE MODULATION TOPOLOGIES

The cascade sigma-delta modulation topology can be used to improve the circuit stability for high order sigma-delta converter implementation. The cascade sigma-delta modulations are also known as the multi-stage sigma-delta (MASH).

In the MASH sigma-delta modulation shown in figure 3.22, each stage re-modulates a signal containing the quantization error generated in the previous stage. In the digital domain, the outputs of the stages are properly processed based on certain DSP algorithm to cancel out the quantization errors of all the stages except the last one in the cascade. This provides a high order noise shaping using multiple stages.

Since the feedback loops in the MASH modulator are localized, unconditional stability can be achieved for high order modulations constructed based on the first and second-order modulation segments. Such approach offers noise shaping performance similar to high order modulation without stability issue.

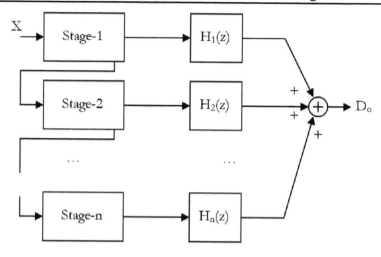

Fig.3.22 MASH Sigma-Delta modulator structure

Shown in figure 3.23 is a MASH modulator example that is based on the cascade of a second-order modulator with a first order modulator. Such a circuit effectively realizes a 3rd order sigma-delta modulation.

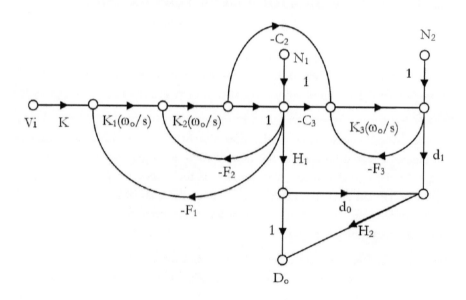

Fig.3.23 3rd order 2-stage 2-1 MASH modulator model

3.2.5 MULTI-BIT MODULATION

The multi-bit quantization structure as shown in figure 3.24 offers an alternative way to improve the sigma-delta modulator dynamic range, where multi-bit quantization reduces the quantization error.

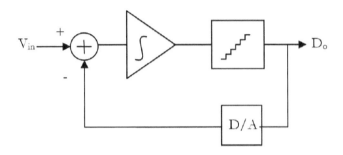

Fig.3.24 Multi-bit Sigma-Delta modulator

3.2.6 OTHER CIRCUIT TECHNIQUES

Note that other circuit techniques, such as the digital correction, the dynamic matching, the dual quantization and the parallel modulation can also be used to improve sigma-delta modulation performances.

3.3 SIGMA-DELTA D/A CONVERSION

In additional to the sigma-delta A/D conversion, the sigma-delta modulation can also be used for the D/A conversion operations.

3.3.1 FIRST-ORDER MODULATION

Show in figure 3.25 is a first-order digital sigma-delta modulator circuit structure.

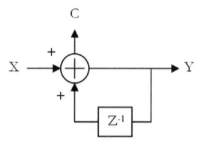

Fig.3.25 First order digital Sigma-Delta modulator

The transfer function of such circuit can be expressed as:

$$C + Y = X + z^{-1}Y \tag{3.52}$$

Such equation can be re-written as

$$C = X - (1 - z^{-1})Y \tag{3.53}$$

Where C is a modulated representation of input X with added noise floor $(z^{-1}-1)$ Y.

Similar to other sigma-delta modulator, this term is small and negligible when the oversampling condition is applied:

$$|(1-z^{-1})Y|_{\omega T \to 0} = |2\sin(\omega T/2)Y|_{\omega T \to 0} \to 0 \tag{3.54}$$

3.3.2 HIGH-ORDER MODULATION

The first-order digital sigma-delta modulation can be extended into high-order modulation circuit as shown in figure 3.26.

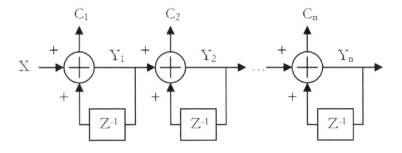

Fig.3.26 High-order digital Sigma-Delta modulator

The relationship of the input to the output parameters for such circuit can be expressed using a group of equations as:

$$\begin{cases} C_1 = X - (1-z^{-1})Y_1 \\ C_2 = Y_1 - (1-z^{-1})Y_2 \\ \dots \\ C_n = Y_{n-1} - (1-z^{-1})Y_n \end{cases} \quad (3.55)$$

Such equations can be re-written as:

$$\begin{cases} C = X - (1-z^{-1})^n Y_n \\ C \equiv C_1 + C_2(1-z^{-1}) + \dots + C_n(1-z^{-1})^{n-1} \end{cases} \quad (3.56)$$

It can be seen that the output signal C is a high-order sigma-delta modulated signal of the input X with the noise floor as:

$$|(1-z^{-1})^n Y_n|_{\omega T \to 0} = |2^n \sin^n(\omega T/2) Y_n|_{\omega T \to 0} \to 0 \qquad (3.57)$$

The above equation can be realized based on the circuit shown in figure 3.27.

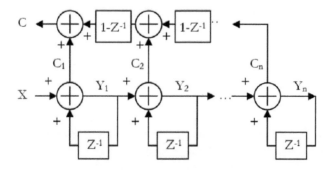

Fig.3.27 Completed high-order digital Sigma-Delta modulator

3.3.3 MODULATOR IMPLEMENTATION

Show in figure 3.28 is a VLSI circuit implementation of the first-order digital sigma-delta modulator circuit structure using a full adder with the feedback.

VLSI Modulation Circuits

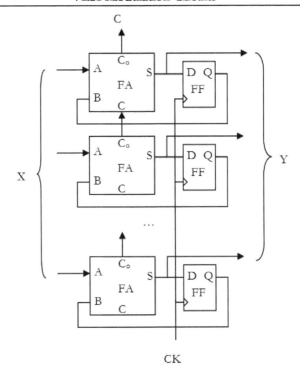

Fig.3.28 VLSI first-order digital Sigma-Delta modulator

In practical applications, multiple first-order modulators can be connected in array to form the high-order digital modulator as shown in figure 3.29.

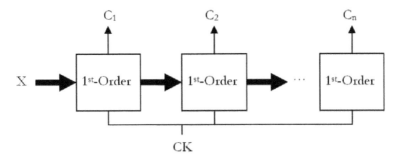

Fig.3.29 VLSI high-order digital Sigma-Delta modulator

Reference:

[1] R. B. Staszewski, J. L. Wallberg, S. Rezeq, Chih-Ming Hung, O. E. Eliezer, S. K. Vemulapalli, C. Fernando, K. Maggio, R. Staszewski, N. Barton, Meng-Chang Lee; P. Cruise, M. Entezari, K. Muhammad, D. Leipold, "All-digital PLL and transmitter for mobile phones," IEEE Journal of Solid-State Circuits, Volume: 40, Issue: 12, Page(s): 2469 - 2482, 2005.

[2] M. H. Perrott, T. L. Tewksbury, C. G. Sodini, "A 27-mW CMOS fractional-N synthesizer using digital compensation for 2.5-Mb/s GFSK modulation," IEEE Journal of Solid-State Circuits, Volume: 32, Issue: 12, Page(s): 2048 - 2060, 1997.

[3] B. E. Boser, B. A. Wooley, "The design of sigma-delta modulation analog-to-digital converters," IEEE Journal of Solid-State Circuits, Volume: 23, Issue: 6, Page(s): 1298 - 1308, 1988.

[4] R. T. Baird, T. S. Fiez, "Linearity enhancement of multibit Δ Sigma; A/D and D/A converters using data weighted averaging," IEEE Transactions on Circuits and Systems II: Analog and Digital Signal Processing, Volume: 42, Issue: 12, Page(s): 753 - 762, 1995.

[5] W. Rhee, B.-S. Song, A. Ali, "A 1.1-GHz CMOS fractional-N frequency synthesizer with a 3-b third-order Δ Sigma; modulator," IEEE Journal of Solid-State Circuits, Volume: 35, Issue: 10, 1453 - 1460. 2000.

[6] S. Rabii, B. A. Wooley, "A 1.8-V digital-audio sigma-delta modulator in 0.8-μm CMOS," IEEE Journal of Solid-State Circuits, Volume: 32, Issue: 6, Page(s): 783 - 796, 1997.

[7] P. Malcovati, S. Brigati, F. Francesconi, F. Maloberti, P. Cusinato, Baschirotto, "A Behavioral modeling of switched-capacitor sigma-delta modulators," IEEE Transactions on circuits and Systems I: Fundamental Theory and Applications, Volume: 50, Issue: 3, Page(s): 352 - 364, 2003.

[8] S. Yan, E. Sanchez-Sinencio, "A continuous-time sigma-delta modulator with 88-dB dynamic range and 1.1-MHz signal bandwidth," IEEE Journal of Solid-State Circuits, Volume: 39, Issue: 1, Page(s): 75 - 86, 2004.

[9] T. L. Brooks, D. H. Robertson, D. F. Kelly, A. Del Muro, S. W. Harston, "A cascaded sigma-delta pipeline A/D converter with 1.25 MHz signal bandwidth and 89 dB SNR," IEEE Journal of Solid-State Circuits, Volume: 32, Issue: 12, Page(s): 1896 - 1906, 1997.

[10] S. A. Jantzi, K. W. Martin, A. S. Sedra, "Quadrature bandpass &Delta & Sigma; modulation for Digital Radio," IEEE Journal of Solid-State Circuits, Volume: 32, Issue: 12, Page(s): 1935 - 1950, 1997.

[11] Keliu Shu; Sanchez-Sinencio, E.; Silva-Martinez, J.; Embabi, S.H.K.; "A 2.4-GHz monolithic fractional-N frequency synthesizer with robust phase-switching prescaler and loop capacitance multiplier," IEEE Journal of Solid-State Circuits, Volume: 38, Issue: 6, Page(s): 866 - 874, 2003.

[12] B. H. Leung, S. Sutarja, "Multibit Σ-Δ A/D converter incorporating a novel class of dynamic element matching techniques," IEEE Transactions on Circuits and Systems II: Analog and Digital Signal Processing, Volume: 39, Issue: 1, Page(s): 35 - 51, 1992.

[13] V. Peluso, P. Vancorenland, A. M. Marques, M. S. J. Steyaert, W. Sansen, "A 900-mV low-power &Delta Sigma; A/D converter with 77-dB dynamic range," IEEE Journal of Solid-State Circuits, Volume: 33, Issue: 12, Page(s): 1887 - 1897, 1998.

[14] R. Schreier, B. Zhang, "Delta-sigma modulators employing continuous-time circuitry," *IEEE* Transactions on Circuits and Systems I: Fundamental Theory and Applications, Volume: 43, Issue: 4, Page(s): 324 - 332, 1996.

[15] Libin Yao, M. S. J. Steyaert, W. Sansen, "A 1-V 140-μW 88-dB audio sigma-delta modulator in 90-nm CMOS," IEEE Journal of Solid-State Circuits, Volume: 39, Issue: 11, Page(s): 1809 - 1818, 2004.

[16] Li Yu, W. M. Snelgrove, "A novel adaptive mismatch cancellation system for quadrature IF radio receivers," IEEE Transactions on Circuits and Systems II: Analog and Digital Signal Processing, Volume: 46, Issue: 6, Page(s): 789 - 801, 1999.

[17] A. Marques, V. Peluso, M. S. Steyaert, W. M. Sansen, "Optimal parameters for Delta Sigma; Modulator Topologies," IEEE Transactions

on Circuits and Systems II: Analog and Digital Signal Processing, Volume: 45, Issue: 9, Page(s): 1232 - 1241, 1998.

[18] B. P. Brandt, B. A. Wooley, "A 50-MHz multibit sigma-delta modulator for 12-b 2-MHz A/D Conversion," IEEE Journal of Solid-State Circuits, Volume: 26, Issue: 12, Page(s): 1746 - 1756, 1991.

[19] E. J. Vander Zwan, E. C. Dijkmans, "A 0.2-mW CMOS sigma-deltaModulator for Speech Coding with 80 dB Dynamic Range," IEEE Journal of Solid-State Circuits, Volume: 31, Issue: 12, Page(s): 1873 - 1880, 1996.

[20] M. R. Miller, C. S. Petrie, "A multibit sigma-delta ADC for multimode receivers," IEEE Journal of Solid-State Circuits, Volume: 38, Issue: 3, Page(s): 475 - 482, 2003.

[21] A. Jayaraman, P. F. Chen, G. Hanington, L. Larson, P. Asbeck, "Linear high-efficiency microwave power amplifiers using bandpass delta-sigma modulators," IEEE Microwave and Guided Wave Letters, Volume: 8, Issue: 3, Page(s): 121 - 123, 1998.

[22] L. J. Breems, R. Rutten, G. Wetzker, "A Cascaded Continuous-time sigma-deltaModulator with 67-dB Dynamic Range in 10-MHz bandwidth," IEEE Journal of Solid-State Circuits, Volume: 39, Issue: 12, Page(s): 2152 - 2160, 2004.

[23] B. Dufort, G. W. Roberts, "On-chip analog signal generation for mixed-signal built-in self-test," IEEE Journal of Solid-State Circuits, Volume: 34, Issue: 3, Page(s): 318 - 330, 1999.

[24] R. Adams, K. Q. Nguyen, "A 113-dB SNR oversampling DAC with segmented noise-shaped scrambling," IEEE Journal of Solid-State Circuits, Volume: 33, Issue: 12, Page(s): 1871 - 1878, 1998.

[25] Ki Young Nam; Sang-Min Lee; D. K. Su, Wooley, B.A.; "A low-voltage low-power sigma-delta modulator for broadband analog-to-digital conversion," IEEE Journal of Solid-State Circuits, Volume: 40, Issue: 9, Page(s): 1855 - 1864, 2005.

[26] E. Roza, "Analog-to-digital conversion via duty-cycle modulation," IEEE Transactions on Circuits and Systems II: Analog and Digital Signal Processing, Volume: 44, Issue: 11 Page(s): 907 - 914, 1997.

[27] R. H. M. Van Veldhoven, "A triple-mode continuous-time sigma-deltamodulator with switched-capacitor feedback DAC for a GSM-EDGE/CDMA2000/UMTS receiver," IEEE Journal of Solid-State Circuits, Volume: 38, Issue: 12, Page(s): 2069 - 2076, 2003.

[28] O. Shoaei, W. M. Snelgrove, "Design and implementation of a tunable 40 MHz-70 MHz Gm-C bandpass Delta Sigma modulator," IEEE Transactions on Circuits and Systems II: Analog and Digital Signal Processing, Volume: 44, Issue: 7, Page(s): 521 - 530, 1997.

[29] M. Sarhang-Nejad, G. C. Temes, "A high-resolution multibit Σ Δ ADC with digital correction and relaxed amplifier requirements," IEEE Journal of Solid-State Circuits, Volume: 28, Issue: 6, Page(s): 648 - 660, 1993.

[30] Ong, A.K.; Wooley, B.A.; "A two-path bandpass sigma-deltaModulator for digital IF extraction at 20 MHz," IEEE Journal of Solid-State Circuits, Volume: 32, Issue: 12, Page(s): 1920 - 1934, 1997.

[31] J. Candy, O. Benjamin, "The Structure of Quantization Noise from sigma-delta Modulation," IEEE Transactions on Communications, Volume: 29, Issue: 9, Page(s): 1316 - 1323, 1981.

[32] Van Engelen, J.A.E.P.; Van De Plassche, R.J.; Stikvoort, E.; Venes, A.G.; "A sixth-order continuous-time bandpass sigma-delta modulator for digital radio IF," IEEE Journal of Solid-State Circuits, Volume: 34, Issue: 12, Page(s): 1753 - 1764, 1999.

[33] Breems, L.J.; van der Zwan, E.J.; Huijsing, J.H.; "A 1.8-mW CMOS sigma-deltaModulator with integrated mixer for A/D conversion of IF signals," IEEE Journal of Solid-State Circuits, Volume: 35, Issue: 4, Page(s): 468 - 475, 2000.

[34] Leslie, T.C.; Singh, B.; "An improved sigma-delta modulator architecture, "1990 IEEE International Symposium on Circuits and Systems, Page(s): 372 - 375 vol.1

[35] Keskin, M.; Un-Ku Moon; Temes, G.C.; "A 1-V 10-MHz clock-rate 13-bit CMOS Delta Sigma; modulator using unity-gain-reset op amps," IEEE Journal of Solid-State Circuits, Volume: 37, Issue: 7, Page(s): 817 - 824, 2002.

[36] Jantzi, S.A.; Snelgrove, W.M.; Ferguson, P.F., Jr.; "A fourth-order bandpass sigma-delta modulator," IEEE Journal of Solid-State Circuits, Volume: 28, Issue: 3, Page(s): 282 - 291, 1993.

[37] Tai-Haur Kuo; Kuan-Dar Chen; Horng-Ru Yeng; "A wideband CMOS sigma-delta modulator with incremental data weighted averaging," IEEE Journal of Solid-State Circuits, Volume: 37, Issue: 1, Page(s): 11 - 17, 2002.

[38] Filiol, N.M.; Riley, T.A.D.; Plett, C.; Copeland, M.A.; "An agile ISM band frequency synthesizer with built-in GMSK data modulation," IEEE Journal of Solid-State Circuits, Volume: 33, Issue: 7, Page(s): 998 - 1008, 1998.

[39] Williams, L.A., III; Wooley, B.A.; "A third-order sigma-delta modulator with extended dynamic range," IEEE Journal of Solid-State Circuits, Volume: 29, Issue: 3, Page(s): 193 - 202, 1994.

[40] Galton, I.; Jensen, H.T.; "Oversampling parallel delta-sigma modulator A/D conversion," IEEE Transactions on Circuits and Systems II: Analog and Digital Signal Processing, Volume: 43, Issue: 12, Page(s): 801 - 810, 1996.

[41] Sauerbrey, J.; Tille, T.; Schmitt-Landsiedel, D.; Thewes, R.; "A 0.7-V MOSFET-only switched-opamp sigma-deltaModulator in standard digital CMOS technology," IEEE Journal of Solid-State Circuits, Volume: 37, Issue: 12, Page(s): 1662 - 1669, 2002.

[42] Gray, R.; "Oversampled sigma-delta Modulation," IEEE Transactions on Communications, Volume: 35, Issue: 5, Page(s): 481 - 489, 1987.

[43] Candy, J.; "A Use of Double Integration in sigma-delta Modulation," IEEE Transactions on Communications, Volume: 33, Issue: 3, Page(s): 249 - 258, 1985

[44] Galton, I.; Jensen, H.T.; "Delta-Sigma modulator based A/D conversion without oversampling," IEEE Transactions on Circuits and Systems II: Analog and Digital Signal Processing, Volume: 42, Issue: 12, Page(s): 773 - 784, 1995.

[45] He, N.; Kuhlman, F.; Buzo, A.; "Multi-loop sigma-delta quantization," IEEE Transactions on Information Theory, Volume: 38, Issue: 3, Page(s): 1015 - 1028, 1992.

[46] Ribner, D.B.; "A comparison of modulator networks for high-order oversampled Σ Δ analog-to-digital converters," IEEE Transactions on Circuits and Systems, Volume: 38, Issue: 2, Page(s): 145 - 159, 1991.

[47] Gray, R.M.; Chou, W.; Wong, P.W.; "Quantization noise in single-loop sigma-delta modulation with sinusoidal inputs," IEEE Transactions on Communications, Volume: 37, Issue: 9, Page(s): 956 - 968, 1989.

[48] Bazarjani, S.; Snelgrove, W.M.; "A 160-MHz fourth-order double-sampled SC bandpass sigma-delta modulator," IEEE Transactions on Circuits and Systems II: Analog and Digital Signal Processing, Volume: 45, Issue: 5, Page(s): 547 - 555, 1998.

[49] Reddy, K.; Pavan, S.; "Fundamental Limitations of Continuous-Time Delta–Sigma Modulators Due to Clock Jitter," IEEE Transactions on Circuits and Systems I: Regular Papers, Volume: 54, Issue: 10, Page(s): 2184 - 2194, 2007.

[50] Gerfers, F.; Ortmanns, M.; Manoli, Y.; "A 1.5-V 12-bit power-efficient continuous-time third-order sigma-deltaModulator," IEEE Journal of Solid-State Circuits, Volume: 38, Issue: 8, Page(s): 1343 - 1352, 2003.

[51] Benabes, P.; Keramat, M.; Kielbasa, R.; "A methodology for designing continuous-time sigma-delta modulators," 1997 ED&TC 97. Proceedings European Design and Test Conference, Page(s): 46 - 50.

[52] Brandt, B.P.; Wingard, D.E.; Wooley, B.A.; "Second-order sigma-delta modulation for digital-audio signal acquisition," IEEE Journal of Solid-State Circuits, Volume: 26, Issue: 4, Page(s): 618 - 627, 1991.

[53] Hongjiang Song, "Current Mode Approaches to Low Voltage/Power VLSI Design for Portable Mixed Signal System," Ph.D. Thesis, 1996.

CHAPTER 4
VLSI CLASS-D AMPLIFIER CIRCUITS

The class-D amplifiers represent the analog signal voltage by fast switching the output signal among two or more discrete signal levels using a PWM or PCM modulation where the switching rates are much faster than that of the signal bands. The modulated analog signal can then be amplified and delivered to the load device (such as the loudspeaker) in full digital form. A filter network will be used to filter-out the carrier and recover the original signal with very high signal quality and power efficiency. The class-D amplifiers may offer better than 90% efficiency at compact size and light weight for large power amplification factors, where high efficiency is achieved by operating the output stages in switching modes that are either fully on or fully off, thereby minimizing the power dissipation. The switching (or digital) operation of the class-D output stage also avoids the requirement of linear characteristic of power device to achieve high amplification linearity. Such a signal amplification approach may provide all of the advantages of class-AB amplifiers (i.e., good linearity and minimal board-space requirements) with the added bonus of higher power efficiency thus making them very suitable for numerous VLSI circuit applications.

4.1 CLASS-D AMPLIFIER PRINCIPLE

A class-D amplifier offers superior efficiency for power amplification that can be described by the conceptual amplify model shown in figure 4.1a where R_L is the load resistor and Z_C the equivalent impedance of the power amplifier.

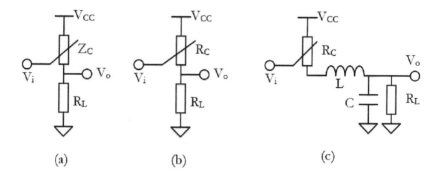

Fig.4.1 Conceptual power amplifier model

For purely resistive driver output impedance as shown in figure 4.1b, the output voltage of the amplifier can be expressed in terms of the load resistance, the power amplifier equivalent resistance and the supply voltage as:

$$V_o(t) = \frac{R_L}{R_L + R_C(t)} V_{cc} \tag{4.1}$$

As the output voltage is controlled by the input signal through varying the impedance R_C of the circuit, the power efficiency of this amplifier can be derived as:

$$\eta = \frac{V_o \cdot I_L}{V_{cc} \cdot I_L} = \frac{R_L}{R_L + R_C(t)} \tag{4.2}$$

Since the output voltage will vary around the common-mode voltage (e.g. Vcc/2), the power efficiency of this conceptual amplifier will typically be less than 50%. A method to improve this amplifier power efficiency is to add the energy storage components to the circuit as show in figure 4.1c. The voltage drop across the lossy part of the amplifier impedance Zc can be reduced to a fraction of total voltage drop in this amplifier operation. The power efficiency of this amplifier can then be significantly improved:

$$\eta = \frac{V_o \cdot I_L}{V_{CC} \cdot I_L} = \frac{R_L}{R_L + R_C(t)} \Big|_{R_C \to 0} \to 100\% \quad (4.3)$$

In this operation, the resistance Rc works in the switching mode with very low on resistance and very high off resistance. Such an operation is the basis of a typical class-D amplifier where the driver devices operate in the switching mode to provide high power efficiency and high signal amplification quality.

The class-D amplifiers commonly use the pulse width modulation (PWM) or the single-bit pulse coded modulation (PCM) as shown in figure 4.2 to control the on/off phases of the amplifier equivalent resistors that allows the pulse density of the output voltage signal before the LC filter proportional to the input signal magnitude.

The PWM allows the pulse density of the signal continuously varied through the changing pulse width of the signal as shown in figure 4.2b. On the other hand, a PCM signal can only provide the discretely adjustable pulse density synchronized to a control clock.

The PWM signal involves the modulation of its duty-cycle, to either convey information over a communication channel or to control the amount of power sent to a load. The PWM uses a rectangular pulse wave whose pulse width represents the analog signal value.

If we consider a pulse waveform f(t) with a low value V_{min}, a high value V_{max} and a duty-cycle k, the average value of the waveform is given by:

$$\overline{f(t)} = V_{min} + kV_{max} \quad (4.4)$$

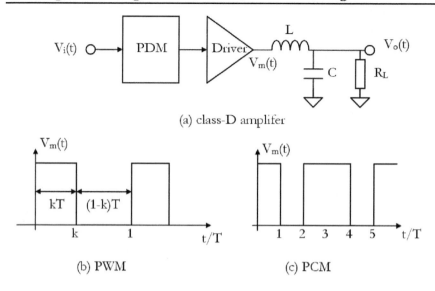

Fig.4.2 PDM for class-D amplification

It is obvious that the average value of the signal f(t) is directly related to the duty-cycle k of the PWM signal.

In the PCM signal case, the analog signal can be represented by the number of "1" pulses divided by the total clock period that can be extracted using a lowpass filter.

Due to their ideally perfect efficiency and linearity, the class-D amplifiers have become a very attractive solution to implement power amplification for audio band signals and now even RF signals. Shown in figure 4.3 are normalized power efficiency curves for the typical power amplifier families reported in literatures. It can be seen that class-D amplifiers may offer significantly higher power efficiency compared with the class-A and class-B amplifier counterparts.

VLSI Modulation Circuits

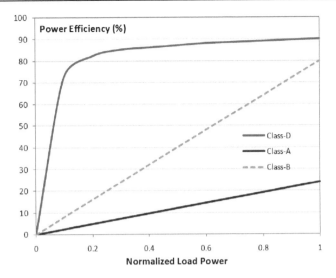

Fig. 4.3 Efficiency of typical power amplifier families

4.2 ANALOG CLASS-D AMPLIFIER ARCHITECTURES

The class-D amplifier circuits offer a very efficient way of power amplification by switching between fully on and fully off at frequency much faster than the bandwidth of the signal and the load devices. The class-D amplifiers with analog inputs can be implemented in either the open-loop or the closed-loop configurations. The signal sources to the class-D amplifiers can be either in the analog or the digital signal forms.

4.2.1 OPEN-LOOP ANALOG CLASS-D AMPLIFIERS

While there are a variety of modulator topologies for the class-D amplifiers, the most basic topology utilizes the pulse-width modulation (PWM) with a triangle-wave (or sawtooth) oscillator. Figure 4.4 shows a simplified block diagram of a PWM-based, half-bridge class-D amplifier. It consists of a pulse-width modulator, two output MOS devices, and an external lowpass filter to recover the amplified audio signal. The driver PMOS and NMOS devices operate as the current-steering switches by alternately connecting the output node to Vcc and to the ground. Because the output is either forced to Vcc or ground, the resulting output of a class-D amplifier is a high-frequency square wave.

The switching frequency (f_{SW}) for most class-D amplifiers is typically between 250 kHz to 1.5MHz for a typical 20 kHz audio signal band. The output square wave is pulse-width modulated by the input audio signal. The PWM is accomplished by comparing the input audio signal to an internally generated triangle-wave (or sawtooth) oscillator. This type of modulation is also often referred to as "natural sampling" where the triangle-wave oscillator acts as the sampling clock. The resulting duty-cycle of the square wave is proportional to the level of the input signal.

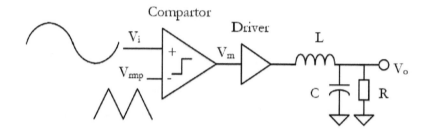

Fig.4.4 Open-loop analog class-D amplifier circuit

The generation of PWM signal using a sawtooth signal and a comparator is shown in figure 4.5.

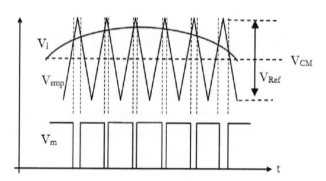

Fig.4.5 Analog PWM modulation

The duty-cycle for the analog signal input V(t) can be derived in term of the ramp signal parameter V_{CM} and V_{Ref} as:

$$k = \frac{V(t) - V_{CM}}{V_{Ref}} \quad (4.5)$$

The analog voltage output at the load after the lowpass filtering is given as:

$$V_o(t) = \frac{V_{cc}}{V_{Ref}}(V(t) - V_{CM}) \quad (4.6)$$

Where Vcc is the power supply voltage of the driver.

The advantage of the above analog class-D amplifier is the circuit simplicity since only one comparator is needed. However such open-loop class-D amplifier suffers from poor PSRR performance since the output analog signal magnitude is directly proportional to the power supply voltage Vcc. The PSRR of such open-loop class-D amplifier circuit can be improved by making the modulation gain (i.e. the ratio of power supply voltage to the magnitude of the ramp signal) relatively constant:

$$V_{cc}/V_{Ref} = cons\tan t \quad (4.7)$$

Shown in figure 4.6 is a linear s-domain SFG model of the open-loop class-D amplifier circuit.

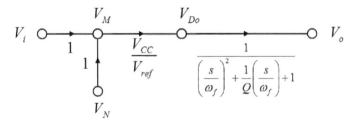

Fig. 4.6 SFG model of open-loop class-D amplifier

Where V_N is the equivalent quantization noise due to PWM modulation. V_{CC}/V_{ref} represents the modulation gain of the class-D amplifier. ω_f and Q are the equivalent resonant frequency and the quality factor of the output lowpass filter that are given approximately as:

$$\omega_f = 1/\sqrt{LC} \qquad (4.8)$$

$$Q = \frac{1}{R}\sqrt{\frac{L}{C}} \qquad (4.9)$$

The signal and noise transfer functions of such open-loop class-D amplifier are given respectively as:

$$H_s(s) \equiv \frac{V_o}{V_i}\bigg|_{V_N=0} = (\frac{V_{CC}}{V_{ref}})\frac{1}{\left(\frac{s}{\omega_f}\right)^2 + \frac{1}{Q}\left(\frac{s}{\omega_f}\right) + 1} \qquad (4.10)$$

$$H_N(s) \equiv \frac{V_o}{V_N}\bigg|_{V_i=0} = (\frac{V_{CC}}{V_{ref}})\frac{1}{\left(\frac{s}{\omega_f}\right)^2 + \frac{1}{Q}\left(\frac{s}{\omega_f}\right) + 1} \qquad (4.11)$$

The signal and noise transfer functions of the circuit are basically the same (no noise shaping), implying that the signal and modulation noise band must be separated before the amplifier to ensure good THD+N at the load. In addition the resonant frequency and the quality factor of the post lowpass filter should be selected such that the useful signal will be able to pass the circuit without significant attenuation and distortion. Since the signal transfer function is directly proportional to the power supply voltage. Such class-D amplifier has poor PSRR as described in early section.

A fully differential circuit implementation of the open-loop class-D amplifier is shown in figure 4.7. The modulator output waveforms of such differential PWM modulation are shown in figure 4.8.

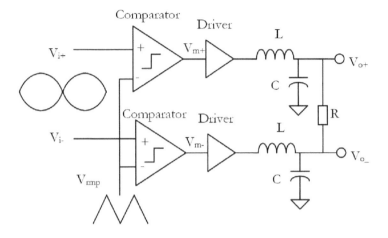

Fig.4.7 Fully differential class-D amplifier circuit

The positive, the negative and the differential duty-cycles of modulation can be derived respectively as:

$$k_+ = \frac{V_{i+}(t) - V_{CM}}{V_{Ref}} \quad (4.12)$$

$$k_- = \frac{V_{i-}(t) - V_{CM}}{V_{Ref}} \quad (4.13)$$

$$k_d = k_+ - k_- = \frac{V_{i+}(t) - V_{i-}(t)}{V_{Ref}} \quad (4.14)$$

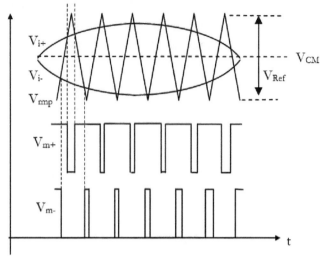

Fig.4.8 Differential analog PWM modulation

Since the load resistor in this circuit is connected in the BTL configuration, this circuit structure can be used to provide much higher output power than the single-end class-D amplifier circuit configuration.

There are four commonly used ways to generate a PWM signal for class-D amplification:

- The pulse center is fixed to the center of the time window and both edges of the pulse moved to compress or expand the width.

- The lead edge is fixed and the tail edge is modulated.

- The tail edge is fixed and the lead edge is modulated.

- The pulse repetition frequency is varied by the signal with the pulse width fixed. Note that this method has a more-restricted range of average output than the other three.

4.2.2 CLOSED-LOOP ANALOG CLASS-D AMPLIFIERS

The closed-loop class-D amplifiers utilize negative feedback from the PWM output back to the input. Such a closed-loop approach not only improves the linearity of the device, but also allows the device to have high power-supply noise rejection. Because the output waveform is sensed and fed back to the input of the amplifier in a closed-loop topology, deviations in the supply rail are detected at the output and corrected by the control loop.

Typical class-D amplifiers operate with a noise-shaping type of feedback loop, which greatly reduces inband noise due to the nonlinearities of the pulse-width modulator, the output stage, and the supply-voltage deviations. This circuit topology is similar to the noise-shaping operation used in sigma-delta modulators. Figure 4.9 shows a simplified block diagram of a first-order noise shaper to illustrate this noise-shaping function. The feedback network typically consists of a resistive network. The transfer function for the integrator has been simplified to equal $1/s$ because the gain of an ideal integrator is inversely proportional to frequency. It is also assumed that the PWM block has a unity-gain and zero-phase-shift contribution to the control loop. Using basic control-block analysis, the following expression can be derived for the output closed-loop class-D amplifier circuit configurations.

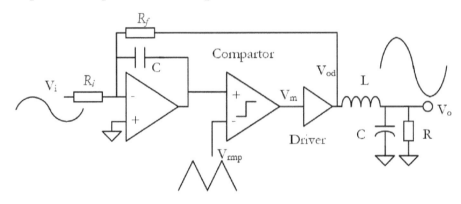

Fig.4.9 First-order closed-loop analog class-D amplifier circuit

The linear SFG model of a closed-loop class-D amplifier circuit is shown in figure 4.10.

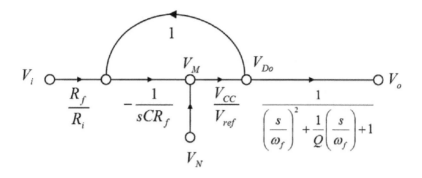

Fig.4.10 SFG model of closed-loop class-D amplifier

The signal and noise transfer functions of this closed-loop class-D amplifier are given respectively as:

$$H_s(s) = \frac{(\frac{R_f}{R_i})}{1+sCR_f(\frac{V_{ref}}{V_{cc}})} \cdot \frac{1}{\left(\frac{s}{\omega_f}\right)^2 + \frac{1}{Q}\left(\frac{s}{\omega_f}\right) + 1} \qquad (4.15)$$

$$H_N(s) = \frac{sCR_f}{1+sCR_f(\frac{V_{ref}}{V_{cc}})} \cdot \frac{1}{\left(\frac{s}{\omega_f}\right)^2 + \frac{1}{Q}\left(\frac{s}{\omega_f}\right) + 1} \qquad (4.16)$$

Based on the above transfer function expressions, we can see that

- The power supply voltage is eliminated from the gain parameter of the signal transfer function, implying an improved PSRR;
- The amplifier gain programming capability using the feedback versus the input resistance ratio;
- The lowpass frequency response of the signal transfer function that allows the useful signal to pass without attenuation and distortion;
- The highpass frequency response of the noise transfer function to shape the modulation noise to higher frequency band for effective filtering.

Shown in figure 4.11 is a fully differential circuit implementation of the first-order closed-loop class-D amplifier that employs a BTL load configuration.

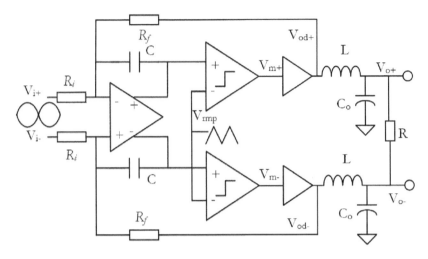

Fig.4.11 Fully differential first order closed-loop class-D amplifier circuit

The closed-loop class-D amplifier circuit can also be implemented using a high order feedback loop as shown in figure 4.12.

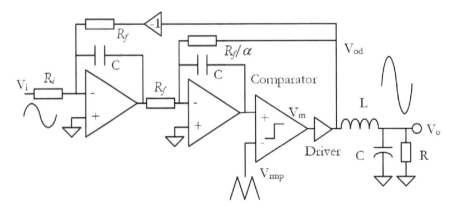

Fig.4.12 Second-order closed-loop analog class-D amplifier circuit

Such a circuit consists of two integrators to provide second-order noise shaping operation. A SFG model of such second-order class-D amplifier is shown in figure 4.13.

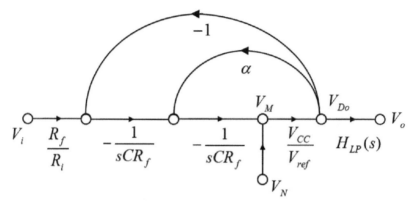

Fig.4.13 SFG model of closed-loop class-D amplifier

Where $H_{LP}(s)$ is the post lowpass filter transfer function given as

$$H_{LP}(s) = \frac{1}{\left(\frac{s}{\omega_f}\right)^2 + \frac{1}{Q}\left(\frac{s}{\omega_f}\right) + 1} \quad (4.17)$$

The signal and noise transfer functions of such second-order class-D amplifier are given respectively as:

$$H_s(s) = \frac{\left(\frac{R_f}{R_i}\right)}{[sCR_f]^2 \left(\frac{V_{ref}}{V_{CC}}\right) + \alpha[sCR_f] + 1} \cdot H_{LP}(s) \quad (4.18)$$

$$H_N(s) = \frac{[sCR_f]^2}{[sCR_f]^2 \left(\frac{V_{ref}}{V_{CC}}\right) + \alpha[sCR_f] + 1} \cdot H_{LP}(s) \quad (4.19)$$

This class-D amplifier circuit offers a few important properties:

- A higher PSRR since the gain parameter of the signal transfer function is independent of power supply voltage;
- More flexible gain programming capability using the feedback versus the input resistance ratio;
- A minimized inband signal loss and distortion with lowpass frequency response of the signal transfer function;
- A second-order noise shaping with highpass frequency response of the noise transfer function for effective filtering.

The second-order class-D amplifier circuit can also be implemented in the circuit form as shown in figure 4.14.

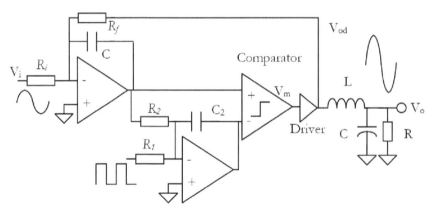

Fig.4.14 Alternative second-order closed-loop analog class-D amplifier

Shown in figure 4.15 is a fully differential second-order closed-loop class-D amplifier circuit.

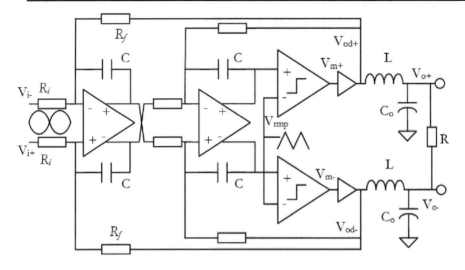

Fig.4.15 Fully differential 2nd-order closed-loop class-D amplifier circuit

4.2.3 SELF-OSCILLATION CLASS-D AMPLIFIERS

The ramp signals of class-D amplifier can be eliminated using a delay based self-oscillation circuit structure as shown in figure 4.16.

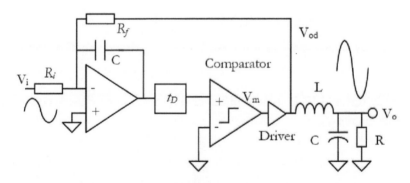

Fig.4.16 Delay based self-oscillating class-D amplifier circuit

The self-oscillation frequency of this class-D amplifier can be derived as:

$$f_{PWM} = \frac{1-M^2}{4t_D} = \frac{1-(2\cdot k-1)^2}{4t_D} \qquad (4.20)$$

Where M is the modulation depth.

The self-oscillation class-D amplifier can also be constructed using the hysteretic in comparator as shown in figure 4.17.

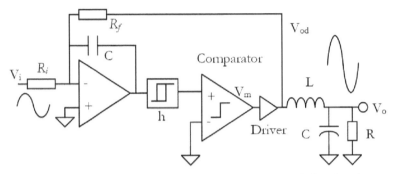

Fig.4.17 Hysteretic self-oscillating class-D amplifier circuit

The self-oscillation frequency of such amplifier can be derived as:

$$f_{PWM} = \frac{1-M^2}{4hRC} = \frac{1-(2k-1)^2}{4hRC} \qquad (4.21)$$

Shown in figure 4.18 is a differential hysteretic self-oscillation class-D amplifier structure.

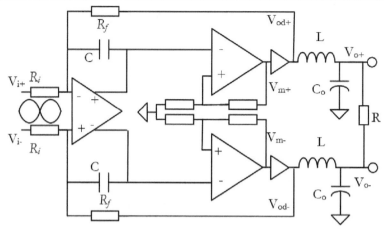

Fig.4.18 Fully differential hysteretic closed-loop class-D amplifier circuit

4.2.4 ALTERNATIVE ANALOG CLASS-D AMPLIFIERS

The ramp signal in class-D amplifier circuit can be replaced by using the square signal as shown in figure 4.19.

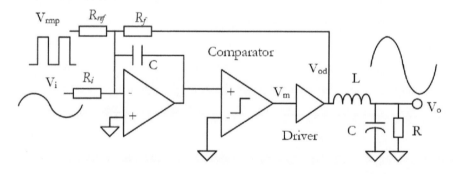

Fig.4.19 Alternative first-order closed-loop class-D amplifier

4.3 DIGITAL CLASS-D AMPLIFIER ARCHITECTURES

The class-D amplifiers based on the digital input are commonly used in the DVD, the iPOD and the notebook PCs applications that store and process the signals in the digital forms.

One conventional implementation for digital class-D amplification is to convert the digital signal into the analog first and then apply the analog class-D amplification as shown in figure 4.20. However this approach may suffer from the penalty in the power efficiency. The alternative methods for digital class-D amplification use the digital PCM to digital PWM conversion as shown in figure 4.21.

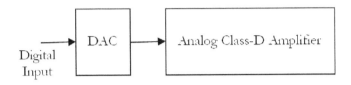

Fig.4.20 Hybrid digital class-D amplifier

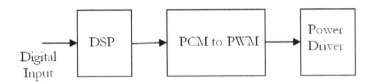

Fig.4.21 Direct PCM-PWM digital class-D amplifier

A method for digital class-D amplifier as shown in figure 4.22 employs the sigma-delta modulation to generate bit stream.

Fig.4.22 Sigma-delta digital class-D amplifier

115

4.3.1 HYBRID CLASS-D AMPLIFIER

The digital audio data at the input of the digital class-D amplifier is usually a pulse code modulation (PCM) signal from sources such as CD and DVD. A conventional way shown in figure 4.23 uses a D/A converter to convert the signal into analog form and then to use the analog class-D amplifier for power amplification.

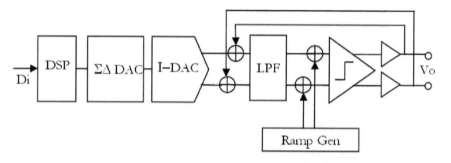

Fig.4.23 Sigma-delta digital class-D amplifier

Such approach can be realized using a digital sigma-delta D/A converter to boost the frequency of the multiple bit digital input to single-bit digital streams and then feed it to the circuit as shown in figure 4.24 and figure 4.25.

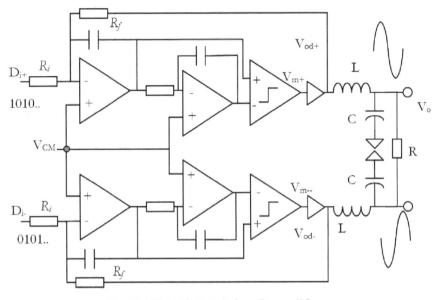

Fig.4.24 Hybrid digital class-D amplifier

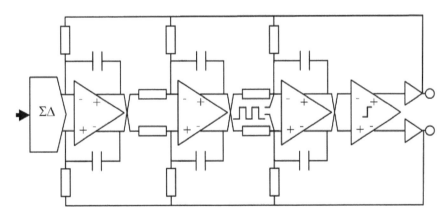

Fig.4.25 Fully differential 3rd-order digital class-D amplifier circuit

4.3.2 DIRECT PCM-PWM CLASS-D AMPLIFIER

The direct stream digital sound encoding method uses a generalized form of pulse-width modulation called pulse density modulation (PDM), at a high enough sampling rate (typically in the order of MHz) to cover the whole acoustic frequencies range with sufficient fidelity. The reproduction of the encoded audio signal is essentially similar to the method used in the class-D amplifiers.

Many digital circuits can generate PWM signals. They normally use a counter that increments periodically (it is connected directly or indirectly to the clock of the circuit) and is reset at the end of every period of the PWM. When the counter value is more than the reference value, the PWM output changes state from high to low (or low to high).

The incremented and periodically reset counter is the discrete version of the intersecting method's sawtooth. The analog comparator of the intersecting method becomes a simple integer comparison between the current counter value and the digital (possibly digitized) reference value. The duty-cycle can only be varied in discrete steps, as a function of the counter resolution. However, a high-resolution counter can provide quite satisfactory performance.

Shown in figure 4.26 are two open-loop PWM class-D amplifier circuits with digital inputs. In these class-D amplifier circuit structures, the PWM modulations are directly realized in the digital domain without using D/A conversion.

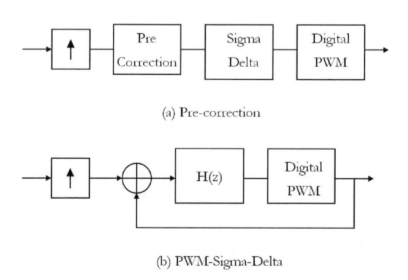

(a) Pre-correction

(b) PWM-Sigma-Delta

Fig.4.26 Digital open loop PWM circuit structures

In the sigma-delta modulation PWM scheme shown in figure 4.27, the waveform is the reference signal, on which the output signal is subtracted to form the error signal and this error is integrated when the integral of the error exceeds the limits of the output changes state.

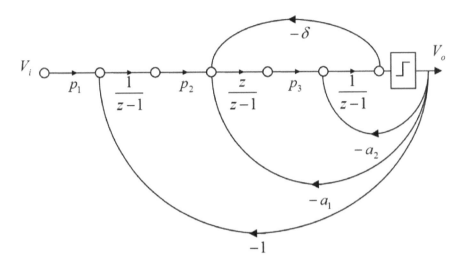

Fig. 4.27 Single-bit 3rd-order Sigma-Delta modulator

4.3.3 VLSI CLASS-D AMPLIFIER TOPOLOGIES

Class-D amplifier can be realized in different circuit topologies such as the half bridge and the full bridge. A full bridge uses two half-bridge stages to drive the load differentially. This configuration is often referred to as a bridge-tied load (BTL). The full-bridge configuration operates by alternating the conduction path through the load. This allows bi-directional current to flow through the load without the need of a negative supply or a DC-blocking capacitor.

A full-bridge class-D amplifier offers similar advantages of a class-AB BTL amplifier with higher power efficiency. The first advantage of BTL amplifiers is that they do not require DC-blocking capacitors on the output when operating from a single supply. The same is not true for a half-bridge amplifier as its output swings between V_{CC} and ground and idles at 50% duty-cycle. This means that its output has a $V_{CC}/2$ dc offset. With a full-bridge amplifier, this offset appears on each side of the load, which means that zero dc current flows at the output. The second advantage is that they can achieve twice the output signal swing when compared to a half-bridge amplifier with the same supply voltage because the load is driven differentially. This results in a theoretical 4x increase in the maximum output power over a half-bridge amplifier operating from the same supply.

A full-bridge class-D amplifier, however, requires twice as many MOS switches as a half-bridge topology. Someone consider this to be a disadvantage, because more switches typically mean more conduction and switching losses. However, this generally is only true with high-output power amplifiers (e.g. > 10W) due to the higher output currents and supply voltages involved. For this reason, half-bridge amplifiers are typically used for high-power applications for their slight efficiency advantage. Most high-power full-bridge amplifiers exhibit power efficiencies in the range beyond 80% with 8Ω loads. However, half-bridge amplifiers achieve power efficiencies greater than 90% while delivering more than 14W per channel into 8Ω.

4.4 VLSI PWM CIRCUIT IMPLEMENTATIONS

The class-D amplifier circuits can be realized in VLSI circuit form using the VLSI manufacture process technologies that are compatible to most digital circuits for SOC applications.

4.4.1 TRIANGULAR RAMP SIGNAL GENERATION

A sawtooth or triangle wave as shown in figure 4.28 is used in PWM circuit to convert the audio signal into a pulse width modulated (PWM) signal. The triangle wave generator typically consists of an integrator and a hysteresis comparator. The circuit integrates a square wave that was created by the hysteresis comparator. The frequency of the sawtooth oscillator impacts the performance of the class-D amplifier. A balance between performance and component size can usually be achieved with an oscillator frequency at least 10 times the maximum signal frequency that is typically 250kH to 1.5 MHz for audio band signal of 10 to 20 kHz. Such selection allows attenuating the switching frequency sufficiently while allowing most of the audio signal to pass to the load without attenuation. Too low oscillator frequency can introduce distortion because of the amplifier's lower sampling rate. A lower oscillator frequency also requires a lower filter-cutoff frequency to sufficiently attenuate the switching frequency that requires larger circuit component values and increasing the overall cost. A higher oscillator frequency allows the filter cutoff to be higher, thus attenuating the switching noise sufficiently while allowing smaller component values.

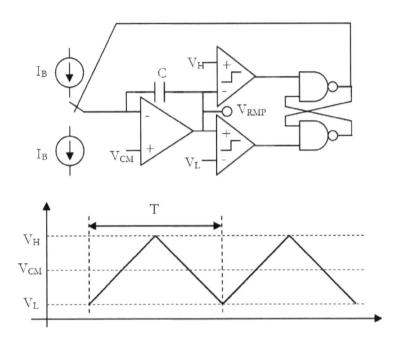

Fig.4.28 VLSI Triangular voltage signal generation circuit

Reasonably high oscillator frequency also reduces the cost and the component sizes for the output filter while extending the amplifier's frequency range. However, higher sampling rate reduces distortion to a certain extent, but slew rates for the comparator and bridge drivers become a limiting factor at high frequencies.

4.4.2 PWM SIGNAL GENERATION CIRCUITS

A typical pulse width modulation (PWM) circuit shown in figure 4.29 and 4.30 uses the natural sampling operation where the signal is compared to a triangular (sawtooth) waveform to generate the digital signal.

The slope of the triangular waveform can be expressed as:

$$Slope = \frac{2V_M}{T} \qquad (4.22)$$

The pulse widths of the two comparator outputs can be expressed respectively as

$$t_+ = \frac{T}{2}\left(1 + \frac{(V_+ - V_-)}{V_M}\right) \qquad (4.23)$$

$$t_- = \frac{T}{2}\left(1 + \frac{(V_- - V_+)}{V_M}\right) \qquad (4.24)$$

The differential pulse width is given as

$$t_+ - t_- = T\left(\frac{(V_+ - V_-)}{V_M}\right) \qquad (4.25)$$

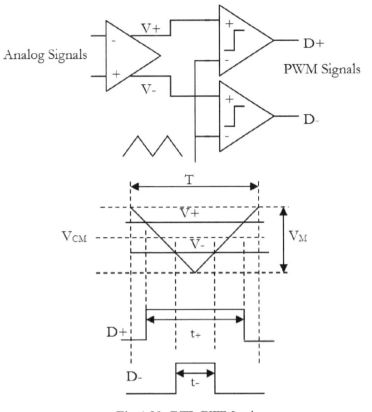

Fig.4.29 BTL PWM scheme

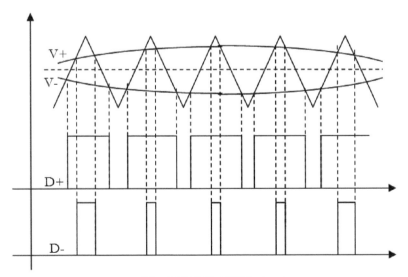

Fig.4.30 BTL PWM circuit

An alternative PWM modulation generation circuit is shown in figure 4.31 and 4.32.

The pulse widths of the two comparator outputs can be expressed respectively as

$$t_+ = \frac{T}{2}\left(1 + \frac{(V_+ - V_-)}{2V_M}\right) \qquad (4.26)$$

$$t_- = \frac{T}{2}\left(1 + \frac{(V_- - V_+)}{2V_M}\right) \qquad (4.27)$$

Finally the differential pulse width is given as

$$t_+ - t_- = T\left(\frac{(V_+ - V_-)}{2V_M}\right) \qquad (4.28)$$

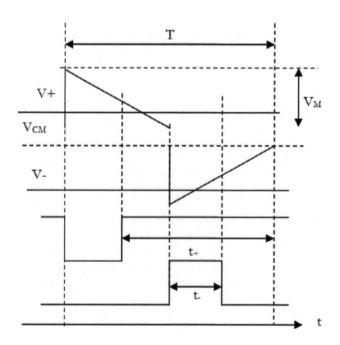

Fig.4.31 Alternative BTL PWM scheme

VLSI Modulation Circuits

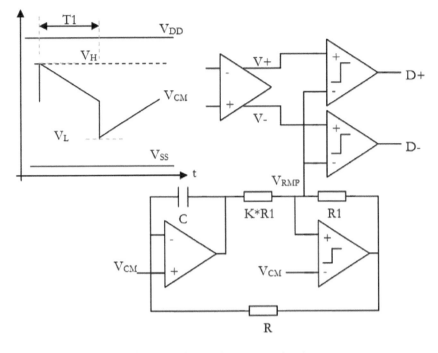

Fig.4.32 Alternative PWM circuit

The slope of the triangular waveform in this modulation scheme can be expressed as:

$$Slope = \frac{2V_M}{T} \qquad (4.29)$$

4.4.3 RAMP-LESS PWM CIRCUIT

Ramp signal generation circuit can be eliminated using a self-oscillation PWM modulation circuit as shown in figure 4.33.

The hysteresis voltage in this PWM circuit can be expressed as:

$$\Delta V = k(V_{MAX} - V_{MIN}) = 2kV_M \qquad (4.30)$$

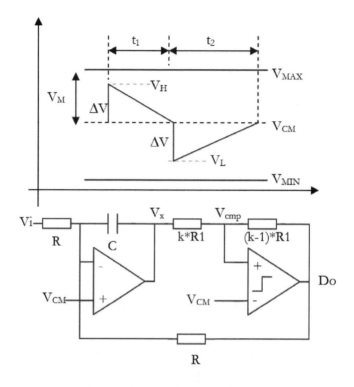

Fig.4.33 Self oscillation PWM circuit

The time durations in this circuit can be derived as:

$$t_1 = \frac{(k-1)kRC}{1+\dfrac{V_i}{V_M}} \qquad (4.31)$$

$$t_2 = \frac{(k-1)kRC}{1-\dfrac{V_i}{V_M}} \qquad (4.32)$$

The period of the PWM signal that is given as the sum of the above time parameters is given as:

$$T = t_1 + t_2 = \frac{2(k-1)kRC}{1-(\frac{V_i}{V_M})^2} \tag{4.33}$$

The duty-cycle of the modulation can then be expressed as function of the input signal magnitude as:

$$k = \frac{t_2 - t_1}{t_1 + t_2} = \frac{V_i}{V_M} \tag{4.34}$$

It can be seen that there is a linear relation between the duty-cycle k and the signal voltage, implying the high linearity amplification of the circuit.

4.4.4 DIGITAL INPUT PWM SIGNAL GENERATION CIRCUITS

Shown in figure 4.34 and 4.35 are two typical digital PWM modulator implementations, where the analog PWM circuits are realized in the digital circuit forms. Such circuits offer the advantages of fully digital technology compatibility and simpler interfaces with the digital signal sources.

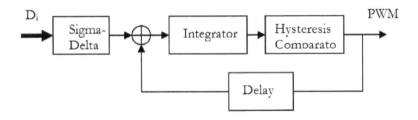

Fig.4.34 First-order PWM modulator

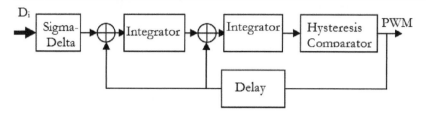

Fig.4.35 Second-order PWM modulator

4.5 PWM SIGNAL POWER SPECTRA

The spectra of various PWM signals are similar, and each usually contains a dc component, a base sideband containing the modulating signal and phase modulated carriers at each harmonic of the frequency of the pulse. The amplitudes of the harmonic groups are restricted by a sinx/x envelope (sinc function) and extend to infinity.

On the other hand, the pulse slew rate and the modulation modes will impact the power spectrum of the PWM modulated signals. As shown in figure 4.36, the pulse slew rate has impact similar to a lowpass FIR filter that impacts the high frequency contents of the modulation output. In practical applications, this behavior can be used to minimize the EMI effect of the class-D amplifier circuits. The PWM modulation modes, such as the BD and AD modulation modes shown in figure 4.37 and 4.38 will also impact the spectrum content of the signal.

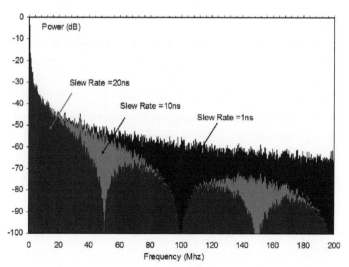

Fig.4.36 Power spectrum versus amplifier driver output slew rate

In order to extract the amplified audio signal from this PWM waveform, the output of the class-D amplifier can be fed to a lowpass filter with cut off frequency at least one order of magnitude lower than the switching frequency. As the result, the output of the filter is equal to the average value of the square wave. In addition, the lowpass filter prevents high-frequency switching energy from being dissipated in the resistive load. Assume that the filtered output voltage remain constant during a single switching period. This assumption is fairly accurate because switching frequency is much higher than the highest input audio signal frequency. Therefore, the relationship between the duty-cycle and resulting filtered output voltage can be derived using a simple time-domain analysis of the inductor voltage and current.

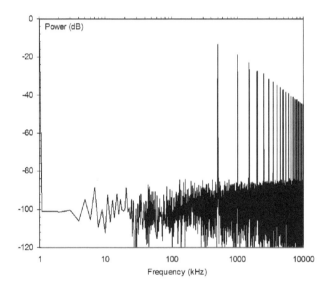

Fig.4.37 Power spectrum of 1 kHz signal with 250 kHz BD model PWM

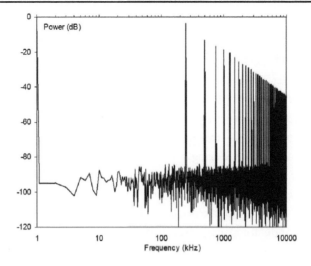

Fig.4.38 Power spectrum of 1 kHz signal with 250 kHz AD model PWM

The feedback loop in a close-loop class-D amplifier helps to shape the quantization noise that greatly reduces in-band noise due to the nonlinearities of the pulse-width modulator, the output stage, and the supply-voltage deviations. Such topology is similar to the noise shaping in the sigma-delta modulators.

One of the major drawbacks of traditional class-D amplifiers is the need for an external LC filter. This need not only increases a solution's cost and board space requirements, but also introduces the possibility of additional distortion due to filter component nonlinearities. Fortunately, many modern class-D amplifiers utilize advanced "filterless" modulation schemes to eliminate, or at least minimize, external filter requirements. One disadvantage of filterless operation is the possibility of radiated EMI from the speaker cables. Because the class-D amplifier output waveforms are high-frequency square waves with fast-moving transition edges, the output spectrum contains a large amount of spectral energy at the switching frequency and integer multiples of the switching frequency. Without an external output filter located within close proximity of the device, the speaker cables can radiate this high-frequency energy. In such case, spread spectrum modulation can help to mitigate possible EMI problems in filterless class-D amplifiers.

Reference:

[1] Bah-Hwee Gwee, et al, "A Micro-power Low-Distortion Digital class-D Amplifier Based on an Algorithmic Pulse width Modulation," IEEE Tran. Circuits and Systems-I: Regular Papers. Vol. 52. No. 10. Oct. 005.

[2] Miguel Angel Roja-Gonzalez, et al, "Low-Power High-Efficiency class-D Audio Power Amplifiers," IEEE JSSC. Vol. 44, No. 12. Dec 2009.

[3] Kyoungsik, et al, "class-D Audio Amplifier using 1-Bit Fourth-Order Delta-Sigma Modulation," IEEE Tran. Circuits and Systems-II: Express Briefs. Vol. 55. No. 8. August 2008.

[4] F. nyboe, et al, "A 240W Monolithic class-D Audio Amplifier Output Stage," ISSCC 2006.

[5] Mikkel C. W. Hoyerby, et al, "Carrier Distortion in Hysteretic Self-Oscillating class-D Audio Power Amplifiers: Analysis and Optimization," IEEE Tran. Power Electronics. Vol. 24. No.3. March 2009.

[6] Vahid M. Tousi, et al, "A 3.3V/1W class-D Audio Power Amplifier with 103dB DR and 90% Efficiency," Proc. 23rd International Conference on Microelectronics. Vol. 2. NIS Yugoslavia. 12-15 May 2002.

[7] Patrick P. Siniscalchi, et al, "A 20W/Channel class-D Amplifier with Significantly Reduced Common-Mode Radiated Emission," ISSCC2009.

[8] Mykhaylo Teplechuk, et al, "Filterless Integrated class-D Audio Amplifier Achieving 0.0012% THD+N and 96dB PSRR When Supplying 1.2W," ISSCC 2011.

[9] Marco Berkhout, "A 460W class-D Output Stage with Adaptive Gate Drive," ISSCC 2009.

[10] Toru IDO, et al, "A Digital Input Controlller for Audio class-D Amplifier with 100W 0.004% THD+N and 113dB DR," ISSCC2006.

[11] Miguel A. Rojas-Gonzalez, et al, "Two class-D Audio Amplifiers with 89/90% Efficiency and 0.02/0.03% THD+N consuming Less than 1mW of Quiescent Power," ISSCC 2009.

[12] Marco Berkhout, "An Integrated 200-W class-D Audio Amplifier," IEEE JSSC. Vol. 38. No.7. July 2003.

[13] Simon Cimin Li, et al, "New High-Efficiency 2.5V/0.45W RWDM class-D Audio Amplifier for Portable Consumer Electronics," IEEE Tran. Circuits and Systems-I: Regular Papers. Vol. 52. No.9. Sept. 2005.

[14] Mei-Ling Yeh, et al, "An Electromagnetic Interference (EMI) Reduced High-Efficiency Switching Power Amplifier," IEEE Tran. Power Electronics. Vol. 25. No.3. March 2010.

[15] Wei Shu, et al, "Power Supply Noise in Analog Audio class-D Amplifier," IEEE Tran. Circuits and Systems-I: Regular Papers. Vol. 56. No. 1. Jan. 2009.

[16] Kwang-Chan Lee, et al, "A High-Performance Fast Switching Charge Dump assist anted class-K Audio Amplifier," IEEE Tran. Circuits and Systems-I: Regular Papers. Vol. 57. No. 6. June 2010.

[17] Sokal, N.O.; Sokal, A.D.; "Class E-A new class of high-efficiency tuned single-ended switching power amplifiers" IEEE Journal of Solid-State Circuits, Volume: 10, Issue: 3, Page(s): 168 - 176, 1975

[18] El-Hamamsy, S.-A.; "Design of high-efficiency RF class-D power amplifier," IEEE Transactions on Power Electronics, Volume: 9, Issue: 3, Page(s): 297 - 308, 1994.

[19] Berkhout, M.; "An integrated 200-W class-D audio amplifier," IEEE Journal of Solid-State Circuits, Volume: 38, Issue: 7. Page(s): 1198 - 1206, 2003.

[20] Koizumi, H.; Suetsugu, T.; Fujii, M.; Shinoda, K.; Mori, S.; Iked, K.; "class-DE high-efficiency tuned power amplifier," IEEE Transactions on Circuits and Systems I: Fundamental Theory and Applications, Volume: 43, Issue: 1, Page(s): 51 - 60, 1996.

[21] Chang, J.S.; Meng-Tong Tan; Zhihong Cheng; Yit-Chow Tong; "Analysis and design of power efficient class-D amplifier output stages," IEEE Transactions on Circuits and Systems I: Fundamental Theory and Applications, Volume: 47, Issue: 6 Page(s): 897 - 902, 2000.

[22] Kyu Min Cho; Won Seok Oh; Young Tae Kim; Hee Jun Kim; "A New Switching Strategy for Pulse Width Modulation (PWM) Power Converters," IEEE Transactions on Industrial Electronics, Volume: 54, Issue: 1, Page(s): 330 - 337, 2007.

[23] Ertl, H.; Kolar, J.W.; Zach, F.C.; "Basic considerations and topologies of switched-mode assisted linear power amplifiers," IEEE Transactions

on Industrial Electronics, Volume: 44, Issue: 1, Publication Year: 1997, Page(s): 116 - 123, 1997.

[24] Kobayashi, H.; Hinrichs, J.M.; Asbeck, P.M.; "Current-mode class-D power amplifiers for high-efficiency RF applications," IEEE Transactions on Microwave Theory and Techniques, Volume: 49, Issue: 12, Page(s): 2480 - 2485, 2001.

[25] Pascual, C.; Zukui Song; Krein, P.T.; Sarwate, D.V.; Midya, P.; Roeckner, W.J.; "High-fidelity PWM inverter for digital audio amplification: Spectral analysis, real-time DSP implementation, and results," IEEE Transactions on Power Electronics, Volume: 18, Issue: 1, Part: 2, Page(s): 473 - 485, 2003.

[26] Forejt, B.; Rentala, V.; Arteaga, J.D.; Burra, G.; "A 700+-mW class-D design with direct battery hookup in a 90-nm process," IEEE Journal of Solid-State Circuits, Volume: 40, Issue: 9, Page(s): 1880 - 1887, 2005.

[27] van der Zee, R.A.R.; van Tuijl, E.A.J.M.; "A power-efficient audio amplifier combining switching and linear techniques," IEEE Journal of Solid-State Circuits, Volume: 34, Issue: 7, Page(s): 985 - 991, 1999.

[28] Johnson, T.; Stapleton, S. P.; "RF class-D Amplification With Bandpass Sigma–Delta Modulator Drive Signals," IEEE Transactions on Circuits and Systems I: Regular Papers, Volume: 53, Issue: 12, Page(s): 2507 - 2520, 2006.

[29] Kazimierczuk, M.K.; Szaraniev, W.; "class-D zero-voltage-switching inverter with only one shunt capacitor," Electric Power Applications, IEE Proceedings B Volume: 139, Issue: 5, Page(s): 449 - 456, 1992.

[30] Bah-Hwee Gwee; Chang, J.S.; Adrian, V.; "A micropower low-distortion digital class-D amplifier based on an algorithmic pulsewidth modulator," Circuits and Systems I: Regular Papers, IEEE Transactions on Volume: 52, Issue: 10, Page(s): 2007 - 2022, 2005.

[31] Oliva, A.R.; Ang, S.S.; Vo, T.V.; "A multi-loop voltage feedback filterless class-D switching audio amplifier using unipolar pulse-width-modulation," Consumer Electronics, IEEE Transactions on Volume: 50, Issue: 1, Page(s): 312 - 319, 2004.

[32] Ertl, H.; Kolar, J.W.; Zach, F.C.; "Analysis of a multilevel multicell switch-mode power amplifier employing the "flying-battery" concept," IEEE Transactions on Industrial Electronics, Volume: 49, Issue: 4, Page(s): 816 - 823, 2002.

[33] Walker, G.R.; "A class-B switch-mode assisted linear amplifier," IEEE Transactions on Power Electronics, Volume: 18, Issue: 6, Page(s): 1278 - 1285, 2003

[34] Li, S.C.; Lin, V.C.-C.; Nandhasri, K.; Ngarmnil, J.; "New high-efficiency 2.5 V/0.45 W RWDM class-D audio amplifier for portable consumer electronics," IEEE Transactions on Circuits and Systems I: Regular Papers, Volume: 52, Issue: 9, Publication Year: 2005, Page(s): 1767 - 1774, 2005

[35] Putzeys, B.; " Digital audio's final frontier," IEEE Spectrum, Volume: 40, Issue: 3, Page(s): 34 - 41, 2003.

[36] Rojas-Gonzalez, M.A.; Sanchez-Sinencio, E.; "Design of a class-D Audio Amplifier IC Using Sliding Mode Control and Negative Feedback," IEEE Transactions on Consumer Electronics, Volume: 53, Issue: 2, 609 - 617, 2007.

[37] Chudobiak, W.J.; Page, D.F.; "Frequency and power limitations of class-D transistor amplifiers," IEEE Journal of Solid-State Circuits, Volume: 4, Issue: 1, Page(s): 25 - 37, 1969.

[38] Mosely, I.D.; Mellor, P.H.; Bingham, C.M.; "Effect of dead time on harmonic distortion in class-D audio power amplifiers," Electronics Letters. Volume: 35, Issue: 12, Page(s): 950 - 952, 1999.

[39] Midya, P.; Roeckner, B.; Bergstedt, S.; "Digital correction of PWM switching amplifiers," IEEE Power Electronics Letters, Volume: 2, Issue: 2, Page(s): 68 - 72, 2004.

[40] Sang-Hwa Jung; Nam-In Kim; Gyu-Hyeong Cho; "class-D audio power amplifier with fine hysteresis control," Electronics Letters Volume: 38, Issue: 22, Page(s): 1302 - 1303, 2002.

[41] Gaalaas, E.; Liu, B.Y.; Nishimura, N.; Adams, R.; Sweetland, K.; " Integrated stereo $\Delta\Sigma$ class-D amplifier," IEEE Journal of Solid-State Circuits, Volume: 40, Issue: 12, Page(s): 2388 - 2397, 2005.Kazimierczuk, M.K.; "class-D voltage-switching MOSFET power amplifier," IEE Proceedings B Electric Power Applications, Volume: 138, Issue: 6, 285 - 296, 1991.

[42] Wei Shu; Chang, J.S.; "THD of Closed-Loop Analog PWM class-D Amplifiers," IEEE Transactions on Circuits and Systems I: Regular Papers, Volume: 55, Issue: 6, 1769 - 1777, 2008.

[43] Koizumi, H.; Kurokawa, K.; Mori, S.; "Analysis of class-D inverter with irregular driving patterns," IEEE Transactions on Circuits and Systems I: Regular Papers, Volume: 53, Issue: 3, Page(s): 677 - 687, 2006.

[44] Tong Ge; Chang, J.S.; "Modeling and Technique to Improve PSRR and PS-IMD in Analog PWM class-D Amplifiers," IEEE Transactions on Circuits and Systems II: Express Briefs, Volume: 55, Issue: 6, Page(s): 512 - 516, 2008.

[45] Meng Tong Tan; Chang, J.S.; Hock Chuan Chua; Bah Hwee Gwee; "An investigation into the parameters affecting total harmonic distortion in low-voltage low-power class-D amplifiers," IEEE Transactions on Circuits and Systems I: Fundamental Theory and Applications, Volume: 50, Issue: 10 Page(s): 1304 - 1315, 2003.

[46] Himmelstoss, F.A.; Edelmoser, K.H.; "High dynamic class-D power amplifier," IEEE Transactions on Consumer Electronics, Volume: 44, Issue: 4, Page(s): 1329 - 1333, 1998.

[47] Morrow, P.; Gaalaas, E.; McCarthy, O.; "A 20-W stereo class-D audio output power stage in 0.6-μm BCDMOS technology," IEEE Journal of Solid-State Circuits, Volume: 39, Issue: 11, Page(s): 1948 - 1958, 2004.

[48] Mellor, P.H.; Leigh, S.P.; Cheetham, B.M.G.; "Reduction of spectral distortion in class-D amplifiers by an enhanced pulse width modulation sampling process," IEE Proceedings G Circuits, Devices and Systems, Volume: 138, Issue: 4, Page(s): 441 - 448, 1991.

[49] Burrow, S.; Grant, D.; "Efficiency of low power audio amplifiers and loudspeakers," IEEE Transactions on Consumer Electronics, Volume: 47, Issue: 3, Page(s): 622 - 630, 2001.

[50] Tsai-Pi Hung; Metzger, A.G.; Zampardi, P.J.; Iwamoto, M.; Asbeck, P.M.; "Design of high-efficiency current-mode class-D amplifiers for wireless handsets," IEEE Transactions on Microwave Theory and Techniques, Volume: 53, Issue: 1, Page(s): 144 - 151, 2005.

CHAPTER 5
VLSI PHASE (DELAY) INTERPOLATION CIRCUITS

A VLSI phase interpolation (PI) circuit belongs to the phase shift key modulation circuit that can be used to adjust the phase of the output reference clock signal. A PI offers a finite number of tunable discrete output clock phases. Each phase is represented by a particular digital input code. In the VLSI high-speed I/O circuit applications, the PI circuit can be used to generate an optimal reference clock to sample the receiver incoming data stream.

There are two basic types of VLSI phase interpolation schemes that can be used to provide perfectly linear phase interpolation operation, including the sinusoidal phase interpolation and the triangular (or linear) phase interpolation. A VLSI phase interpolation operation is typically realized based on the weighted summation (called mixing) of two or more signals of same frequency and different phase (commonly known as the I/Q signals if the signals are 90 degree apart).

The practical VLSI phase interpolations are usually based on the combination of the basic phase interpolation algorithms that employs the single tone-like pre-conditioning using band-limiting circuits and the linear weighting coefficients. Such an approach provides a workable solution for the trade-off between the design simplicity and the circuit performance.

5.1 PHASE INTERPOLATION PRINCIPLE

There are two basic types of VLSI phase interpolation schemes that can provide perfectly linear phase interpolation, including the sinusoidal phase interpolation and the triangular (or linear) phase interpolation. The sinusoidal interpolation algorithm is based on the weighted summation of two single-tone sinusoidal signals of same frequency and 90 degrees phase difference (known as I/Q signals). The linear phase interpolation, on the other hand, is based on the weighted summation of the two or more triangular (linear ramping) clock signals.

A VLSI sinusoidal phase interpolation circuit usually offers the advantage of implementation simplicity in the signal pre-condition that relies on the single-tone signals, which can be generated using a simple lowpass or bandpass filter to eliminate the harmonic contents of the input I/O clock references. However the sinusoidal phase interpolation may suffer from the constraint in the VLSI circuit implementation since it usually requires nonlinear weighting coefficients for perfectly linear phase interpolation. The VLSI triangular phase interpolation circuit, on the other hand, offers the advantage of high VLSI circuit implementation simplicity in the weighting coefficient generation circuit. However, a triangular phase interpolation circuit typically suffers from the design complexity in the VLSI signal precondition circuit implementations for highly linear triangular signal generation for the perfectly linear phase interpolation.

5.1.1 SINUSOIDAL PHASE INTERPOLATION

A phase interpolation based on the I/Q sinusoidal clock signals can be realized by weighted addition of two sinusoidal clock signals of same frequency with $\pi/2$ (i.e. 90º) phase difference as shown in figure 5.1.

The sinusoidal phase interpolation operation can be mathematically expressed by the phase interpolation equation as

$$V_o(t) = A\sin(\omega t - \phi) = A[\cos(\phi)\sin(\omega t) - \sin(\phi)\cos(\omega t)] \quad (5.1)$$

This equation implies that an arbitrary phase sinusoidal signal output can be realized by weighted combination of the I/Q signal sinusoidal clock input as:

$$\begin{cases} V_o(t) = A\sin(\omega t + \phi) = \alpha_I \cdot V_I(t) + \alpha_Q \cdot V_Q(t) \\ V_I(t) = A\sin(\omega t) \\ V_Q(t) = -A\cos(\omega t) \\ \alpha_I = \cos(\phi) \\ \alpha_Q = \sin(\phi) \end{cases} \quad (5.2)$$

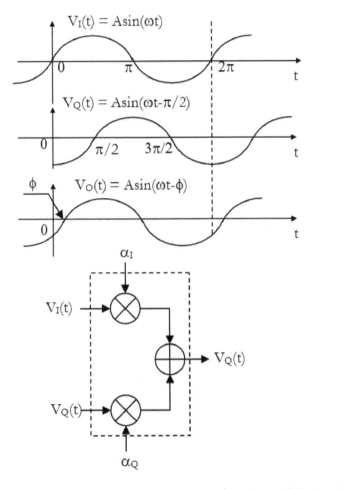

Fig.5.1 Phase interpolation based on I/Q sinusoidal signals

The operation of the sinusoidal phase interpolation can also be expressed using a phase diagram as shown in figure 5.2 that represents the relation between the interpolation weights (or coefficients) of the input I/Q clock signals and the interpolated output clock phase.

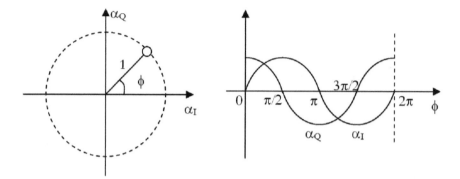

Fig.5.2 Phase diagram for sinusoidal phase interpolation

It can be seen that for a sinusoidal phase interpolation the relationship between the output clock phase and the interpolation weighting coefficient is non-linear. In addition, we can also see that the sum of the two weighting coefficient is not constant:

$$\begin{cases} \phi = \tan^{-1}(\dfrac{\alpha_Q}{\alpha_I}) \\ \alpha_I^2 + \alpha_Q^2 = 1 \end{cases} \quad (5.3)$$

Such nonlinear properties have a major drawback in VLSI circuit implementation, which may significantly increase the complexity for the interpolation circuit implementation.

5.1.2 TRIANGULAR PHASE INTERPOLATION

The nonlinear phase relation and the non-constant weighting coefficient sum in the sinusoidal phase interpolation can be avoided using the triangular phase interpolation technique as shown in figure 5.3. The triangular phase interpolation offers linear relation between the interpolation phase and the interpolation weight coefficients. The constant sum of the weighting coefficients is very suitable for the VLSI circuit implementation.

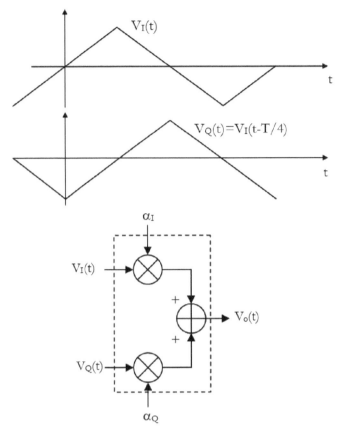

Fig.5.3 Linear signal based phase interpolation algorithm

The operation of the triangular phase interpolation can be expressed by the following equation:

$$\begin{cases} \alpha_I \cdot V_m(\frac{\phi}{T}) + \alpha_Q \cdot V_m(\frac{\phi - T/4}{T}) = 0 \\ \alpha_I + \alpha_Q = 1 \end{cases} \quad (5.4)$$

$$\Rightarrow \phi = \alpha_Q \cdot \frac{T}{4} \quad (5.5)$$

It is important to see that the zero-crossing time of the weighted I/Q triangular waveforms are linear with respect to the interpolation coefficient.

The phase diagrams for the above triangular phase interpolation scheme is shown in figure 5.4

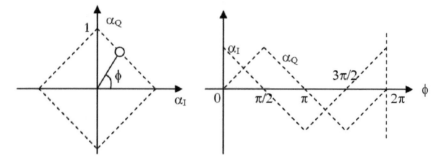

Fig.5.4 Phase diagram for linear signal based phase interpolation

It can be seen that for the triangular phase interpolation the interpolated phase is linearly related to the interpolation weighting coefficients and that the sum of the weighting coefficients is constant. Such relation can also be demonstrated based on the circuit simulation results as shown in figure 5.5 for various weighting coefficients.

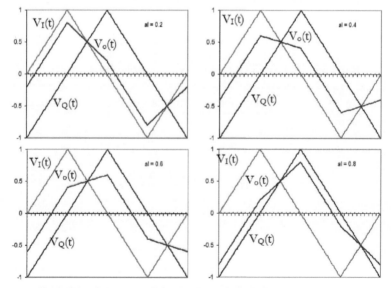

Fig.5.5 Simulation waveform for linear based phase interpolation

5.2 VLSI PI CIRCUIT IMPLEMENTATIONS

A typical VLSI phase interpolator circuit usually consists of four major building components as shown in figure 5.6, including the polyphase clock generator, the conditioner, the mixer and the control circuit.

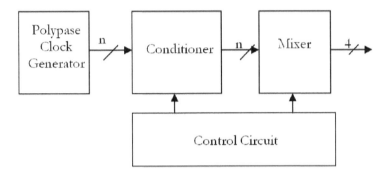

Fig.5.6 VLSI phase interpolation circuit block diagram

A VLSI phase interpolation circuit typically includes the following functions:

- The polyphase clocks are generated in the polyphase clock generation circuit, employing either a VCO, a VCDL or other methods;

- The polyphase clocks are pre-conditioned either through a lowpass or bandpass filter or through a linear ramps generation circuit based on the sinusoidal or triangular method, that shapes the harmonic contents in the clocks;

- The pre-conditioned polyphase clocks are mixed based on programmable weighted addition operation. The mixed clock is amplified and limited for output;

- The phase interpolation control circuit for output clock phase tuning controls the weighting coefficients.

Shown in figure 5.7 are two VLSI digital phase interpolation circuit implementations that are based on the direct phase interpolator circuit structure, where one additional phase is interpolated from the two phases (0° and 90°).

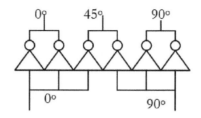

(a) 1-to-2 interpolation

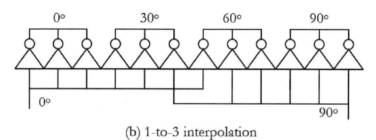

(b) 1-to-3 interpolation

Fig. 5.7 VLSI direct phase interpolation circuits

Shown in figure 5.8 is a VLSI R-ring phase interpolation circuit structure, where the signal conditioning is realized using a simple RC lowpass filter. The I/Q phase mixing coefficients in this phase interpolation circuit structure are implemented through the weighted addition of the conditioned I/Q clock phase signal based on the ratio of the resistors as shown figure 5.9:

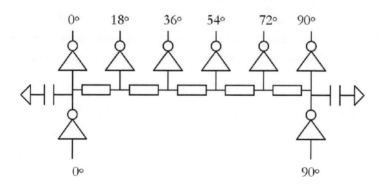

Fig. 5.8 VLSI R-ring 1-to-5 phase interpolation circuit

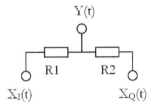

Fig. 5.9 VLSI resistor ratio phase interpolation circuit

Based on the weighted summation of two clock signals:

$$Y(t)(\frac{1}{R_1} + \frac{1}{R_2}) = X_I(t)\frac{1}{R_1} + X_Q(t)\frac{1}{R_2} \tag{5.6}$$

we have that

$$\Rightarrow \begin{cases} Y(t) = \alpha_I \cdot X_I(t) + \alpha_Q \cdot X_Q(t) \\ \alpha_I \equiv \dfrac{R_2}{R_1 + R_2} \\ \alpha_Q \equiv \dfrac{R_1}{R_1 + R_2} \end{cases} \tag{5.7}$$

It can be seen that

$$\alpha_I + \alpha_Q = 1 \tag{5.8}$$

Therefore such PI circuit belongs to the linear coefficient phase interpolation type and it is preferred to have triangular input I/Q clock signals for the phase interpolation.

Shown in figure 5.10 is a VLSI phase interpolation circuit based on the current I/Q clock mixing operation. This circuit can be used to approximate the sinusoidal phase interpolation circuit operations.

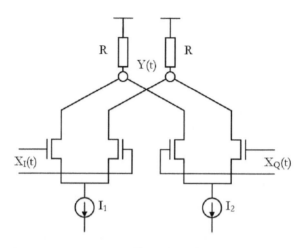

Fig. 5.10 VLSI current steeling phase interpolation circuits

For this current steeling PI mixer circuit, the differential input and output signal relation can be expressed as

$$\begin{cases} Y(t) = R \cdot g_{m1} \cdot X_I(t) + R \cdot g_{m2} \cdot X_Q(t) \\ g_{m1} = \sqrt{2\beta I_1} \\ g_{m2} = \sqrt{2\beta I_2} \end{cases} \quad (5.9)$$

By letting

$$\begin{cases} Y(t) = A[\alpha_I \cdot X_I(t) + \alpha_Q \cdot X_Q(t)] \\ \alpha_I = \sqrt{I_1/(I_1 + I_2)} \\ \alpha_Q = \sqrt{I_2/(I_1 + I_2)} \end{cases} \quad (5.10)$$

We can see that

$$\alpha_I^2 + \alpha_Q^2 = \frac{I_1}{I_1+I_2} + \frac{I_2}{I_1+I_2} = 1 \qquad (5.11)$$

It implies that such a phase interpolation circuit is suitable for the sinusoidal I/Q clock signals.

The differential VLSI circuit shown in figure 5.11 can usually be used to control the weight current for the VLSI phase interpolation operation.

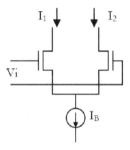

Fig. 5.11 VLSI phase interpolation coefficient generation circuit

For such circuit structure we have that

$$\begin{cases} I_1 = \dfrac{I_B}{2} + \dfrac{g_m}{2}V_i \\ I_2 = \dfrac{I_B}{2} - \dfrac{g_m}{2}V_i \end{cases} \qquad (5.12)$$

And

$$I_1 + I_2 = I_B \qquad (5.13)$$

Shown in figure 5.12, figure 5.13 and figure 5.14 are several VLSI phase interpolation mixer circuit implementations that are based on the symmetrical load differential buffer circuit structures. Such circuits provide regulated signal voltage swing for the mixer circuit core and the pre-conditioner circuit employing a replica biasing circuit structure.

Above circuit structures can be used to implement the PI mixer circuit by including the digital control circuits that provide the digitally controlled weights to support the VLSI PI mixing operations.

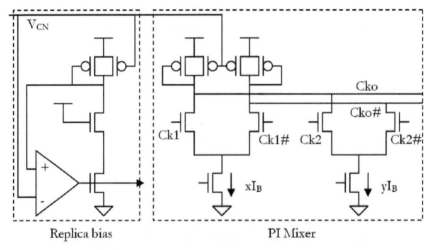

Fig. 5.12 VLSI phase interpolation based on CML circuit

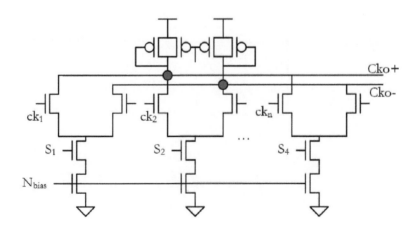

Fig. 5.13 VLSI PI mixer circuit employing the replica biasing circuit

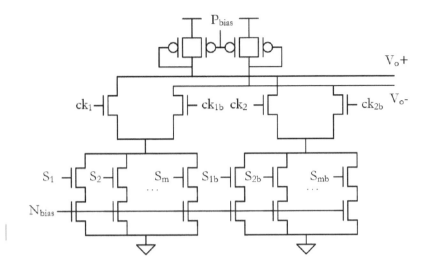

Fig. 5.14 Alternative VLSI PI mixer circuit

5.3 VLSI PI CIRCUIT PHASE NOISE MODEL

The VLSI PI circuit structures are similar to the basic delay buffer circuit structures, where the random phase noise of the PI circuit can be approximately expressed as:

$$\frac{\Delta t_{RMS}}{t_d} \approx \sqrt{\frac{2kT}{C_L}} \sqrt{1 + \frac{2}{3} a_V} \frac{1}{V_{PP}} \qquad (5.14)$$

Or

$$\Delta t_{RMS} \propto \frac{\sqrt{TC_L}}{V_{PP}} \qquad (5.15)$$

Reference:

[1] F. Yang, et al, "A CMOS Low-Power Multiple 2.5-3.125 Gb/s Serial Link Macrocell for High IO Bandwidth Network ICs", IEEE JSSC, Vol 37, No. 12, Dec. 2002.

[2] T. Kim, et al, "Phase Interpolator Using Delay Locked Loop", SSMSD 2003.

[3] L. Yang, et al, "An Arbitrarily Skewable Multiphase Clock Generator Combing Direct Interpolation with Phase Error Averaging", 2003 IEEE.

[4] J-M Chou, et al, "A 125Mhz 8b Digital-to-Phase Converter," ISSCC 2003.

[5] R. Kreienkamp, et al, "A 10-Gb/s CMOS Clock and Data Recovery Circuit with an Analog Phase Interpolator," IEEE2003 Custom Integrated Circuits Conference.

[6] A. Maxim, "A 0.16-2.55Ghz CMOS Active Clock Deskewing PLL Using Analog Interpolator," IEEE JSSC. Vol. 40, No. 1, Jan. 2006.

[7] J-M. Chou, et al, "Phase Averaging and Interpolation Using Resistor Strings or Resistor Rings for Multi-Phase Clock Generation," IEEE Trans. Circuits & Systems, Vol.53, No. 5, May 2006.

[8] W. Rhee, et al, "A Semi-Digital Delay-Locked Loop Using an Analog-Based Finite State Machine", IEEE Trans. Circuits & Systems, Vol.51, No.11, Dec. 2004.

[9] B. W. Garlepp, et al, "A Portable Digital DLL for High-Speed CMOS Interface Circuits," IEEE JSSC Vol.34, No. 5, May 1999.

[10] G. Jovannovic, et al, "Delay Locked Loop with Linear Delay Element," Serbia and Montenegro, Nis, September 28-30, 2006.

[12] J. G. Maneatis, "Low-Jitter Process-Independent DLL and PLL Based on Self-Biased Techniques," IEEE JSSC, Vol. 31, No. 11, Nov. 1996.

[13] B. W. Garlepp, et al, "A Portable Digital DLL for High-speed CMOS Interface Circuits", IEEE JSSC, Vol. 34, No. 5, May 1999.

[14] S-S Hwang, et al, "A DLL Based 10-320 Mhz Clock Synchronizer," ISCAS 2000, May 28-31, Geneva, Switzerland.

[15] C. Kim, et al, "A Low-Power Small-Area +/- 7.28ps Jitter 1-Ghz DLL-Base Clock Generator", IEEE JSSC, Vol. 37, No. 11, Nov. 2002.

[16] Y Arai, "A High-Resolution Time Digitizer Utilizing Dual PLL Circuits," IEEE2004.

[17] H. J. Song, "Digital Delay Locked Loop for Adaptive De-skew Clock Generation", US Patent, No. 6275555, 2001.

[18] H. J. Song, "Programmable De-Skew Clock Generation based on Dual Digital Delay Locked Loop Structure", SOCC2001.

[19] G.C. Hsieh, et al, "Phase-Locked Loop Techniques-A Survey", IEEE Transactions on Industrial Electronics, Vol.43, No.6, Dec.1996.

[20] P. Heydari, et al, "Analysis of the PLL Jitter Due to Power/Ground and Substrate Noise," IEEE Transactions on Circuits and Systems-I: Regular Papers, Vol 51, No. 12, Dec. 2004.

[21] T. M. Almeida, et al, "High Performance Analog and Digital PLL Design," IEEE 1999.

[22] V. F. Kroupa, "Noise Properties of PLL Systems," IEEE Transactions on Communications, Vol COM-30, No. 10, Oct. 1982.

[23] A. Maxim, "A 0.16-2.55-Ghz CMOS Active Clock De-Skewing PLL Using Analog Phase Interpolation," IEEE, JSSC, Vol. 40, No.1, Jan. 2006.

[24] M. Inoue, et al, "Over-Sampling PLL for Low-Jitter and Responsive Clock Synchronization," IEEE2006.

[25] T. Toifl, et al, "A 0.94ps-RMS-Jitter 0.016-mm2 2.5Ghz Multiphase Generation PLL with 360o Digitally Programmable Phase Shift for 10-Gb/s Serial Links," IEEE JSSC Vol. 40, No.12, December 2005.

[26] J. F. Bulzacchelli, et al, "A 10-Gb/s 5-Tap DFE/4-Tap FFE Transceiver in 90nm CMOS Technology," IEEE JSSC Vol. 41, No. 12, December 2006.

[27] B. G. Kim, et al, "A 250 Mhz-2Ghz Wide Range Delay-Locked Loop," IEEE JSSC Vol. 40, No.6, June 2005.

[28] T. Matano, et, al, "A 1-Gb/s/pin 512-Mb DDRII SDRAM Using a Digital DLL and a Slew-Rate-Controlled Output Buffer," IEEE JSSC Vol.38, No. 5, May 2003.

[29] R. C. H. Van de Beek, et, al, "Low-Jitter Clock Multiplication: A Comparison Between PLLs and DLLs," IEEE Tran. Circuits and System-II: Analog and Digital Signal Processing, Vol.49, No. 8, Aug. 2002.

[30] T. C. Weigandt, et, al, "Timing Jitter Analysis for High-frequency Low-power CMOS Ring-Oscillator Design," Proc. Int. Symp., Circuits and Systems. London, UK., June 1994.

[31] P. Vancorenland, et, al, "A Wideband IMRR Improved Quadrature Mixer/LO Generator," IEEE.

[32] P. Minami, et, al, "A 1Ghz Portable Digital Delay-Locked Loop with Infinite Phase Capture Range," 2000 IEEE ISSCC.

[33] J. G. Maneatis, et, al, "Precise Delay Time Generation Using Coupled Oscillators," 1993 IEEE ISSCC.

[34] M. Kokubo, et, al, "Spread spectrum Clock Generator for Serial ATA using Fractional PLL Controlled by $\Delta\Sigma$ Modulator with Level Shifter" 2005 IEEE ISSCC.

[35] P. Lu, et, al, "A Low-Jitter Frequency Synthesizer with Dynamic Phase Interpolation for High-Speed Ethernet," 2000 IEEE ISSCC.

[36] T. Matsumoto, "High-Resolution On-Chip Propagation Delay Detector for Measuring Within-Chip and Chip-to-Chip Variation," 2004 Symposium on VLSI Circuits Digest of Technique Papers.

[37] M. Bazes, et, al, "An Interpolating Clock Synthesizer," IEEE JSSC, vol. 31, no. 9, Sept. 1996.

[38] Y. J. Jung, et, al, "A Dual-Loop Delay-Locked Loop Using Multiple Voltage-Controlled Delay Lines," IEEE 2001.

[39] M. G. Johnson, et, al, "A Variable Delay Line PLL for CPU-Coprocessor Synchronization," IEEE JSSC, vol. 23, no. 5, Oct. 1988.

[40] A. Ghaffari, et, al, "A Novel Wide-Range Delay Cell for DLLs," ICECE 2006.

[41] M. T. Hsieh, et, al, "Clock and Data Recovery with Adaptive Loop Gain for Spread Spectrum SerDes Applications," IEEE 2005.

[42] J. M. Chou, et, al, "Phase Averaging and Interpolation Using Resistor Strings or Resistor Rings for Multi-Phase Clock Generation," IEEE Tran. Circuits and Systems-I: Regular Papers, vol. 53, No. 5, May 2006.

[43] C. C. Wang, et, al, "Clock-and-Data Recovery Design for LVDS Transceiver Used in LCD Panels," IEEE Tran. Circuits and Systems-II: Express Briefs, vol. 53, No. 11, Nov. 2006.

[44] U. Yodprasit, *et, al*, "Realization of a Low-Voltage and Low-Power Colpitts Quadrature Oscillator," IEEE Tran. Circuits and Systems-II: Express Briefs, vol. 53, No. 11, Nov. 2006.

CHAPTER 6
VLSI AUTO-ZERO AND CHOPPER STABLIZATION CIRCUITS

Static voltage offset and ac noise effects (especially the 1/f noise and thermal drift) are among the major performance limitations of VLSI analog circuits. Various VLSI circuit techniques have been developed to mitigate these effects. A static voltage offset can usually be minimized by using circuit techniques such as by employing symmetric circuit structure, by using matched device layout and by adopting the laser or fuse trimming techniques. On the other hand, the ac noises can usually be minimized employing two major circuit techniques such as the auto-zero and the chopper stabilization circuit techniques.

A VLSI auto-zero circuit is based on the correlative double sampling (CDS) method to cancel out the dc offset and the low frequency noises. Alternatively a chopper circuit modulates the static offset and low frequency noise to higher frequency band such that they can be filtered out using a lowpass filter. These circuit techniques can also be combined for better offset and ac noise cancellations.

6.1 VLSI NOISE EFFECTS AND COMPENSATION

The VLSI circuits suffer from various noise effects such as the dc offset, the drifts, the 1/f noise and the white noise as shown in figure 6.1. The typical dc offset in a CMOS circuit technology is in the order of 1~10mV that is mainly caused by the component mismatch in VLSI circuit. The VLSI circuit offset also depends on the circuit structure, the device size and the circuit operation conditions. The drift of the circuit is usually caused by time varying VLSI devices non-idealities such as temperature variation, aging and package stress. VLSI circuits also suffer from the ac noise as such the 1/f noise, white noise, the power supply noise, and the cross-talk noise effects.

In VLSI CMOS process technologies, the dc offset and the 1/f noise are usually among the dominant noise effects since they are typically a few orders of magnitude higher than that of a circuit fabricated using bipolar technology.

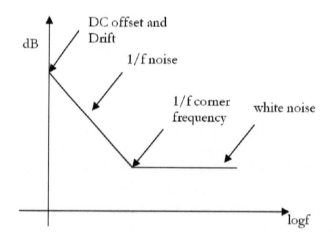

Fig.6.1 Typical VLSI circuit noise power spectrum

For a VLSI amplifier circuit, the input referred differential noise can typically be modeled as:

$$E_i = \frac{E_o}{A} + (I_{B+}R_{s+} + I_{B-}R_{s-}) + \frac{E_{CM}}{CMRR} + \frac{E_s}{PSRR} \qquad (6.1)$$

These noise effects can be used to calculate the input referred noise by dividing the open-loop gain of the amplifier circuit. The second and the third terms are due to the loading effects of the amplifier to the signal source. The 4th and 5th terms are noise contents due to the common-mode noise and the supply noise effects.

The offset voltage of a differential amplifier circuit is defined as the voltage difference between input pins of an amplifier that causes the output to be the same as the plus input pin as shown in figure 6.2.

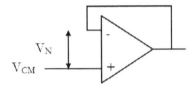

Fig.6.2 VLSI amplifier input referred noise

The sources of the amplifier offset include the device V_T and I_d mismatches of input differential MOS pair and the non-symmetry in the circuit structure. In most VLSI amplifier circuits, the dc offset, the drift and the 1/f noise are the major contributors of the circuit noises. The input offset voltage of an amplifier becomes important when a small signal is amplified with a high gain.

There are a few major static circuit methods to mitigate the VLSI circuit dc offset, the drift and the low frequency noise effects such as (a) circuit design optimization practices using symmetric circuit structure (e.g. fully differential circuit), (b) the use of large device sizes and careful device physical layout for better device parameter matching and low 1/f noise, and (c) the trimming based circuit techniques employing the laser trimming or fuse.

On the other hand, dynamic circuit techniques such as the auto-zero and the chopper stabilization circuit techniques can be used to minimize both the dc and ac noise effects. These dedicated techniques or combination of both techniques can be used to achieve very low (in the nV-level) input offset and voltage drifts therefore to address the circuit issues such as the temperature drift and the 1/f noise. In addition, such zero-drift amplifiers can offer higher open-loop gain, power-supply rejection, and common-mode rejection than the standard amplifiers.

The VLSI amplifiers based on the auto-zero circuit techniques can be used to achieve very low dc offset. Such circuits also offer wide bandwidth operation with very low ripple noise (especially at the sampling frequency). However, these amplifiers may have degraded low frequency noise due to the aliasing effect. In addition, they usually have higher power dissipation.

VLSI amplifiers based on the synchronous chopper circuit technique may offer very low dc offset at low power dissipation. However, these amplifiers have narrower bandwidth and higher ripple noise (especially at the chopping frequency).

The VLSI amplifiers may combine both the auto-zero and the chopper-stabilization circuit techniques to achieve very low dc offset through the noise shaped over frequency. Such amplifiers offer the widest operation bandwidth with lower ripple at the chopper frequency.

6.2 VLSI AUTO-ZERO CIRCUITS

VLSI auto-zero circuits are based on the correlative double sampling (CDS) principle that works in two or more clock phases, where the noise signal is first sampled and stored. The sampled noise signal is then subtracted from the sampled signal in the later phase either at the input, the output or at some intermediate nodes of the circuit. As the result, the noise signal can be eliminated from the circuit output.

This CDS operation can be used together with the signal amplification operation. An auto-zero amplifier based on such technique corrects the input offset in the similar way. However this sample-and-hold operation turns auto-zero amplifiers into sampled-data circuits, making them prone to aliasing and fold-back effects. At low frequency, the noise changes slowly, so the subtraction of the two consecutive noise samples results in true cancellation. This correlation diminishes at higher frequencies, with subtraction errors causing wideband components to fold back into the baseband. Thus, the auto-zero amplifiers usually have more in-band noise than standard amplifiers.

6.2.1 AUTO-ZERO CIRCUIT PRINCIPLE

Shown in figure 6.3 is a simple VLSI auto-zero amplifier circuit structure. This circuit works with a noise sampling phase (ϕ1 phase) and a noise cancellation phase (ϕ2 phase).

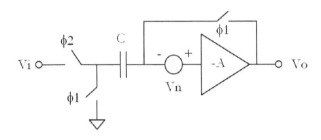

Fig.6.3 CDS based VLSI auto-zero technique

Shown in figure 6.4 is the equivalent circuit of the auto-zeroing phase where V_N is the effective input referred low frequency noises (e.g. the dc offset and the 1/f noise) of the amplifier circuit. The parameter A is the gain of the noise free ideal amplifier. In this circuit configuration, the input is connecting to a ground (or reference) and the amplifier input and output terminals are shorted together.

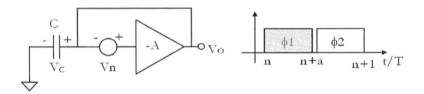

Fig.6.4 Auto-zeroing circuit

The effective noise voltage in this noise sampling phase is sampled and stored in the sampling capacitor as:

$$V_C(nT + aT) = -(1 - \frac{1}{A})V_n(nT + aT) \qquad (6.2)$$

During the signal amplification phase, the sampling capacitor is re-connected to the input as shown in figure 6.5. The sampled noise voltage can be subtracted from the output voltage as:

$$V_o(t) = -A(V_i(t) + V_C(nT + aT) + V_n(t)) \qquad (6.3)$$

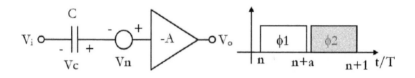

Fig.6.5 Amplification phase of CDS technique

The output voltage of the amplifier at the end of the $\phi 2$ phase is given as:

$$V_o(nT+T) = -A(V_i(nT+T) + V_n(nT+T) - (1-\frac{1}{A})V_n(nT+aT)) \quad (6.4)$$

The effective input referred offset after the CDS compensation is given as:

$$V_n'(nT+T) = V_n(nT+T) - (1-\frac{1}{A})V_n(nT+aT) \qquad (6.5)$$

Especially, if the input referred noise is a dc offset, we have that

$$(V_n')_{DC} = \frac{1}{A}(V_n)_{DC} \qquad (6.6)$$

It can be seen that the dc offset is reduced by a factor equal to the dc gain of the amplifier.

For low frequency noise, the input referred noise can be expressed as a function of the original noise and the amplifier gain A by applying the z-transform to above equation as:

$$V_n'(z) = V_n(z)[1-(1-\frac{1}{A})z^{a-1}] \tag{6.7}$$

The frequency response of this effective input referred noise as a function of the original amplifier noise, the amplifier gain, and the frequency can be expressed as:

$$\frac{V_n'(z)}{V_n(z)}\bigg|_{z=e^{j\omega T}} = [1-(1-\frac{1}{A})e^{j\omega T(a-1)}] \tag{6.8}$$

For signal frequency much lower than the clock rate, we have that

$$\omega T \ll 1 \tag{6.9}$$

The noise transfer function of the circuit can be simplified as:

$$\frac{V_n'(z)}{V_n(z)}\bigg|_{z=e^{j\omega T}} \approx [1-(1-\frac{1}{A})(1+j\omega T(a-1))] \tag{6.10}$$

It can be seen that such noise transfer function has a highpass noise shaping that can be used to minimize the dc offset, the low frequency noise (such as 1/f noise) if the gain of the amplifier is high enough.

6.2.2 AUTO-ZERO CIRCUIT IMPLEMENTATION

Shown in figure 6.6 is a basic auto-zero amplifier circuit structure.

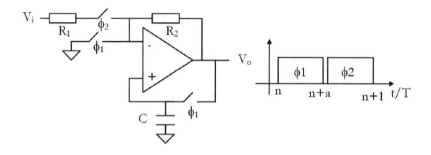

Fig.6.6 Basic auto-zero amplifier structure

During the auto-zero phase ($\phi1$ phase) as shown in figure 6.7, the input referred noise of the amplifier is sampled and stored on the capacitor C that is connected to the positive input terminal of the amplifier.

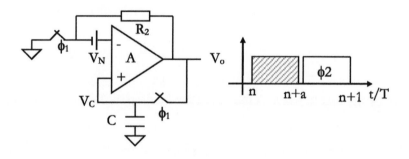

Fig.6.7 Equivalent circuit in $\phi1$ phase

The voltage of capacitor at the end of the $\phi1$ phase is given as:

$$V_C(n+a) = \frac{A}{A-1}V_N(n+a) \approx V_N(n+a) \qquad (6.11)$$

Where A is the gain of the amplifier and V_N is the equivalent input referred noise of the amplifier.

During the amplification phase ($\phi2$ phase) shown in figure 6.8, the stored offset voltage is then subtracted from the amplifier output since the positive terminal of the amplifier is connected to the sampled offset voltage.

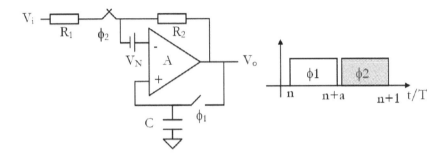

Fig.6.8 Equivalent circuit in $\phi2$ phase

The output of the amplifier at the end of $\phi2$ phase is given as:

$$V_o(n+1) = A[\frac{R_1 V_o(n+1) + R_2 V_i(n+1)}{R_1 + R_2} + \frac{A V_N(n+a)}{A-1} - V_N(n+1)] \quad (6.12)$$

For large gain A we have that

$$V_o(n+1) = -\frac{R_2}{R_1}V_i(n+1) + \frac{R_1 + R_2}{R_1}[V_N(n+1) - V_N(n+a)] \quad (6.13)$$

The first term in the equation represents the normal inverted gain operation of the circuit. The second term is an additive noise introduced by the input referred noise V_N of the amplifier.

$$V_{No}(n+1) = \frac{R_1 + R_2}{R_1}[V_N(n+1) - V_N(n+a)] \quad (6.14)$$

By applying the z-transform to above equation we have that z-domain noise transfer function of the circuit as

$$H_N(z) \equiv \frac{V_{No}(z)}{V_N(z)} = \frac{R_1 + R_2}{R_1}[1 - z^{-(1-a)}] \quad (6.15)$$

The frequency response of such output noise transfer function is given as:

$$H_N(j\omega) = H_N(z)|_{z=\exp(j\omega T)} = \frac{R_1 + R_2}{R_1}[1 - e^{-j\omega T(1-a)}] \propto \sin(\frac{\omega T(1-a)}{2}) \quad (6.16)$$

It is important to see that

- The noise transfer function as shown in figure 6.9 has a highpass frequency response. Therefore such a circuit can be used to minimize low frequency noise (such as 1/f noise and the drift) effects that are much lower than the control clock frequency (i.e. 1/T).

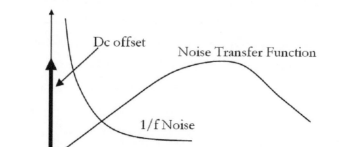

Fig.6.9 Noise transfer function of auto-zero circuit

- The dc offset represents the noise content at dc that will be totally filtered

6.2.3 ALIASING EFFECT OF AUTO-ZERO CIRCUITS

Since the auto-zero circuit is basically a sampled data circuit, the wideband noise will be aliased down into the baseband, resulting in increased inband noise unless the system is already a sampled data circuit or there is no noise at high frequency. The noise aliasing effect can be analyzed as follows.

By assuming the source noise voltage corresponds to a PSD SN(f), the PSD of the auto-zero voltage across the switch can be decomposed into two terms including the inband noise that will be reduced by the auto-zero operation and the fold over components caused by the aliasing effect:

$$S_{No}(f) = |H_0(f)|^2 S_N(f) + S_{Fold}(f) \qquad (6.17)$$

where

$$S_{Fold}(f) = \sum_{\substack{n=-\infty \\ n \neq 0}}^{n=+\infty} |H_n(f)|^2 S_N(f - \frac{n}{T}) \qquad (6.18)$$

The transfer function is given as

$$|H_0(f)| = k^2 \{[1 - \frac{\sin(2\pi f T_h)}{2\pi f T_h}]^2 + [1 - \frac{\cos(2\pi f T_h)}{2\pi f T_h}]^2\} \qquad (6.19)$$

where k is the duty-cycle of the sampling clock (k = T_h/T). For $fT_h \ll 1$, we have that

$$|H_0(f)| \approx \pi f T_h \qquad (6.20)$$

It has a highpass frequency response (i.e. noise shaping of auto-zero circuits) that compensate for the dc and low frequency noises.

The PSD at the output of the auto-zero circuit depends on the source noise spectrum. For a typical VLSI circuit contain the 1/f and the white noise with corner frequency f_k as:

$$S_N(f) = S_0(1 + \frac{f_k}{|f|}) \qquad (6.21)$$

The foldover noise of an ideal amplifier with bandwidth B can be estimated as

$$\sum_{-\infty}^{+\infty} S_N(f - \frac{n}{T}) \approx 2BTS_0 \qquad (6.22)$$

and

$$S_{Fold-White}(f) \approx (2BT - 1)S_0[\frac{\sin(\pi f T)}{\pi f T}]^2 \qquad (6.23)$$

The first-order lowpass filtered white noise and 1/f noise can be given respectively as:

$$S_{White}(f) = \frac{S_0}{1 + (\frac{f}{f_c})^2} \qquad (6.24)$$

$$S_{1/f}(f) = \frac{S_0 f_k}{|f|[1 + (\frac{f}{f_c})^2]} \qquad (6.25)$$

The foldover noises are given respectively as:

$$S_{Fold-White}(f) \approx (2f_cT-1)S_0[\frac{\sin(\pi fT)}{\pi fT}]^2 \qquad (6.26)$$

and

$$S_{Fold-1/f}(f) \approx (1+\ln(\frac{2}{3}f_cT))2f_kS_0[\frac{\sin(\pi fT)}{\pi fT}]^2 \qquad (6.27)$$

The total foldover noise for both white and 1/f noises is given as:

$$S_{Fold}(f) \approx [(2f_cT-1)+(1+\ln(\frac{2}{3}f_cT))2f_k]S_0[\frac{\sin(\pi fT)}{\pi fT}]^2 \qquad (6.28)$$

6.3 VLSI CHOPPER STABLIZATION CIRCUITS

Unlike the auto-zero circuits, the VLSI chopper stabilization circuits modulate the dc and low frequency noise to the chopper frequency such that they can be filtered out using a lowpass filter. The chopper stabilization circuit techniques can be used very effectively to compensate for the dc offset, the drift and the 1/f noises in the VLSI circuits.

VLSI chopper amplifier techniques are widely used in high precision amplifier design for low noise and offset characteristics where static offset cancellation circuit techniques, such as trimming, cannot be used to reduce the time dependent noises such as the 1/f noise.

A classic application of the chopper circuit is the dc amplification, where the signal input to be amplified can be so small that an incredibly high gain is required. However very high gain dc is difficult to achieve without low offset and low 1/f noise, and reasonable stability and bandwidth. A chopper circuit is used to break up the input signal so that it can be processed as if it were an ac signal. In this way, extremely small dc signals can be amplified.

The chopper amplifiers were originally invented years ago to solve the dc drift issue of amplifiers by converting the dc voltage to an ac signal. The initial circuit implementations used ac coupling of the switched input signal and synchronous demodulation of the ac signal to recover the dc signal at the output. These amplifiers had limited bandwidth and required post-filtering to remove the large ripple voltages generated by the chopping action.

6.3.1 CHOPPER CIRCUIT PRINCIPLE

A conceptual chopper amplifier circuit is shown in figure 6.10. The signal and noise power spectrum along the signal path of the circuit are shown in figure 6.11. It can be seen that the band-limited signal at input is chopped such that its frequency is shifted to the chopper clock frequency and its harmonics at location V_A. The dc and the $1/f$ noise will be added at location V_B after the chopped signal goes through the signal processing circuit K. The chopped signal with added noise is then converted back to the low frequency with the second chopper stage. Such an operation also shifts the added dc offset and low frequency noise to the chopper frequency and its harmonics as shown at location V_C. This noise can be separated from the signal in frequency domain and then be filtered using a lowpass filter.

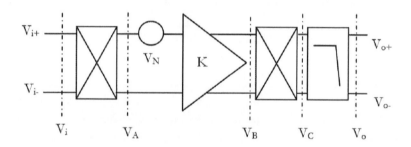

Fig.6.10 Chopper amplifier principle

An s-domain equivalent circuit model of the chopper amplifier is shown in figure 6.12. In such a chopper amplifier circuit structure, the single-tone input signal VI(t) is first chopped using a clock of frequency f_m. Mathematically, the chopped single-tone signal can be expressed as:

$$V_A(t) = A\sin(\omega t) \sum_{\substack{k=1 \\ k=odd}}^{\infty} \frac{4}{k\pi} \sin(k\omega_m t)$$

$$= A \sum_{\substack{k=1 \\ k=odd}}^{\infty} \frac{2}{k\pi} [\cos((k\omega_m - \omega)t) - \cos((k\omega_m + \omega)t)]$$

(6.29)

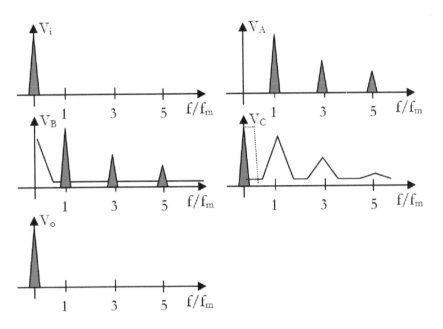

Fig.6.11. Chopper amplifier signal spectrum

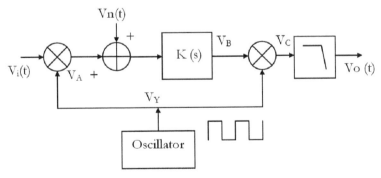

Fig.6.12 Chopper amplifier s-domain model

Shown in figure 6.13 is the simulated power spectrum density of a chopped single-tone input.

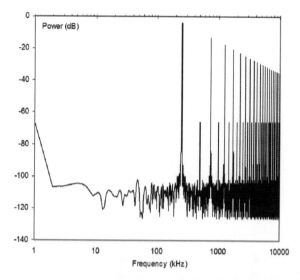

Fig. 6.13 Power spectrum of 1 kHz signal with 250 kHz chopper

It can be seen that after the chopping operation, the signal is transposed around the odd harmonic frequency of the modulation frequency. It is then amplified and de-modulated back to the original band. Because of the finite bandwidth of the amplifier, the output signal contains spectral components around the even harmonics of the chopper frequency. The output signal that is related to the input signal is given as:

$$V_o(s) = K(s)V_i(s) \qquad (6.30)$$

On the other hand, the noise and the dc offset are only chopped once and translated to the odd harmonics of the chopper modulation frequency. The output spectrum is given as:

$$P_{no}(f) = (\frac{2}{\pi})^2 \cdot \sum_{n=-\infty}^{+\infty} \frac{1}{(2n+1)^2} |K(f - \frac{2n+1}{T})|^2 \, P_{ni}(f - \frac{2n+1}{T}) \qquad (6.31)$$

Where $P_{ni}(f)$ is the equivalent input noise power density.

It can be seen that at higher frequency beyond the chopping frequency, the equivalent low-frequency input noise of the chopper amplifier is equal to the original amplifier white-noise components. Contrary to the auto zero technique, where the white noise is not aliased because there is no sample/hold process. This effect suggests that an auto-zero circuit technique is more suitable for sampled data circuits, such as the switched-capacitor circuit where the under-sampling process is unavoidable and that the chopper technique is better used in continuous-time applications.

6.3.2 VLSI CHOPPER CIRCUIT STRUCTURES

Shown in figure 6.14 is a simple VLSI chopper amplifier circuit where the chopper (or commutating) amplifier modulates the input to create an ac signal. This chopped signal is then amplified using an ac amplifier to increase the signal magnitude. The amplified ac signal is then synchronously demodulated back to the dc at the output of the circuit.

Since the switching operation of the chopper may introduce noise at the fundamental and harmonic of the chopper frequency, a lowpass filter is usually needed to attenuate such switching noise.

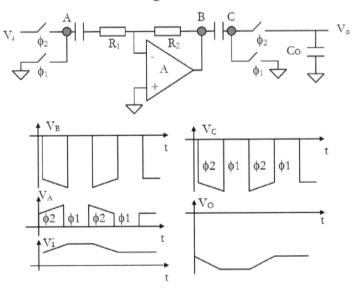

Fig.6.14 Classic chopper amplifier structure

Another classic implementation of the chopper amplifier is shown in figure 6.15 where the low frequency input signal is chopped into an ac signal using a chopping clock that is significantly higher than the signal band. The amplifier A_1 then amplifies the chopped signal. The amplified signal is synchronous demodulated back to low (base band) frequency. The demodulated signal is lowpass filter constructed using amplifier A_2 to filter the chopping noise and modulated dc offset, drift and 1/f noises.

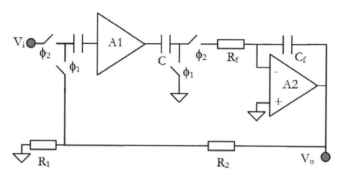

Fig.6.15 Classic chopper amplifier circuit implementation

This chopper amplifier has a high dc gain that is equal to the product of the open-loop gain of the two amplifiers. The output offset, the drift and the 1/f noise can be significantly minimized when it is referred back to the input. The dc offset, drift and 1/f of the first amplifier will be modulated to a band closed to chopping frequency that will be filtered by the second amplifier stage.

For the simple chopper amplifier circuit shown in figure 6.16 the dc or low frequency input signal can be amplified by chopped into an ac signal with frequency range between 100 to 10 kHz and recovered using a lowpass filter at the circuit output.

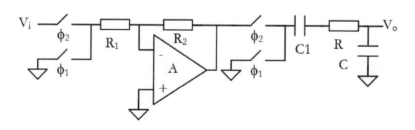

Fig.6.16 Simple chopper amplifier structure

The node voltage waveforms of this chopper amplifier are shown in figure 6.17.

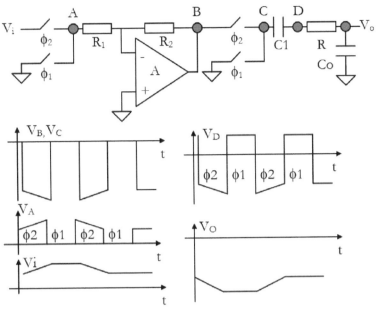

Fig.6.17 Simple chopper amplifier operation

An alternative chopper amplifier implementation is shown in figure 6.18.

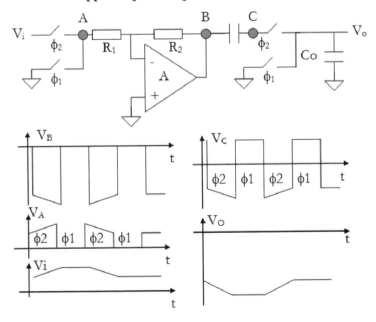

Fig.6.18 Alternative simple chopper amplifier

6.4 VLSI CIRCUIT IMPLEMENTATIONS

The VLSI auto-zero and the VLSI chopper circuit techniques can be used together to compensate for the dc offset and the low frequency noises.

These VLSI input offset and noise cancellation circuits are based on the following circuit techniques:

- The output of the circuit is reset to V_N. The input capacitor stores the noise and then subtracted from the input signal;
- The output of the circuit is held during the reset. The input and the feedback capacitors are referred to the previous value of virtual ground voltage;
- The output of the circuit during reset anticipates the next output. Signal path capacitors are referred to the anticipated value with or without the use of extra offset-storage capacitor;
- An auxiliary input signal is established during reset then it is compensated for slow varying noise effects.

6.4.1 INVERTED GAIN CHOPPER STAGE

Shown in figure 6.19 is a VLSI inverted chopper amplifier circuit. In this circuit the differential input is chopped such that the output will take two signal paths either directly from the first Opamp output or from the inverted output of the second Opamp.

In the $\phi 1$ phase, the equivalent circuit is similar to the basic circuit shown in Fig.6.19a.

In the $\phi 2$ phase, an inversion in the opamp output is introduced to reverse the chopper path in the input signal.

Such a chopper circuit provides a modulation of the input referred noise and offset of the input opamp circuit. The noise of the amplifier circuit is shaped away from the signal frequency band into higher frequency.

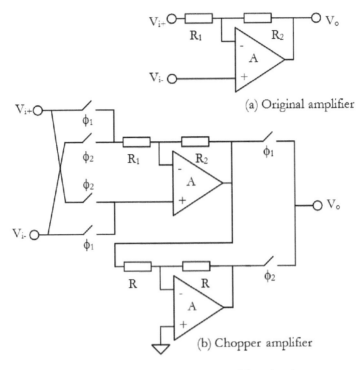

Fig.6.19 Simple VLSI chopper amplifier circuit

6.4.2 OPEN-LOOP CDS CIRCUITS

Shown in figure 6.20 is a CDS circuit structure employing the VLSI open-loop offset cancellation circuit technique.

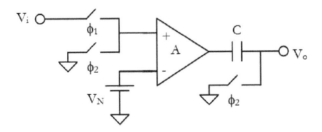

Fig.6.20 CDS based open loop offset cancellation circuit

During the φ2 phase the input referred noise V_N is sampled and stored in the capacitor connected to the output of the amplifier with the value given as:

$$V_C(t_1) = AV_N(t_1) \quad (6.32)$$

where t_1 is the time point at the end of the noise-sampling phase.

During the φ1 phase (signal amplification phase), the amplifier amplifies the input signal and the input noise voltage with the output voltage at the amplifier output given as:

$$V_x(t) = A[V_i(t) - V_N(t)] \quad (6.33)$$

The output voltage in this phase can be further given as:

$$V_o(t) = A[V_i(t)] - A[V_N(t) - V_N(t_1)] \quad (6.34)$$

It can be seen that the amplifier input noise is highpass shaped with the offset being eliminated from the output signal.

The above circuit can be extended into multiple gain stage as shown in figure 6.21.

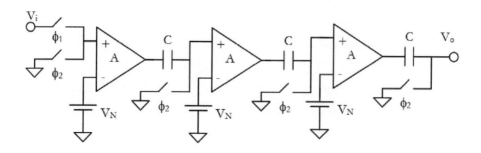

Fig. 6.21 Multi-stage CDS based open loop offset cancellation circuit

6.4.3 CLOSED-LOOP CDS CIRCUITS

Shown in figure 6.22 is a VLSI closed-loop CDS offset cancellation circuit structure.

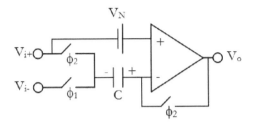

Fig.6.22 Closed-loop CDS offset cancellation circuit

In the $\phi 2$ phase (noise sampling phase), the input referred noise V_N is sampled in the capacitor connected to the input of the amplifier with value given as:

$$V_C(t_1) = AV_N(t_1) \tag{6.35}$$

where t_1 is the time point at the end of the noise-sampling phase.

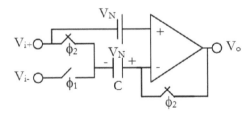

Fig.6.23 $\phi 2$ phase equivalent of offset cancellation circuit

During the $\phi 1$ phase (signal amplification phase), the stored offset and low frequency noise are subtracted from the amplified signal at amplifier output as:

$$V_o(t) = A[V_{i+}(t) - V_{i-}(t)] - A[V_N(t) - V_N(t_1)] \qquad (6.36)$$

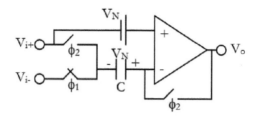

Fig.6.24 φ1 phase equivalent of offset cancellation circuit

Such an operation effectively provides a highpass noise shaping to the input offset and noise.

Similar to the open-loop offset cancellation circuits, the closed-loop offset cancellation circuits can also be cascaded to form the multi-stage offset cancellation circuit as shown in figure 6.25.

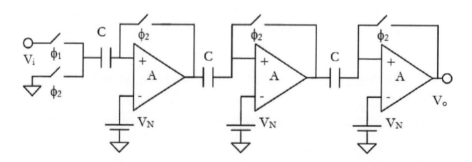

Fig. 6.25 Multi-stage offset cancellation circuit

An alternative offset cancellation circuit structure shown in figure 6.26 uses an addition circuit to store the output offset of the main amplifier in the sampling phase (φ1 phase) and to cancel it out in the operation phase (φ2 phase). In this circuit implementation, the output offset and low frequency noise are detected by the noise amplifier A_N and are sampled into holding capacitor C during the

ϕ_1 phase. These noises are then subtracted from the amplifier output during the ϕ_2 phase.

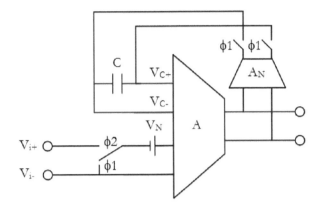

Fig. 6.26 Offset cancellation circuit using additional mulling terminals

Shown in figure 6.27 are two VLSI implementations of the input offset compensation circuit structure.

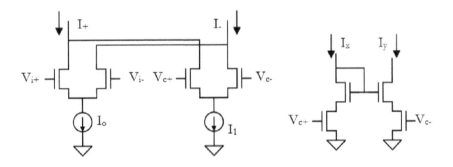

Fig.6.27 VLSI programmable offset circuits

6.4.4 INVERTER CDS COMPARATOR CIRCUITS

Shown in figure 6.28 is a simple VLSI offset cancellation circuit employing the CMOS inverter comparator for high-speed A/D conversion.

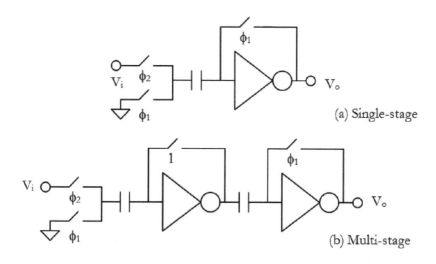

Fig. 6.28 VLSI inverter-based offset cancelled comparators

In the $\phi 1$ phase the inverter's output is shorted to its input. It generates a voltage bias of circuit that equals to the threshold of the inverter as shown in figure 6.29. This bias voltage (including the input referred noise and offset) is then stored to the sampling capacitor C as:

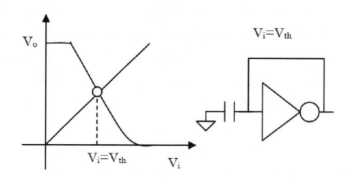

Fig. 6.29 Storage of the comparator threshold in $\phi 1$ phase

$$V_C(t_1) = V_{th}(t_1) + V_N(t_1) \qquad (6.37)$$

In the φ2 phase, the input voltage of the inverter circuit is given as:

$$V_i(t) = V_i(t) + [V_{th}(t_1) + V_N(t_1)] \qquad (6.38)$$

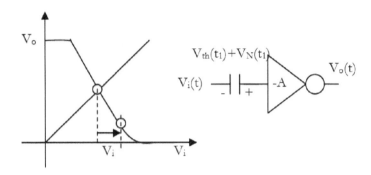

Fig. 6.30 Noise and threshold cancellation in φ2 phase

The output of the circuit can be derived as:

$$V_o(t) = -AV_o(t) + -A[V_{th}(t) + V_N(t) - V_{th}(t_1) - V_N(t_1)] \qquad (6.39)$$

It can be seen that such a circuit offers a highpass frequency shaping operation for the input noise and the threshold voltage variations of the inverter circuit.

6.4.5 CDS SAMPLE/HOLD CIRCUIT

Shown in figure 6.31 is a VLSI SC sample/hold circuit. This circuit offers the offset cancellation capability based on the CDS circuit technique.

At the end of φ1 phase the opamp is connected as the unity gain circuit and the opamp input noise and the input voltage are sampled into the capacitor C as:

$$V_C(t_1) = V_i(t_1) - V_N(t_1) \tag{6.40}$$

In the φ2 phase, the output of the circuit is given as:

$$V_o(t) = V_i(t_1) + V_N(t) - V_N(t_1) \tag{6.41}$$

It can be seen that this circuit also offers a highpass noise shaping for the opamp input noise.

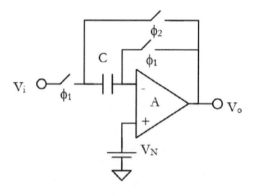

Fig.6.31 SC CDS S/H circuit

6.4.6 CDS AMPLIFIER CIRCUITS

Shown in figure 6.30 is a SC amplifier circuit that offers the CDS offset cancellation capability.

This circuit works the similar way as previously discussed circuit. At the end of φ1 phase the opamp is connected as the unity gain circuit as shown in figure 6.33. The opamp input noise and the input voltage are sampled into the capacitor C_1 and C_2 respectively as:

$$V_{C_1}(t_1) = V_i(t_1) - V_N(t_1) \tag{6.42}$$

$$V_{C_2}(t_1) = V_N(t_1) \tag{6.43}$$

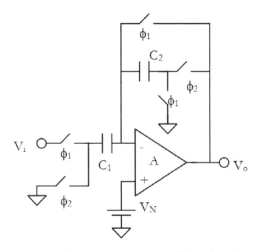

Fig.6.32 SC CDS amplifier circuit

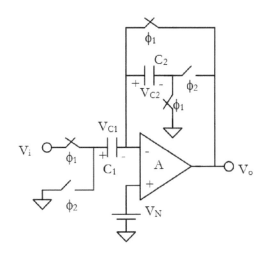

Fig.6.33 Sampling of the noise

In the φ2 phase, the charge in capacitor C_1 is transferred to capacitor C_2 as result of charge redistribution effect and the output of the circuit is given as:

$$V_o(t) = \frac{C_1}{C_2}V_i(t_1) + \frac{C_1+C_2}{C_2}[V_N(t)-V_N(t_1)] \qquad (6.44)$$

It can be seen that such circuit offers a scaling to the input voltage based on the ratio of the two capacitances in the circuit. At the same time it also offers a highpass shaping for the input noise of the opamp circuit.

Shown in figure 6.34 is an alternative SC amplifier circuit that offers the CDS offset cancellation and finite gain compensation capability.

In this amplifier circuit, the feedback reset CDS switch is replaced by an S/H circuit structure using the capacitor C_3.

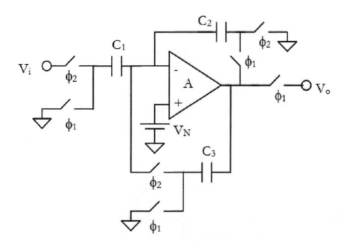

Fig.6.34 SC CDS amplifier circuit with gain compensation

In the $\phi 1$ phase as shown in figure 6.35, capacitor C_1 is charged to the input referred noise $V_N(t_1)$. The output voltage is stored in capacitor C_2 and C_3 as:

$$\begin{cases} Q_1(t_1) = -C_1 V_N(t_1) \\ Q_2(t_1) = -C_2[V_N(t_1)-V_o(t_1)] \\ Q_3(t_1) = C_3 V_o(t_1) \end{cases} \qquad (6.45)$$

In the φ2 phase as shown in figure 6.36, the charges are transferred among the capacitor C_2 as result of the charge redistribution effect and the output of the circuit is given as:

$$\begin{cases} Q_1(t) = C_1[V_i(t) - V_N(t)] \\ Q_2(t) = -C_2 V_N(t) \\ Q_3(t) = C_3[V_o(t) - V_N(t)] \end{cases} \quad (6.46)$$

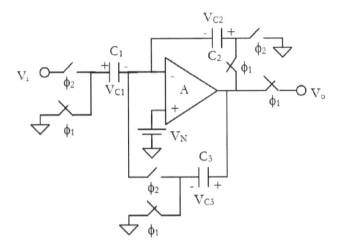

Fig.6.35 φ1 phase equivalent circuit

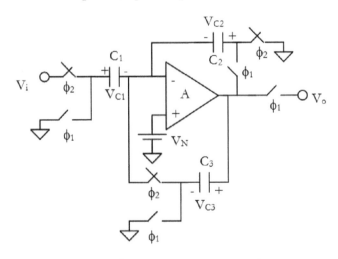

Fig.6.36 φ2 phase equivalent circuit

Combining above equations we have that

$$V_o(t) = \frac{C_1}{C_2}V_i(t_1) + \frac{C_1+C_2}{C_2}[V_N(t) - V_N(t_1)] \qquad (6.47)$$

It can be seen that such a circuit offers a scaling to the input voltage determined by the ratio of the two capacitances. This circuit also offers a highpass noise-shaping operation for the input noise of the opamp circuit.

A wideband VLSI implementation of the SC CDS circuit is shown in figure 6.37. Such circuit shares similar working principle as other circuits discussed above.

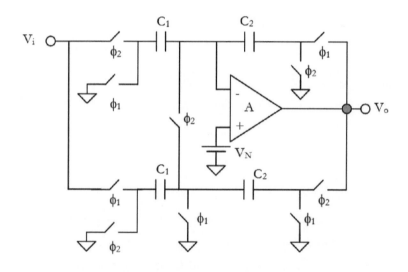

Fig.6.37 SC wideband CDS amplifier circuit

6.5 VLSI COMPOUND ZERO-DRIFT AMPLIFIERS

VLSI chopper circuit techniques can be used to compensate for the dc offset, the drift and the low frequency (such as 1/f) noises in the circuits. However, the signal bandwidths of such circuits are limited by the clock frequency of the chopper clock. Therefore such circuit techniques are only suitable for low frequency signal processing applications, such as audio signal band.

One solution is to use the compound amplifier circuit techniques to separate the normal high bandwidth signal amplification path from the low bandwidth offset and drift compensation path. VLSI compound amplifiers usually use a secondary auxiliary amplifier to correct the input offset voltage of a main amplifier.

VLSI compound amplifier solves the bandwidth limitation problem of chopper amplifier by using the chopper amplifier to stabilize a conventional wideband amplifier that remained in the signal path. A typical compound amplifier employs an auto-zero approach using a two-or-more-stage composite amplifier structure similar to the chopper-stabilized scheme. The difference is that the stabilizing amplifier signals are connected to the wide-band or main amplifier through an additional "nulling" input terminal, rather than one of the differential inputs. Higher frequency signals bypass the nulling stage by direct connection to the main amplifier or through the use of feed-forward techniques, maintaining a stable zero in wide-bandwidth operation.

This technique thus combines dc stability and good frequency response with the accessibility of both inverting and non-inverting configurations. However it may produce interfering signals consisting of high levels of digital switching noise that limit the usefulness of the wider available bandwidth. It also causes intermodulation distortion (IMD), which looks like aliasing between the clock signal and the input signal, producing error signals at the sum and difference frequencies.

6.5.1 COMPOUND CIRCUIT STRUCTURES

Shown in figure 6.38 is a VLSI compound amplifier circuit structure, where A is the main wideband amplifier that may suffer from the dc offset, drift and 1/f noise limitations. A_1 is a low bandwidth auxiliary amplifier with compensated dc offset, drift and 1/f noise employing certain auto-zero circuit technique.

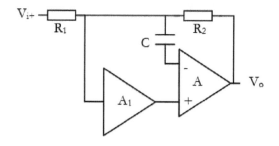

Fig.6.38 VLSI compound amplifier circuit structure (I)

In this compound auto-zero amplifier circuit, the fast signal will go to the main amplifier through the ac coupling capacitor to ensure the wide bandwidth circuit operation. The low frequency and dc signal content will go through the chopper stabilized amplifier signal path. Since the low frequency noise effect of the main amplifier will be attenuated by the high gain of the chopper stabilized amplifier circuit. The dc offset, the drift and the 1/f noise will be eliminated in such circuit.

A VLSI compound opamp circuit structure with similar concept and slightly different implementation is shown in figure 6.39. Such a circuit is based on a trimmable amplifier circuit as shown in figure 6.40 where A is the main wideband amplifier and A_1 is a nulling amplifier employing the chopper circuit techniques.

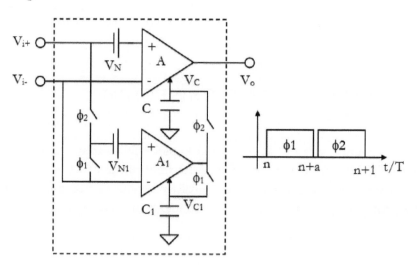

Figure 6.39 VLSI compound amplifier circuit structure (II)

VLSI Modulation Circuits

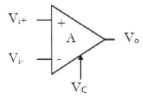

Fig.6.40 trimmable opamp circuit model

For the trimmable opamp circuit the V_o is given as inputs of the opamp and the signal gain A and trim gain B as:

$$V_o = A(V_{i+} - V_{i-}) + BV_C \qquad (6.48)$$

The equivalent circuit in phase 1 (ϕ1 phase) is shown in figure 6.41 where V_N and V_{N1} are used to model the input referred low frequency noise of the main and the nulling amplifiers respectively.

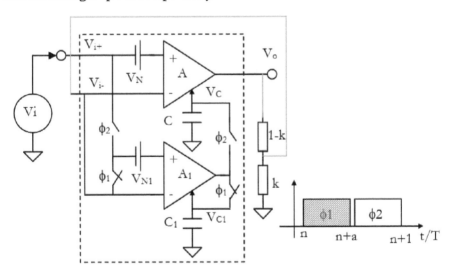

Fig.6.41 Auto-zero phases (ϕ1 phase)

During this phase, a nulling voltage is generated at the output of the nulling amplifier to charge the storage capacitor C_1 with value given at the end of this phase as (ignoring T for simplicity):

$$V_{C1}(n+a) = A_1 V_{N1}(n+a) + B_1 V_{N1}(n+a) = \frac{A_1 V_{N1}(n+a)}{1-B_1} \quad (6.49)$$

The equivalent circuit in φ2 phase (normal operation phase) is shown in figure 6.42, where the input is applied to the positive input terminal of the main amplifier and a negative feedback is applied to the negative terminal of the main amplifier.

In addition, a nulling voltage is applied to the nulling input of the main amplifier as:

$$V_C(t) = A_1(V_i(t) + V_{N1}(t) - kV_o(t)) + B_1 V_N(n+a) \quad (6.50)$$

Or

$$V_C(t) = A_1(V_i(t) + V_{N1}(t) - kV_o(t)) + A_1 \frac{B_1 V_{N1}(n+a)}{1-B_1} \quad (6.51)$$

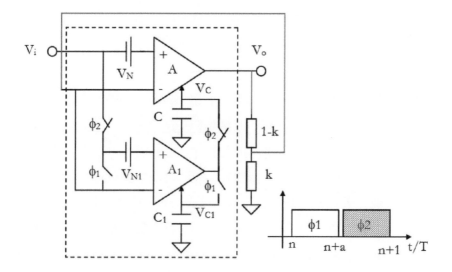

Fig. 6.42 Amplification phase (φ2 phase)

The output voltage of the main amplifier is given in this phase as:

$$V_O(t) = A(V_i(t) + V_N(t) - kV_O(t)) + BV_C(t) \quad (6.52)$$

Solving for Vo from the above equations, we have that

$$V_O(t) = \frac{(A+BA_1)}{1+k(A+BA_1)}V_i(t) + V_{OS}(t) \quad (6.53)$$

Where the output voltage noise Vos(t) of the circuit is given as:

$$V_{OS}(t) \equiv \frac{AV_N(t) + BA_1[V_{N1}(t) - \frac{B_1}{B_1-1}V_{N1}(n+a)]}{1+k(A+BA_1)} \quad (6.54)$$

For high gain amplifier A and A_1 with ignoring dc offset voltage this circuit serves as a wide bandwidth positive gain amplifier circuit with the voltage gain given as

$$\frac{V_O}{V_i} = \frac{1}{k} \quad (6.55)$$

The output noise for the high amplifier gain is given as:

$$V_{OS}(t) \approx \frac{V_N(t)}{k(1+BA_1/A)} \frac{[V_{N1}(t) - V_{N1}(n+a)]}{k(A/(BA_1)+1)} \quad (6.56)$$

The first term is in continuous-time form that will be compensated by the high nulling amplifier signal gain and the nulling gain. The second term, on the other hand, is in the discrete-time that will be shaped by a highpass frequency response similar to the CDS circuit. Therefore these low frequency noises with the chopping clock frequency (dc offset, drift and 1/f) will be highly attenuated.

A third VLSI compound amplifier implementation is shown in figure 6.43 that consists of a two-stage main opamp and a chopper auxiliary amplifier. A negative feedback is used to reduce the dc gain of the main opamp.

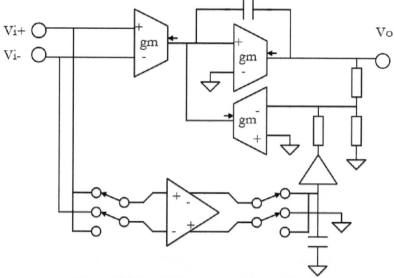

Fig.6.43 VLSI chopper stabilized opamp circuit

Highlighted in figure 6.44 is the main wideband opamp sub-circuit structure. Such a circuit features a high dc gain determined by the Gm and the effective output resistance of the first and second amplifier stages.

$$(A_{DC})_{MAIN} = G_{m1} \cdot R_{o1} \cdot G_{m2} \cdot R_{o2} \qquad (6.57)$$

Highlighted in figure 6.45 is a dc feedback path of the circuit to reduce the dc gain of the first opamp stage. Such a feature can be used to minimize the dc offset, drift and low frequency noise of the main opamp and to improve the phase margin of the opamp circuit.

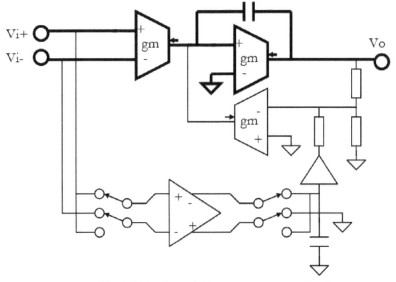

Fig.6.44 Main wideband opamp sub circuit

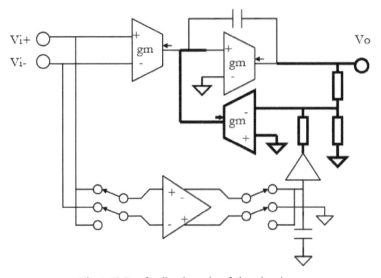

Fig.6.45 Dc feedback path of the circuit

Highlighted in figure 6.46 is the chopper stabilized nulling amplifier sub-circuit structure.

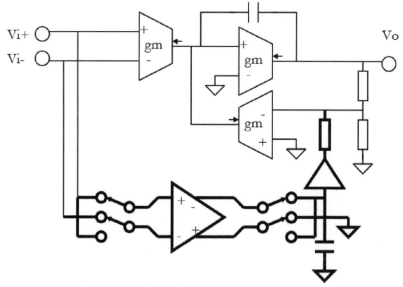

Fig.6.46 VLSI chopper stabilized opamp circuit

Alternatively a compound amplifier can also be implemented based on the dual input scheme as shown in figure 6.47.

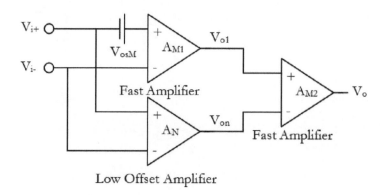

Fig. 6.47. Dual input compound amplifier

Since one drawback of the basic auto-zero amplifier circuit is that phase $\phi 1$ of such circuit is not available for signal amplification. This issue can be resolved by introducing a second amplifier (called nulling amplifier) for the offset cancellation such that the main amplifier can be operated without the influence of the auto-zero operation

Shown in figure 6.48 is a VLSI compound amplifier based on the bias adjustment technique. In this compound amplifier a chopper amplifier (i.e. nulling amplifier) is used to compensate for the dc offset and the low frequency noise of the main amplifier. Since the chopper circuit is only implemented in the nulling amplifier, the impact of the switching effect of the chopper to the main amplifier can be minimized.

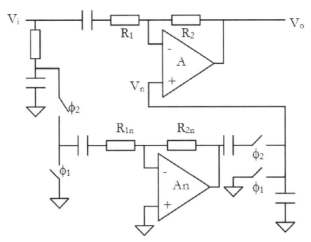

Fig.6.48 Bias adjustment compound amplifier structure

Shown in figure 6.49 is a VLSI compound zero-drift opamp circuit structure that is based on a chopper feedback loop for compensating the dc and 1/f noise.

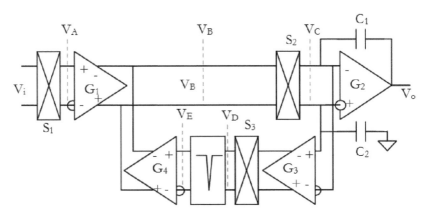

Fig.6.49 compound amplifier with local chopper feedback loop

6.5.2 COMPOUND AMPLIFIER CHARACTERISTICS

VLSI compound auto-zero amplifiers offer a few attractive characteristics. The product of the gains of two amplifiers determines the dc open-loop voltage gain that can be extremely high, typically more than 140dB. The offset voltage after the auto-zero will be very low due to the effect of the large nulling-terminal gain on the uncompensated amplifier offsets. Typical output offset voltages of compound amplifiers could be as low as a few uV. The low effective offset voltage also helps to improve other circuit parameters, such as the dc CMRR and the PSRR, which can typically be better than 140dB. Since the offset voltage is continuously compensated, the shift in voltage offset over time will be extremely small (e.g. as low as $40\text{-}50nV/(month)^{0.5}$). The temperature induced offset can be as low as a few nV/C.

The compound auto-zero amplifiers also minimize the low-frequency noise effects such as the 1/f noise. For the uncompensated circuit, the input voltage noise spectral density increases exponentially inversely with frequency below the corner frequency, which is typically in the order of a $10^0\text{-}10^2$ Hz. The chopper-stabilized amplifier or the auto-zero amplifiers will minimize this low-frequency noise. The auto-correction action becomes more efficient as the frequency approaches dc. As a result of the high-speed chopper action in the auto-zero amplifiers, the low-frequency noise is relatively flat down to dc (i.e. no 1/f noise). The elimination of 1/f noise is very attractive in low-frequency applications where long sampling intervals are common.

In the chopper circuits, the switching induced charge injection effects generate both voltage and current transient noises at the chopping clock frequency and its harmonics. These noises are significant compared to the wideband noise floor of the amplifier that can be a significant error source if they fall within the frequency band of interest for the signal path. Even worse, this switching causes intermodulation distortion of the output signal, generating additional error signals at sum and difference frequencies. As the result, the spread spectrum clocking technique sometimes is used for essentially pseudorandom chopper-related noise. Since there is no longer a peak at a single frequency in either the intrinsic switching noise or aliases signals, these devices can be used at signal bandwidths beyond the nominal chopping frequency without a large error signal showing up in-band.

In certain applications, higher chopping frequency can be used to extend the useful bandwidth. However, this approach can degrade V_{OS} performance and increase the input bias current.

Due to the presence of storage capacitors, the auto-zero amplifiers may require longer time to recover from output saturation (commonly referred to as overload recovery) if large storage capacitors such as external capacitors are used.

Finally, as a consequence of the complex additional circuit required for the auto-correction function, compound auto-zero amplifiers require more quiescent current for the same level of ac performance (bandwidth, slew rate, voltage noise and settling time) than a comparable non-chopped amplifiers.

6.5.3 DESIGN CONSIDERATIONS

The chopper based compound amplifier uses clocked circuit for dynamical compensation of analog offset and low frequency noises. Such design techniques greatly improve the amplifier parameters such as offset voltage, drift and $1/f$ noise. There are some major design considerations in the chopper stabilized amplifier circuits:

- Charge injection effect. Charge injection is caused by the switching action of choppers and auto-zero amplifiers. Charge injection introduces periodic noise pulse at the amplifier inputs that is independent of temperature. Charge injection increases with the size of the switch, the supply voltage, the amplifier gain, the source resistance, and other circuit parameters. Charge injection can usually be reduced by adding a capacitor in the feedback to limit the signal bandwidth, by using lower source and feedback resistors, and by building an active or passive filter after the amplification stage.

- Clock feedthrough effect. Clock feedthrough may occur in poor amplifier design or in circuit using a pure chopping technique.

- Intermodulation distortion effect. The amplifier can provide the signal amplification beyond the auto-zero frequency. An auto-zeroed amplifier's operation speed is related to the gain-bandwidth product of the main amplifier. The amplifier will introduce intermodulation distortion (IMD) as the input approaches the chopping or auto-zero frequency. We will see larger errors, as the input frequency gets closer to the clock frequency. IMD between the high-frequency input signal and the chopping frequency creates tones at frequencies ($f_{Chop}+/-f_{IN}$). IMD can usually be minimized using a pseudorandom auto-zero frequency. Another possibility is to add filters around the amplifier, where the clock noise is shown after filtering. Selecting the right cutoff frequency can enhance circuit response.

- The overload recovery time. The zero-drift amplifiers typically have longer overload recovery time than that of the standard CMOS amplifiers. When the inputs of auto-zero amplifiers are separated by a large amount for any reason, the output will saturate. The nulling amplifier treats this as an offset and tries to null the error. This sends the main amplifier further into saturation and prolongs the recovery time. One way to resolve the overload issue is to use built-in overload detection circuit to force it recover from overload in a short time.

6.6 COMPARSION OF CIRCUIT TECHNIQUES

The classic design approaches to minimize the dc offset and the 1/f noise effects in high precision VLSI amplifier circuits are usually related to the device optimization, the autozero, and the chopper techniques and their combinations. The zero-drift amplifier uses both auto-zeroing and chopping to reduce the energy at the chopping frequency, while keeping the noise very low at lower frequencies. This combined technique also allows for wider bandwidth than is possible with conventional zero-drift amplifiers. The characteristics of these circuit techniques can be states as follows:

- The auto-zeroing circuit techniques use CDS to correct the offset. Because sampling causes noise to fold back into baseband, auto-zero amplifiers have more in-band noise. To suppress noise, more current is used. Therefore, the devices typically dissipate more power. Auto-zero amplifiers are usually better for wideband applications.

- The chopper circuit techniques use modulation and demodulation. Choppers have low-frequency noise consistent with their flat-band noise, but produce a large amount of energy at the chopping frequency and its harmonics. Output filtering may be required, so these amplifiers are most suitable in low-frequency applications. Choppers are a good choice for low-power, low-frequency applications (<100 Hz).

- For applications that require low noise, no switching glitch, and wide bandwidth, the optimal solution is an amplifier that combines auto-zero and chopping techniques.

Shown in figure 6.50 are the noise power spectrums of the typical VLSI amplifier circuit families.

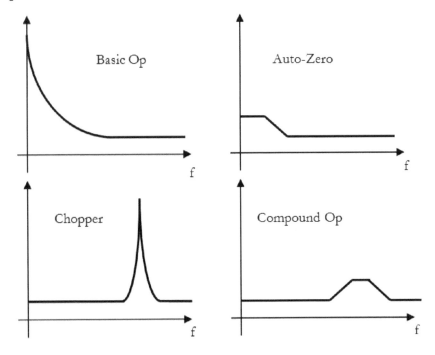

Fig.6.50 Noise power profiles of the zero-draft amplifier families

6.7 BASIC VLSI CHOPPER CIRCUIT ELEMENTS

A combination of a few very basic circuit structures can implement various VLSI chopper circuits.

6.7.1 BASIC CHOPPER MODULATOR CIRCUIT

The basic chopper modulator shown in figure 6.51 consists of four MOS switches that are used to change the connections of the chopper circuit.

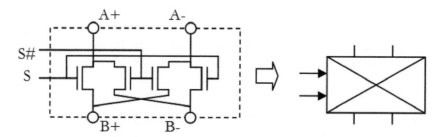

Fig.6.51 Basic chopper modulator circuit

6.7.2 VLSI CHOPPER OPAMP CIRCUITS

The chopper modulator can be implemented into the VLSI Opamp circuit as shown in figure 6.52 and figure 6.53.

VLSI Modulation Circuits

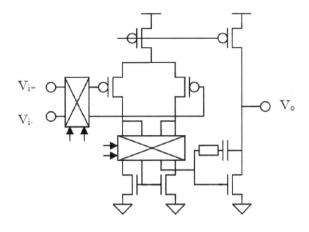

Fig.6.52 Two-stage chopper Opamp circuit

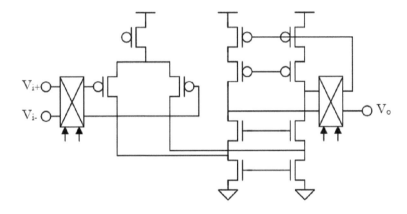

Fig.6.53 Folded cascade chopper Opamp circuit

References:

[1] C. C. Benz, G. C. Temes, "Circuit techniques for reducing the effects of op-amp imperfections: auto zeroing, correlated double sampling, and chopper stabilization," Proceedings of the IEEE Volume: 84, Issue: 11, Page(s): 1584 - 1614, 1996.

[2] A. Bakker, K. Thiele, J.H. Huijsing, "A CMOS nested-chopper instrumentation amplifier with 100-nV offset," IEEE Journal of Solid-State Circuits, Volume: 35, Issue: 12, Page(s): 1877 - 1883, 2000.

[3] H. Van der Ploeg, G. Hoogzaad, H.A.H. Termeer, M. Vertregt, R.L.J. Roovers, "A 2.5-V 12-b 54-Msample/s 0.25-μm CMOS ADC in 1-mm2 with mixed-signal chopping and calibration," IEEE Journal of Solid-State Circuits, Volume: 36, Issue: 12, Page(s): 1859 - 1867, 2001.

[4] C. Menolfi, Qiuting Huang; "A fully integrated, untrimmed CMOS instrumentation amplifier with submicrovolt offset," IEEE Journal of Solid-State Circuits, Volume: 34, Issue: 3, Page(s): 415 - 420, 1999.

[5] C. Menolfi, Qiuting Huang; "A low-noise CMOS instrumentation amplifier for thermoelectric infrared detectors," IEEE Journal of Solid-State Circuits, Volume: 32, Issue: 7, Page(s):

[6] C. C. Enz, E. A. Vittoz, F. Krummenacher, "A CMOS chopper amplifier," IEEE Journal of Solid-State Circuits, Volume: 22, Issue: 3, Page(s): 335 - 342, 1987.

[7] J. E. Witte, K. A. A. Makinwa, J. H. Huijsing, "A CMOS Chopper Offset-Stabilized Opamp," IEEE Journal of Solid-State Circuits, Volume: 42, Issue: 7, Page(s): 1529 - 1535, 2007.

[8] T. Denison, K. Consoer, W. Santa, A-T. Avestruz, J. Cooley, A. Kelly, "A 2 μW 100 nV/rtHz Chopper-Stabilized Instrumentation Amplifier for Chronic Measurement of Neural Field Potentials," IEEE Journal of Solid-State Circuits, Volume: 42, Issue: 12, Page(s): 2934 - 2945, 2007.

[9] L. Toth, Y. P. Tsividis, "Generalization of the principle of chopper stabilization," IEEE Transactions on Circuits and Systems I: Fundamental Theory and Applications, Volume: 50, Issue: 8, Page(s): 975 - 983, 2003.

[10] A. Uranga, X. Navarro, N. Barniol, "Integrated CMOS amplifier for ENG signal recording," IEEE Transactions on Biomedical Engineering, Volume: 51, Issue: 12, Page(s): 2188-

[11] H. Takao, Y. Matsumoto, M. Ishida, "A monolithically integrated three-axis accelerometer using CMOS compatible stress-sensitive differential amplifiers," IEEE Transactions on Electron Devices, Volume: 46, Issue: 1, Page(s): 109 - 116, 1999.

[12] M. A. P. Pertijs, A. Niederkorn, Xu Ma, B. McKillop, A. Bakker, J. H. Huijsing, "A CMOS smart temperature sensor with a 3σ inaccuracy of ±0.5°C from -50°C to 120°C," IEEE Journal of Solid-State Circuits, Volume: 40, Issue: 2, Page(s): 454 - 461, 2005.

[13] C. B. Wang, "A 20-bit 25-kHz delta-sigma A/D converter utilizing a frequency-shaped chopper stabilization scheme," IEEE Journal of Solid-State Circuits, Volume: 36, Issue: 3, Page(s): 566 - 569, 2001.

[14] A. Bilotti, G. Monreal, "Chopper-stabilized amplifiers with a track-and-hold signal demodulator," IEEE Transactions on Circuits and Systems I: Fundamental Theory and Applications, Volume: 46, Issue: 4, Page(s): 490 - 495, 1999.

[15] M. C. W. Coln, "Chopper stabilization of MOS operational amplifiers using feed-forward techniques," IEEE Journal of Solid-State Circuits, Volume: 16, Issue: 6, Page(s): 745 - 748, 1981.

[16] J. M. De la Rosa, S. Escalera, B. Perez-Verdu, F. Medeiro, O. Guerra, R. del Rio, A. Rodriguez-Vazquez, "A CMOS 110-dB@40-kS/s programmable-gain chopper-stabilized third-order 2-1 cascade sigma-delta Modulator for low-power high-linearity automotive sensor ASICs," IEEE Journal of Solid-State Circuits, Volume: 40, Issue: 11, Page(s): 2246 - 2264, 2005.

[17] Jiangfeng Wu, G. K. Fedder, L. R. Carley, "A low-noise low-offset chopper-stabilized capacitive-readout amplifier for CMOS MEMS accelerometers," Digest of Technical Papers. ISSCC. 2002 IEEE International Solid-State Circuits Conference, Volume: 1, Page(s): 428 - 478.

[18] R. Poujois, J. Borel, "A low drift fully integrated MOSFET operational amplifier," IEEE Journal of Solid-State Circuits, Volume: 13, Issue: 4, Page(s): 499 - 503, 1978.

[19] A.-T. Avestruz, W. Santa, D. Carlson, R. Jensen, S. Stanslaski, A. Helfenstine, T. Denison, "A 5 W/Channel Spectral Analysis IC for Chronic Bidirectional Brain–Machine Interfaces," IEEE Journal of Solid-State Circuits, Volume: 43, Issue: 12, Page(s): 3006 - 3024, 2008.

[20] K. Ishida, M. Fujishima, "Chopper-stabilized high-pass sigma-deltamodulator utilizing a resonator structure," IEEE Transactions on Circuits and Systems II: Analog and Digital Signal Processing, Volume: 50, Issue: 9, Page(s): 627 - 631, 2003.

[21] A. Bakker, K. Thiele, J. Huijsing, "A CMOS nested chopper instrumentation amplifier with 100 nV offset," Digest of Technical Papers, 2000 IEEE International Solid-State Circuits Conference, Page(s): 156 - 157

[22] L. Toth, Y. Tsividis, "Generalized chopper stabilization," The 2001 IEEE International Symposium on Circuits and Systems, 2001. Volume: 1, Page(s): 540 - 543 vol. 1

[23] A. Thomsen, D. Kasha, Wai Lee, "A five stage chopper stabilized instrumentation amplifier using feedforward compensation,"1998 Symposium on VLSI Circuits, Page(s): 220 - 223.

[24] E. Rubiola, C. Francese, A. De Marchi, "Long-term behavior of operational amplifiers," IEEE Transactions on Instrumentation and Measurement, Volume: 50, Issue: 1, Page(s): 89 - 94, 2001.

CHAPTER 7
INTRODUCTION TO VLSI LOCK-IN AMPLIFIER CIRCUITS

Lock-in amplifier offers extremely high frequency selectivity, and sensitivity for measuring very weak signals in large noise background.

Lock-in amplifier circuits are based on the principle that if in frequency domain the localized signal to noise ratio (SRN) at a specific narrow frequency band is high enough then this signal can be effectively measured even though in the time domain the noise magnitude is a few orders magnitude higher than the signal to be measured.

Lock-in amplifier can be used to separate narrow band signal from the wideband noise floor (such as white noise background) based on the phase-sensitive detection techniques to achieve extremely high effective quality factor.

7.1 LOCK-IN AMPLIFICATION PRINCIPLE

A lock-in amplification relies on the fact that noise tends to be spread over a wider spectrum, often much wider than the signal. Therefore even though the magnitude of noise background over a significantly wide frequency band is much higher than the magnitude of a signal, the signal magnitude in a highly localized frequency band can overcome the noise magnitude. Such a principle leads to the lock-in amplification concept of highly selective signal measurement using very narrow band circuits.

7.1.1 WEAK SIGNAL DETECTION CONCEPT

The lock-in amplification concept can be explained using the simplest case of a single-tone weak signal in the background of white noise, where the root mean square of noise is much larger (e.g. as large as 10^6 times) than the signal to be measured. If the bandwidth of the amplification circuit can be made very narrow around the signal frequency, then the amplifier circuit can recover the signal from the strong noise background.

For a very weak signal V_s at the frequency f_s that is buried in a random noise floor of V_n $(Hz)^{-1/2}$, if a bandpass filter with bandwidth BW and peak gain A is used to amplify the signal as shown in figure 7.1, the output signal is amplified by a factor A with the magnitude given as:

$$V_{so}(s) = AV_s(s) \qquad (7.1)$$

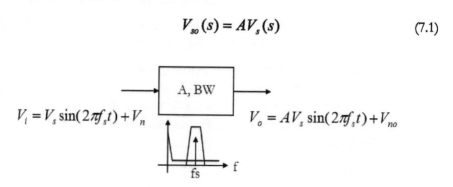

Fig.7.1 Weak signal amplification using narrow bandwidth circuit

Note that the white noise floor will also be amplified. However since the white noise is spread over the wide frequency band, only a small portion of the noise band will be amplified and pass through and amplifier circuit. The output magnitude of the noise can be expressed as a function of the amplifier bandwidth as:

$$V_{no}(s) = AV_n \cdot \sqrt{(BW)} \tag{7.2}$$

The signal-to-noise ratio (SNR) at the amplifier output is inversely related to the bandwidth of the amplifier as:

$$SNR = 20\log(V_s/[V_n \cdot \sqrt{(BW)}]) \tag{7.3}$$

The above equation can be further expressed in term of the BP amplifier center frequency fs and the amplifier quality factor Q as:

$$SNR = 20\log(V_s/[V_n \cdot \sqrt{f_s}]) + 10\log(Q) \tag{7.4}$$

The output SNR of the amplifier versus the amplifier quality factor Q is plotted in figure 7.2.

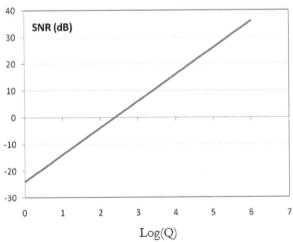

Fig.7.2 Bandpass amplifier output SNR versus amplifier quality factor

It can be seen that the amplifier output SNR increases as the quality factor (log(Q)) of the amplifier increases. For a high enough Q, a significantly high SNR can be achieved.

The SNR values can be evaluated in several special design cases below by assuming a nominal 1kHz weak signal of 10nV magnitude under a typical $5nV/\sqrt{(Hz)}$ white noise floor and a 60dB amplifier gain (1000 times):

- If the bandwidth amplified is 10 kHz, the output signal magnitude and SNR of the amplifier can be calculated as 10uV and -24dB respectively. We can see that the signal magnitude in this case is far below the noise floor therefore it can't be effectively measured.

- If the amplifier has a narrow passband of quality factor Q = 10, the output signal magnitude and the SNR can be calculated as 10uV and -4dB respectively. Such a signal is still below the noise floor therefore it can't be measured effectively.

- If the amplifier has an extremely high equivalent Q of 10^6, the output signal and the SNR in this case can be calculated as 10uV and 26dB respectively. It can be seen that such a high SNR is good enough for accurate measurement of the signal.

Above discussion implies that high amplifier quality factor is the key for the effective narrowband signal amplification and measurement. However since the conventional VLSI amplifier circuit techniques usually will not be able to offer such a high quality factor, we will have to rely on VLSI special signal processing circuit techniques such as the phase-sensitive detector (PSD) circuit to achieve such a high quality factor.

In a typical lock-in amplifier circuit implementation, a PSD can typically be used to realize very high Q (as high as 10^7) that is enough for detecting the signal at 10 kHz with a narrow bandwidth such as 0.01 Hz.

7.1.2 PHASE SENSITIVE SIGNAL DETECTION

The phase sensitive detector circuit is at the center of a VLSI lock-in amplification that provides the extremely high quality factor amplification for weak signals. A PSD is based on the orthogonality signal processing technique by multiplying the signal with a reference and then by passing it through a very lowpass selection filter.

Specifically, when a sinusoidal (single-tone) signal of frequency ω_1 is multiplied by the sinusoidal reference of frequency ω_2 and the output is integrated (averaged) over a time much longer than the period of the two signals. Such a circuit effective realizes a very lowpass filter and the result can be expressed as:

$$V_o = \overline{A\cos(\omega_1 t + \phi_1) \cdot B\cos(\omega_2 t + \phi_2)} \\ = \frac{AB}{2}\cos[(\omega_2 - \omega_1)t + (\phi_2 - \phi_1)] \quad (7.5)$$

We may derive the output of the circuit depends on the property of the two input signals:

- If the frequency of the two signals are different, the average is zero due to the integration effect (i.e. the two signals are orthogonal):

$$V_o = \overline{A\cos(\omega_1 t + \phi_1) \cdot B\cos(\omega_2 t + \phi_2)}\big|_{\omega_1 \neq \omega_2} = 0 \quad (7.6)$$

- If the frequency of the two signals are the same, the output of the averaged output is then a phase sensitive signal:

$$V_o = \overline{A\cos(\omega_1 t + \phi_1) \cdot B\cos(\omega_2 t + \phi_2)}\big|_{\omega_1 = \omega_2} = \frac{AB}{2}\cos(\phi_2 - \phi_1) \quad (7.7)$$

- If the frequency of the two signals are the same and the phase is separated by 90 degree, the output of the averaged output is then zero:

$$V_o = \overline{A\cos(\omega_1 t + \phi_1) \cdot B\cos(\omega_2 t + \phi_2)}|_{\omega_1 = \omega_2} = \frac{AB}{2}\cos(90°) = 0 \quad (7.8)$$

- However, when the frequency and phase of the two signals are the same, the average reaches the maximum:

$$V_o = \overline{A\cos(\omega_1 t + \phi_1) \cdot B\cos(\omega_2 t + \phi_2)}|_{\substack{\omega_1 = \omega_2 \\ \phi_1 = \phi_2}} = \frac{AB}{2} \quad (7.9)$$

The significance of such property is the extremely highly frequency selectivity (i.e. high Q) in such signal amplification.

7.1.3 BASIC LOCK-IN AMPLIFIER MODEL

A lock-in amplifier uses the orthogonality to achieve the high quality factor signal processing. It takes the input signal, multiplies it by the reference signal (internal or external), and integrates it over a specified time, usually on the order of milliseconds to a few seconds. The resulting signal is essentially a dc signal, where the contribution from any signal that is not at the same frequency as the reference signal is attenuated essentially to zero. In addition, the out-of-phase component of the signal that has the same frequency as the reference signal (because sine functions are orthogonal to the cosine functions of the same frequency) is also eliminated.

For a sinusoidal reference signal and a general multi-tone input waveform $V_i(t)$, the dc output signal $V_o(t)$ can be calculated for an analog lock-in amplifier by:

$$V_o(t) = \frac{1}{T}\int_{t-T}^{t} \cos(\omega_{ref}\tau + \phi)V_i(\tau)d\tau \quad (7.10)$$

Where ϕ is a phase that can be set on the lock-in (set to zero by default).

A practical VLSI lock-in amplifier typically uses a set of I/Q (in-phase and quadrant phase) reference signal set generated by the internal or the external PLL circuit to provide low distortion sinusoidal waveform with highly defined frequency and phase. The in-phase and quadrant-phase (I/Q) reference signals can be expressed as:

$$\begin{cases} i(t) = \sin(2\pi f_o) \\ q(t) = \cos(2\pi f_o) \end{cases} \quad (7.11)$$

The input signal to the lock-in amplifier can be expressed as a sinusoidal signal of frequency f_o with added noise and the harmonic distortion n(t) as:

$$v_i(t) = A\sin(2\pi f_o + \theta) + n(t) \quad (7.12)$$

When the input signal is mixed with the I/Q reference signals and averaged over many clock periods we have that

$$\begin{cases} x \equiv 2 \cdot \overline{i(t) \cdot V_i(t)} = A\sin(\theta) \\ y \equiv 2 \cdot \overline{q(t) \cdot V_i(t)} = A\cos(\theta) \end{cases} \quad (7.13)$$

The magnitude and phase of the mixer circuit output can be calculated as:

$$\begin{cases} M = \sqrt{x^2 + y^2} = A \\ ph \equiv \tan^{-1}(\dfrac{y}{x}) = \theta \end{cases} \quad (7.14)$$

Such a set of parameters directly provides the magnitude and phase of the weak signal to be measured insensitive to the phase of the signal.

Note that a random sampling approach can also be used to improve the bandwidth of the amplification without limited by the aliasing effect.

7.2 VLSI LOCK-IN AMPLIFIER IMPLEMENTATIONS

Shown in figure 7.3 is a VLSI lock-in amplifier circuit structure.

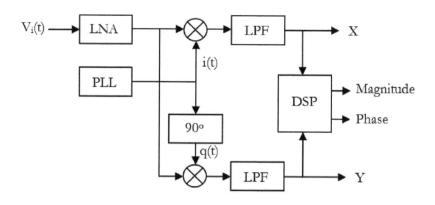

Figure 7.3 Lock-in amplifier blockdiagram

Such a circuit consists of a low noise amplifier (LNA) for pre-conditioning the weak input signal, a reference clock generator for generating the reference, an I/Q mixer for multiplying the I/Q phase reference to the signal and a lowpass filter for averaging the mixer outputs. This circuit also uses a DSP unit to calculate the magnitude and the phase of the detected signal. The mixer and the lowpass filter effectively form the phase sensitive detector in this lock-in amplifier implementation.

7.2.1 LOW NOISE AMPLIFIER (LNA)

A lock-in amplifier can measure signals as small as a few nanovolts where a low-noise signal amplifier is required to precondition (boost) the signal to a level that the A/D converter can digitize the signal without degrading the signal-to-noise. Note that a high gain is usually not necessary in such LNA since it does not improve the signal-to-noise. The LNA gain (ac and dc) is usually determined by the sensitivity requirement of the lock-in amplifier and the

VLSI Modulation Circuits

distribution of the gains (ac versus dc) within the lock-in amplifier that is set by the lock-in amplifier dynamic reserve.

7.2.2 ANALOG PSD

Shown in figure 7.4 is an analog PSD based on the analog multiplier for multiplying the input signal with a sinusoidal reference.

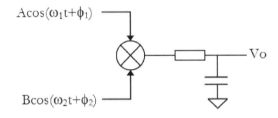

Fig.7.4 Analog multiplication PSD

It can be seen that when the reference is frequency locked to the signal the output signal is sensitive to phase difference between the signal and reference as:

$$V_o = \overline{A\cos(\omega_1 t + \phi_1) \cdot B\cos(\omega_2 t + \phi_2)} = \begin{cases} 0 & \omega_1 \neq \omega_2 \\ \dfrac{AB}{2}\cos(\phi_1 - \phi_2) & \omega_1 = \omega_2 \end{cases}$$

(7.15)

The analog multiplier provides a simple way for realizing the PSD function. However it is difficult for analog multiplier based PSD to operate linearly in the presence of large noise, or other interfering signals. Such a linearity limitation usually results in poor noise rejection and thereby limits the signal recovery capability of the lock-in amplifier.

7.2.3 SWITCHING PSD

A switching PSD uses a simple demodulator consisting of an analog polarity-reversing switch driven at the applied reference frequency as shown in figure 7.5.

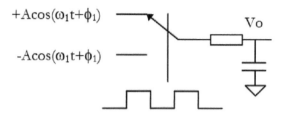

Fig.7.5 Switching PSD

In this type of PSD circuit the reference signal is a square wave at frequency 1/T that might be a sync output from a function generator. Such a circuit can be analyzed using the Fourier series of the switching signal as:

$$V_o(t) = S(t)A\cos(\omega_1 t + \phi_1) \quad (7.16)$$

The square waveform reference can be expanded based on the Fourier transform as:

$$S(t) = \frac{4}{\pi}[\sin(\omega t + \phi_2) + \frac{1}{3}\sin(3\omega t + 3\phi_2) + \frac{1}{5}\sin(5\omega t + 5\phi_2) + ...] \quad (7.17)$$

At the lock-in frequency the lowpass filtered output is given as:

$$V_o(t) = \frac{2A}{\pi}\cos(\phi_1 - \phi_2) \quad (7.18)$$

The switching PSD offers the benefit of very easier for demodulator to operate linearly over a very wide range of input signals. However, the switching PSD not only detects signals at the applied reference frequency, but also at its odd harmonics, where the Fourier analysis of a square wave defines the response at each harmonic relative to the fundamental. Such a response can be used if the signal being detected is also a square wave, but can give problems if the interference (i.e. noise) has strong tone at the harmonics of the reference. This may require using a tuned low-pass or bandpass filter in the signal channel prior to the multiplier to ensure it primarily detects signals at the reference frequency. However, in order to fully reject the 3rd harmonic response, while still offering good performance at the reference frequency, very complicated and expensive filters would be required.

7.2.4 DIGITAL PSD

A digital PSD employing digital multiplication of the signal with the reference can be realized using DSP circuit techniques either in hardware or software form by converting the signal to digital form using an A/D converter. Such approach effectively resolves the harmonic rejection, the output offsets, the limited dynamic reserve, and the gain error problems in the analog PSD circuit implementations. An anti-aliasing filter to prevent higher frequency inputs from aliasing effects usually precedes the A/D converter in such application. The converted digital input data stream is then computed (multiplied) a point at a time, with the computed in-phase and quant rant reference sine waves.

A digital reference signal can typically provide 20 bits of accuracy to achieve very low harmonic distortion (e.g. typically at the -120 dB level). This means that the signal in such PSD is effectively multiplied by a single tone reference (instead of a reference and its many harmonics in the switching PSDs), and only the signal at this reference frequency is detected. As the result, the digital lock-in amplifiers can be completely insensitive to signals at harmonics of the reference. In contrast, a switching lock-in amplifier will detect at all of the odd harmonics of the reference since a square wave contains many odd harmonics.

The digital PSD can also be used to solve the common output offset problem in most analog lock-in amplifier circuits where the signal of interest is a dc output from the PSD that is sensitive to circuit offset and zero drift since there is no erroneous dc output offset from the digital multiplication of the signal and reference, where the actual multiplication is virtually error free.

7.2.5 LOWPASS FILTER

For the input signal made up of signal plus noise, the lock-in amplifier PSD and lowpass filter only detect signals whose frequencies are at the lock-in reference frequency (i.e. high quality factor detection). The lowpass filter rejects noise signal at frequencies different from the reference at the PSD output. Noise at frequencies very close to the reference frequency will result in very low frequency ac at the outputs of the PSD. Their attenuation depends upon the lowpass filter bandwidth and roll off. In general, a narrower filter bandwidth will remove noise very close to the reference frequency; a wider bandwidth allows these signals to pass. Therefore the lowpass filter bandwidth determines the bandwidth of PSD detection. Only the signal at the reference frequency will result in a true dc output and be unaffected by the lowpass filter.

7.2.6 REFERENCE SIGNAL

In the lock-in amplifier operation, the reference must be the same as the signal frequency. In addition, the phase of the signals must be the same as the reference, since any change in the phase will cause the PSD output not a dc signal. In other words, the reference needs to be phase-locked to the signal.
The lock-in amplifiers usually use a phase-locked loop (PLL) to generate the reference, where an external reference signal such as the reference square wave is provided to the lock-in amplifier. The PLL in the lock-in amplifier locks the internal reference oscillator to this external reference, resulting in a reference sine wave at clock frequency with a fixed phase shift of the reference. Since the PLL actively tracks the external reference, changes in the external reference frequency do not affect the measurement.
The PSD output depends on the phase difference between the signal and the lock-in reference. In which case the output of PSD is the signal itself. Conversely, if the phase difference is 90, there will be no output at all. A lock-in amplifier with a single PSD is called a single-phase lock-in. In practical applications, using I/Q PSD employing the I/Q clock references can eliminate this phase dependency.
Based on the Fourier's transformation a given input signal can be represented in frequency domain as the sum of many sine waves of differing amplitudes, frequencies and phases. A lock-in amplifier multiplies the signal by a pure sine wave at the reference. All components of the input signal are multiplied by the

reference simultaneously. Mathematically speaking, sine waves of differing frequencies are orthogonal that has the zero average of the products unless the frequencies are exactly the same. The product of this multiplication produced by the mixer yields a dc output signal proportional to the component of the signal whose frequency is exactly locked to the reference frequency. The lowpass filter (which follows the multiplier) provides the averaging to remove the products of the reference with components at all other frequencies.

7.3 NOISE EFFECTS AND DYNAMIC RESERVE

There are two major types of noises that might impact the VLSI lock-in amplifier circuit performances including the intrinsic noises and the external noises. The intrinsic noises are generated by the intrinsic noise sources in VLSI circuit elements such as resistors. Examples of the intrinsic noise sources are the Johnson noise and the shot noise that are inherent to all physical processes. The external noise sources are noise generated in the application environment such as the power supply di/dt, the IR noises, the substrate noise, the xtalk noise and the EMI effects. In VLSI lock-in amplifier circuits, both intrinsic and external noise effects can be minimized using specific VLSI signal processing circuit techniques, such as the optimal VLSI circuit element selection, the CDS circuit techniques, the grounding, and the shielding.

The dynamic reserve of a lock-in amplifier, at a given full-scale input sensitivity, is defined as the ratio of the largest interfering signal to the full-scale input voltage. The largest interfering signal is defined as the amplitude of the largest signal at any frequency that can be applied to the input before the lock-in cannot measure a signal with its specified accuracy.

7.3.1 INTRINSIC NOISE SOURCES

A resistor in lock-in amplifier circuit will generates a noise voltage across its terminals due to the thermal fluctuations effect in the electron density. An open-circuit noise voltage due to this noise effect is give as:

$$\sqrt{V_n^2} = \sqrt{4kTRB} \qquad (7.19)$$

Where k is the Boltzmann's constant (1.38 x10^{-23} J/K), T is the absolute temperature in Kelvin, R is the resistance in ohms, and B is the bandwidth of the circuit in Hz.

For example, for a typical input signal amplifier of 300kH bandwidth, the effective broadband noise at the amplifier input can be calculated as Vnoise=70$R^{1/2}$ nVrms, or 350$R^{1/2}$ nVpp. Note that the amount of the noise is proportional to the square root of the source impedance and the lock-in amplifier bandwidth.

In a lock-in amplifier, the equivalent noise bandwidth (ENBW) of the lowpass filter (time constant) sets the detection bandwidth.

On the other hand, the shot noise is caused by the electric current due to the finite nature of the charge carriers. The non-uniformity in the electron flow causes noise in the current. Shot noise can result in voltage noise when current is passed through a resistor, or as noise in a current measurement. The shot noise, or current noise, is given as:

$$\sqrt{I_n^2} = \sqrt{2qIB} \qquad (7.20)$$

Where q is the electron charge (1.6x10^{-19} Coulomb), I is the rms ac current or dc current depending upon the circuit, and B is the bandwidth. When the current input of a lock-in amplifier is used to measure an ac signal current, the bandwidth is typically so small that shot noise is not important.

The third type of commonly seen intrinsic noise in lock-in amplifier is the 1/f noise. This type of noise has a 1/f spectrum and makes signal amplification at low frequencies more difficult.

Since all of these noise sources are incoherent. The total random noise is the square root of the sum of the squares of all the incoherent noise sources.

7.3.2 EXTERNAL NOISE SOURCES

There are a variety of external noise sources that might impact the lock-in amplifier circuit performance. Most of these noise sources are asynchronous, i.e. they are not related to the reference, and do not occur at the reference frequency or its harmonics.

These noise sources affect the lock-in amplifier circuit by increasing the required dynamic reserve or the time constant. Some noise sources, however, are related to the reference. Typical sources of synchronous noise are ground loops between the circuit elements and xtalk (capacitive, inductive, resistive, mechanic coupling or Thermocouple effects) from the reference oscillator or other circuits. Many of these noise sources can be minimized with good design practice. There are several ways in which noise sources are coupled into the signal path.

- The capacitance coupling noise. An ac voltage from a nearby piece of circuits can couple to a detector via stray capacitance. Although the coupling capacitance may be very small, the coupled noise may still be larger than a weak signal. This can be very worse if the coupled noise is synchronous (at the reference frequency). If the noise source is at a higher frequency, the coupled noise will be even stronger. If the noise source is at the reference frequency, the problem is even worse. The lock-in amplifier circuit rejects noise at other frequencies, but pick-up at the reference frequency appears as a signal. Commonly used VLSI circuit techniques to minimize the capacitive coupling effects include (1) keeping the noise source far from the lock-in amplifier; (2) minimize the source impedance (noise current generates very little voltage); (3) proper shielding of the signal and circuit.

- The inductance coupling noise. A changing current in a nearby can couple to the lock-in amplifier via a changing magnetic field which induces an emf in the circuit loop in the lock-in amplifier. Commonly used VLSI circuit techniques to minimize the inductive coupling effects include (1) reduce the area of the pick-up loop by using twisted differential signal layout or coaxial shielded route; (2) using magnetic shielding to prevent the magnetic field from crossing the area of the experiment; (3) measuring currents (not voltages) from high impedance detectors.

- The ground bounce noise. Circuits grounded at different places that have different reference potentials can cause power supply noise. VLSI circuit techniques to minimize such noise include (1) single-point ground (i.e. grounding everything to the same physical point); (2) using a heavy ground bus to reduce the resistance of ground connections; (3) removing sources of large ground currents from the ground bus used for small signals.

- The Mechanical noise. Mechanical movement (vibration) in the VLSI circuit or package and PCB trace in VLSI lock-in amplifier can introduce capacitance varies in time to cause a dC/dt noise in lock-in amplifier. Such noise effect can be minimized by several design techniques including (1) eliminating mechanical vibrations near the lock-in amplifier; (2) carrying sensitive signals so they do not move; (3) using a low noise cable that is designed to reduce microphonic effects.

- The thermal couple effects. The emf created by junctions between dissimilar metals can give rise to microvolt level slowly varying voltage noise. This source of noise is typically at very low frequency since the temperature of the lock-in amplifier generally changes slowly. This effect can be a problem for low frequency (especially in the mHz range) lock-in amplifier applications. Commonly used techniques to minimize such noise effects include (1) keeping the temperature of the lock-in amplifier constant; (2) using compensation circuit to cancel the thermal potential of the junction.

7.3.3 DYNAMIC RESERVE

The dynamic reserve can be used to specify the performance of a lock-in amplifier circuit. For lock-in amplifier with input consists of a full-scale signal at reference frequency and noise at some other frequency. The dynamic reserve specifies the ratio of the largest tolerable noise signal to the full-scale signal. For example, if the full scale signal is 1mV, then the 40dB dynamic reserve means a noise as large as 100 mV (40dB greater than full scale) can be tolerated at the input without overload. This means that the noise at the dynamic reserve limit should not cause an overload anywhere (at the input, the PSD circuit, lowpass filter or dc amplifier circuit) in the lock-in amplifier. In practical design cases, this condition can usually be realized by proper distribution of the gain among the circuit blocks of the lock-in amplifier circuit, such as by setting very low input signal gain so that the noise will not overload. This results in very small signal at the PSD input. A lowpass filter can be used to remove the large noise components from the PSD output, which allows the remaining dc component to be amplified (with high gain) to reach the full scale.

Most lock-in amplifiers define tolerable noise as levels, which do not affect the output more than a few percent of full scale. This is more severe than simply not overloading.

The analog lock-in amplifiers usually have lower reserves than the digital lock-in amplifiers that are only limited by the requirement that the analog signal gains before the A/D converters are high enough. Lower reserve means less output error and drift. On the other hand, more reserve can increase the output noise. Since digital phase sensitive detectors usually do not suffer from dc output errors caused by large noise signals. Dynamic reserve of digital lock-in amplifier can be as 100dB without error. In digital lock-in amplifiers large noise signals do not cause output errors from the PSD and large dc gain does not result in increased output drift. The only drawback to using very high dynamic reserves (>60 dB) in digital lock-in amplifier is the increased output noise due to the noise of the A/D converter. This increase in output noise usually only presents when the dynamic reserve is increased above 60 dB and above the minimum reserve.

To set a scale, the digital lock-in amplifier's output noise at 100 dB dynamic reserve is only measurable when the signal input is grounded. In fact, virtually all signal sources will have a noise floor, which will dominate the lock-in amplifier output noise. Of course, noise signals are generally much noisier than pure sine generators and will have much higher broadband noise floors. If the noise does not reach the reserve limit, the digital lock-in amplifiers own output noise may become detectable at ultra-high reserves.

The frequency dependence of dynamic reserve is inherent in the lock-in detection technique. More lowpass filter stages can increase the dynamic reserve close to the reference frequency. The specified reserve applies to noise signals within the operating range of the lock-in, i.e. frequencies below 100 kHz. The reserve at higher frequencies is actually greater but is generally not that useful.

References:

[1] Nakanishi, M.; Sakamoto, Y.; "Analysis of first-order feedback loop with lock-in amplifier," IEEE Transactions on Circuits and Systems II: Analog and Digital Signal Processing, Volume: 43, Issue: 8, Page(s): 570 - 576, 1996.

[2] Lanyon, H.P.D.; Sapega, A.E.; "Measurement of semiconductor junction parameters using lock-in amplifiers," IEEE Transactions on Electron Devices, Volume: 20, Issue: 5, Page(s): 487 - 491, 1973.

[3] Davies, R.; Meuli, G.; "Development of a digital lock-in amplifier for open-path light scattering measurement," 2010 IEEE Symposium on Industrial Electronics & Applications (ISIEA), Page(s): 50 - 55, 2010

[4] An Hu; Chodavarapu, V.P.; "CMOS Optoelectronic Lock-In Amplifier With Integrated Phototransistor Array," IEEE Transactions on Biomedical Circuits and Systems, Volume: 4, Issue: 5, Page(s): 274 - 280, 2010.

[5] Jiawei Xu; Guy Meynants; Merken, P.; "Low-power lock-in amplifier for complex impedance measurement,"3rd International Workshop on Advances in sensors and Interfaces, Page(s): 110 - 114.

[6] Moe, A.E.; Marx, S.R.; Bhinderwala, I.; Wilson, D.M.; "A miniaturized lock-in amplifier design suitable for impedance measurements in cells [biological cells]," Proceedings of IEEE Sensors, 2004. Page(s): 215 - 218 vol.1

[7] Nakanishi, M.; Sakamoto, Y.; "Analysis of first-order feedback loop with lock-in amplifier," IEEE Transactions on Circuits and Systems II: Analog and Digital Signal Processing, Volume: 43, Issue: 8, Page(s): 570 - 576, 1996.

[8] Gabal, Miguel; Medrano, Nicolas; Calvo, Belen; Celma, Santiago; "A Low-Voltage Single-Supply Analog Lock-in Amplifier for Wireless Embedded Applications," 2010 European Workshop on Smart Objects: Systems, Technologies and Applications (RFID Sys Tech), Page(s): 1 - 6.

[9] Galzerano, G.; Bava, E.; ottoboni, R.; Svelto, C.; "Lock-in amplifier up to 530 MHz with phase and amplitude demodulation," Proceedings of the 20th IEEE Instrumentation and Measurement Technology Conference, 2003. Volume: 2, Page(s): 1665 - 1668 vol.2.

[10] Arafat Bin Azidin, F.; Bin Zainudin, Z.; "Detecting Human Pulse in P-SPICE Model Using Lock-In Amplifier," International Conference of Soft Computing and Pattern Recognition, 2009. Page(s): 720 - 723.

[11] De Marcellis, A.; Ferri, G.; Patrizi, M.; Stornelli, V.; D'Amico, A.; Di Natale, C.; Martinelli, E.; Alimelli, A.; Paolesse, R.; "An integrated analog lock-in amplifier for low-voltage low-frequency sensor interface," second International Workshop on Advances in Sensors and Interface, 2007. Page(s): 1 - 5.

[12] Zhu, Y.; Takada, T.; "Measurement technique for electric field distribution in liquid dielectrics by using Kerr effect and a diagnostic image lock-in amplifier," Proceedings of the 5th International Conference on Properties and Applications of Dielectric Materials, 1997. Volume: 2, Page(s): 1103 - 1106 vol.2.

[13] Walker, W.D. "Sub-micro degree phase measurement technique using lock-in amplifiers," 2008 IEEE International Frequency Control Symposium, Page(s): 825 - 828.

[14] ShengQian Ma; Schroder, J.; Hauptmann, P.; "Sensor impedance spectrum measurement interface with lock-in amplifier," Proceedings of IEEE Sensors, 2002, Volume: 2, Page(s): 1313 - 1316.

[15] Sonnaillon, M.O.; Urteaga, R.; Bonetto, F.J.; Ordonez, M.; " Implementation of a high-frequency digital lock-in amplifier," Canadian Conference on Electrical and Computer Engineering, 2005. Page(s): 1229 - 1232

[16] Sonnaillon, M.O.; Urteaga, R.; Bonetto, F. J.; "High-Frequency Digital Lock-In Amplifier Using Random Sampling," IEEE Transactions on Instrumentation and Measurement, Volume: 57, Issue: 3, Page(s): 616 - 621, 2008.

CHAPTER 8
VLSI SWITCHED-CAPACITOR DC/DC CONVERTER CIRCUITS

The VLSI charge pump dc/dc converters offer combination of low power, high simplicity, and low cost power management applications for VLSI circuits or sub-circuits that can't be directly powered by the main supply due to reasons such as the available supply rails are not directly usable, nor is the direct use of battery voltage. In such cases, VLSI charge pump dc/dc converter offers effective small form factor solution options with the flexibility of the number of cells to use, or for resolving the issue of the declining voltage of a discharging battery.

A VLSI charge pump circuits use the capacitors (instead of inductor) as the main energy storage elements to create either higher, lower or inverted power supply sources that are capable of power efficiencies as high as 90-95% while being electrically simple.

8.1 VLSI CHARGE PUMP CIRCUIT PRINCIPLE

VLSI switched-capacitor dc/dc converter circuits use switched-capacitor charge pump to converter voltage sources, where charge is pumped from one source to another.

8.1.1 BASIC CHARGE PUMP CIRCUIT CONCEPT

Shown in figure 8.1 is the basic switched-capacitor charge pump operation model where charge in source E is pumped into storage capacitor C_o through a fly capacitor C and a switched network.

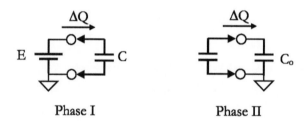

Fig.8.1 Basic switched-capacitor charge pump circuit

This circuit works in two control clock phases:

- In phase I: fly capacitor is charged to the voltage E of the source.

    Where the charge stored on C is given as:

    $$Q = CE \qquad (8.1)$$

- In phase II: the fly capacitor is charge shared with a large load capacitor Co. If the voltage V_o of the C_o is lower than the voltage E of the source. Part of the charge on the fly capacitor C will be transferred to Co as:

$$\Delta Q = C(E - V_o) \tag{8.2}$$

In the ideal case, neither fly capacitor nor the switches consume the energy and total charge (i.e. energy) will be transferred to load C_o with one control clock period.

By slight modifying the basic charge pump circuit structure, we can construct charge pump for voltage conversion as shown in figure 8.2.

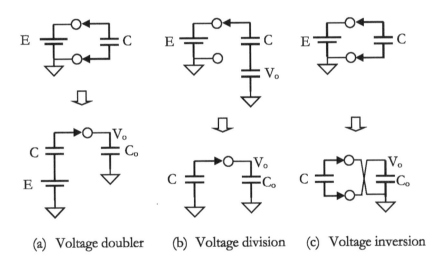

(a) Voltage doubler   (b) Voltage division   (c) Voltage inversion

Fig.8.2 Switched-capacitor charge pump circuit structure

8.1.2 BASIC CHARGE PUMP CIRCUIT COMPONENTS

A VLSI switched-capacitor charge pump circuit is typically based on the following circuit elements as shown in figure 8.3:

- Voltage source: the source is the element the switched-capacitor charge pump circuit get charge from. It is usually a battery, a solar cell or another voltage (energy) source. I can also be a charged high value capacitor.
- Fly capacitors: The fly capacitors are used to transfer charge form the source to the load. The fly capacitors usually have reconfigurable terminals to enable the charge transfer among capacitors.
- Load: the load for switched-capacitor charge pump can be a rechargeable battery, a large value storage capacitor and with a load resistor.
- Switch network: The switched network is used to re-configure the switched-capacitor network for the charge transfer operation. I can be a switch network in the active switch in synchronous switched-capacitor charge pump circuit or unidirectional circuit elements like diode in passive switched-capacitor charge pump circuits.
- Control circuits: Control circuits are used to generate the control for the switched in the charge pump circuit for charge transfer operations. They may include the network configuration control for programmable charge pump circuit. For regulated charge pump circuit, they may include the feedback circuit structure in the control circuit.

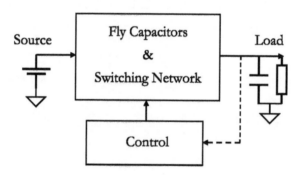

Fig.8.3 Switched-capacitor charge pump circuit elements

8.1.3 CHARGE PUMP CIRCUIT CLASSIFICATIONS

VLSI charge pump circuit can usually defined based on various circuit aspects such as voltage conversion gain, switch device types, conversion stages, and circuit structures.

VLSI Modulation Circuits

Under the classification of the conversion gain, switched capacitor charge pump circuits can be defined as

- The **boost** (i.e. step-up) charge pump circuits that provides higher than one voltage conversion gain;
- The **buck** (i.e. step-down) charge pump circuits that provides lower than one voltage conversion gain;
- The **inversion** charge pump circuits that provides negative voltage compared with the source.

Under the switch device type aspect, VLSI charge pump circuit can be defined as

- The **synchronous** (or active) charge pump circuits that use active switches and controlled with synchronous clock sequence;
- The **diode-based** (or passive) charge pump circuits that use unidirectional circuit elements, such as PN junction diodes or diode-connection MOS switches.

Switched-capacitor charge pump can also implemented based on various stages for better conversion gain:

- Single-stage charge pump uses one fly capacitor stage for transferring charge from the source to the load. Such single stage converter can only offers limited conversion gain in the voltage conversion operation.
- Multi-stage charge pump with linear, exponential or Fibonacci types of gain enhancement.

Based on the circuit structure classification, charge pump circuit can be defined as the following circuit families:

- The H-Bridge charge pump circuits;
- The Cockcroft-Walton (CW) voltage multiplier;

- The Villard charge pump circuits;
- The Greinacher voltage doubler circuit;
- The Delon charge pump circuits;
- The Marx generator charge pump circuits;
- The Dickson charge pump circuits; and
- The cross-coupled charge pump circuit

8.2 CHARGE PUMP CIRCUIT STRUCTURES

A very basic VLSI switched-capacitor level shift circuit structures employing a source and a fly capacitor is shown in figure 8.4, where the switch is ON if Ck ="1" and OFF if CK="0". It can be seen that such circuit realizes a voltage level shift circuit operation. The input signal Ck is voltage shifted by the source voltage E.

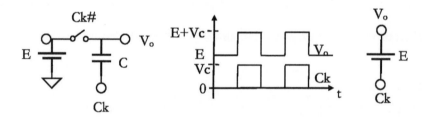

Fig. 8.4 Basic VLSI switched-capacitor level shift circuit structure (I)

However, for the same circuit, if the control phase of one of the switches is modified as shown in figure 8.5, we can see that it realizes a –E voltage shift circuit operation.

When a synchronous rectifying switch is added to the output the voltage shifted signal value can be kept at the output, the output is a voltage shifted value of the input as shown in figure 8.6.

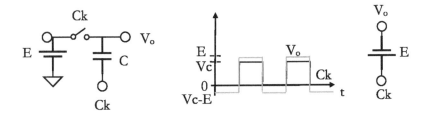

Fig. 8.5 Basic VLSI switched-capacitor level shift circuit structure (II)

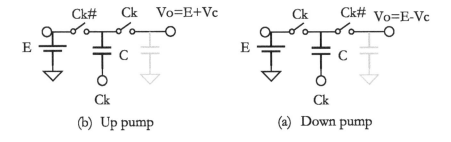

(b) Up pump (a) Down pump

Fig. 8.6 Basic VLSI switched-capacitor charge pump circuit structures

Since the basic switched-capacitor structure as shown in figure 8.4 is a level shifter that can be used to generate the synchronous rectifier control phase Ck as shown in figure 8.7a and figure 8.7b. This circuit structure is equivalent to circuit in figure 8.7c if the forward diode voltage drop is ignored.

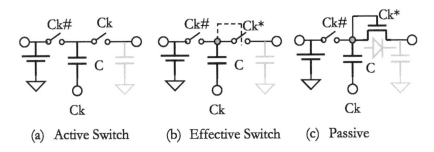

(a) Active Switch   (b) Effective Switch   (c) Passive

Fig. 8.7 Equivalent passive VLSI charge pump circuit structures

231

In addition, for control clock Ck# in figure 8.8a, if X<<E, it creates a natural control for the input switch and leads to the passive circuit implementation as figure 8.8b and figure 8.8c.

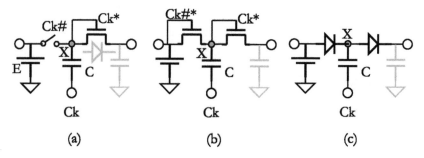

Fig. 8.8 Basic passive VLSI charge pump circuit structures

The above design concept also leads to the passive voltage level shift circuits as shown in figure 8.9.

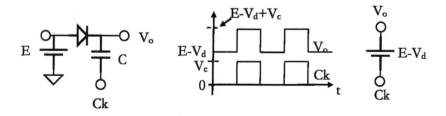

Fig. 8.9 Basic VLSI passive switched-capacitor level shift circuit structure

8.2.1 H-BRIDGE CHARGE PUMP CIRCUITS

VLSI H-bridge charge pump circuit structure provides the simplest charge pump building block to realize complex VLSI charge pump circuits. The basic H-bridge circuit element shown in figure 8.10 consists of one fly capacitor and four switched. These switches can be controlled and connected to other circuit components for various charge pump circuit functions.

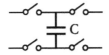

Fig.8.10 the H-bridge circuit

Example 8.1. Shown in figure 8.11 is an H-bridge doubler circuit that can provide 2x unloaded voltage conversion.

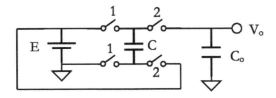

Fig.8.11 H-bridge voltage doubler circuit

This unloaded voltage doubler circuit works in two operation phases:

- Phase I: The fly capacitor is charged to battery voltage in this control phase as shown in figure 8.12.

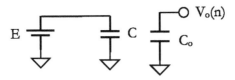

Fig.8.12 H-bridge voltage doubler circuit (phase I)

- Phase II: The fly capacitor is used to charge the load capacitance as shown in figure 8.13.

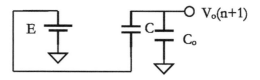

Fig.8.13 H-bridge voltage doubler circuit (phase II)

Assuming the voltage of the load (i.e. storage) capacitor is before this phase is given as $V_o(n)$, the new output voltage value can be expressed based on charge conservation as

$$C_o V_o(n+1) + C(V_o(n+1) - E) = C_o V_o(n) + CE \qquad (8.3)$$

We can get the time domain transfer function of the circuit as:

$$V_o(n+1) = \frac{C_o}{C_o + C} V_o(n) + \frac{2C}{C_o + C} E \equiv \alpha V_o(n) + \beta \qquad (8.4)$$

Assume initially no charge on the storage capacitor, such that $V_o(0)=0$, the output voltage for each clock period can be solved by iterations as:

- n=1

$$V_o(1) = \beta = \frac{1-\alpha^1}{1-\alpha} \beta \qquad (8.5)$$

- n=2

$$V_o(2) = \alpha\beta + \beta = \frac{1-\alpha^2}{1-\alpha} \beta \qquad (8.6)$$

- n=3

$$V_o(3) = \alpha^2\beta + \alpha\beta + \beta = \frac{1-\alpha^3}{1-\alpha}\beta \qquad (8.7)$$

...

- n=k

$$V_o(3) = \alpha^{k-1}\beta + ... + \alpha^2\beta + \alpha\beta + \beta = \frac{1-\alpha^k}{1-\alpha}\beta \qquad (8.9)$$

When k goes to infinite, the final output voltage is given as:

$$\lim_{n\to\infty} V_o(n) = \frac{1}{1-\alpha}\beta = \frac{\frac{2C}{C_o+C}E}{1-\frac{C_o}{C_o+C}} = 2E \qquad (8.10)$$

It converges to 2x of the source voltage.

Example 8.2 Shown in figure 8.14 is an H-bridge voltage division circuit that can provide 1/2 unloaded voltage conversion. Such circuit operates on two clock phases:

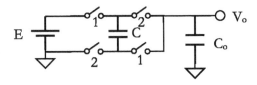

Fig.8.14 H-bridge voltage division circuit

- Phase I: The fly capacitor is charged with the output storage capacitor in series as shown in figure 8.15.

- Phase II: The fly capacitor is used to charge the load capacitance as shown in figure 8.16.

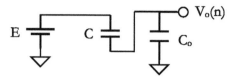

Fig.8.15 H-bridge voltage division circuit (phase I)

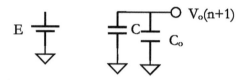

Fig.8.16 H-bridge voltage division circuit (phase II)

The final steady state voltage of the charge pump output can be derived by letting $V_o(n+1)= V_o(n)= V_o$ as follows.

Since the fly capacitor has been charged to Vo in the (n-1) phase, we have that in phase I:

$$E = V_o + V_o = 2V_o \qquad (8.11)$$

Or

$$V_o = \frac{E}{2} \qquad (8.12)$$

Example 8.3 Shown in figure 8.17 is an H-bridge voltage inversion circuit that can provide unloaded voltage inversion. Such circuit operates on two clock phases:

- Phase I: The fly capacitor is charged with the output storage capacitor in series as shown in figure 8.18.

- Phase II: The fly capacitor is used to back charge the load capacitance as shown in figure 8.19.

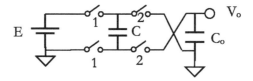

Fig.8.17 H-bridge voltage inversion circuit

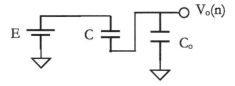

Fig.8.18 H-bridge voltage inversion circuit (phase I)

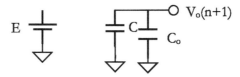

Fig.8.19 H-bridge voltage inversion circuit (phase II)

The final steady state voltage of the charge pump output as described by $V_o(n+1)=V_o(n)=V_o$ can be derived as:

$$V_o = -E \qquad (8.13)$$

Multiple H-bridge elements can be used to construct more complex charge pump circuit structures.

Example 8.4 Shown in figure 8.20 is an H-bridge 3x voltage multiplier circuit that can provide 3x unloaded voltage conversion. Such circuit operates on two clock phases:

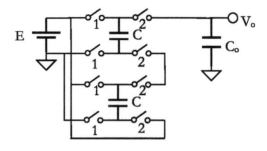

Fig.8.20 H-bridge 3x voltage multiplier circuit

In phase I of the circuit operation, both fly capacitors are charged to E. In phase II and the two fly capacitors are stuck with the source to form 3E voltage source to charge the storage capacitor Co. As the result, the steady state final output voltage for the unloaded charge pump is 3x of the source voltage as:

$$V_o = 3E \tag{8.14}$$

Example 8.5 Similarly the two elements H-bridge can be used to construct 1/3 division circuit as shown in figure 8.21.

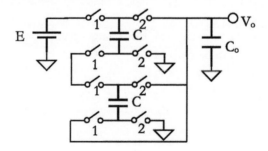

Fig.8.21 H-bridge 1/3 voltage division circuit

Example 8.6 Shown in figure 9.22 is a more complex two elements H-bridge can be used to construct 2/3 voltage division.

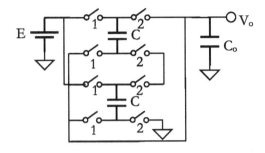

Fig.8.22 H-bridge 1/3 voltage division circuit

This circuit works as shown in figure 8.23:

- In phase I: both fly capacitor is parallels charged in series with the storage capacitor.
- In phase 2: the tow fly capacitors are connected into series to charge the storage capacitor

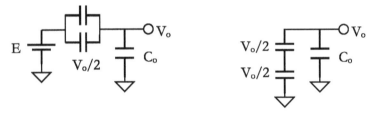

Fig.8.23 H-bridge voltage 2/3 division circuit operation

The steady state output voltage for the unloaded charge pump can be calculated based the figure as:

$$(1+\frac{1}{2})V_o = E \tag{8.15}$$

Or

$$V_o = \frac{2}{3}E \tag{8.16}$$

Example 8.7 Shown in figure 9.24 is a more complex two elements H-bridge can be used to construct (3/2)x voltage boost.

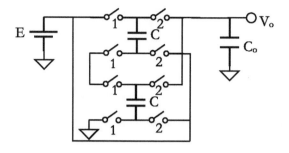

Fig.8.24 H-bridge (3/2)x voltage boost circuit

This circuit works as shown in figure 8.25:

- In phase I: both fly capacitor is parallels charged in series.
- In phase 2: the two fly capacitors are connected in parallel and then in series with the source to charge the storage capacitor

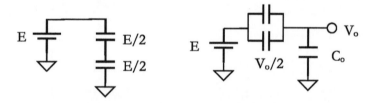

Fig.8.25 H-bridge voltage 3/2 gain boost circuit operation

The steady state output voltage for the unloaded charge pump can be calculated based the figure as:

$$V_o = (1+\frac{1}{2})E = \frac{3}{2}E \qquad (8.17)$$

Shown in figure 8.26 are typical VLSI active charge pump circuit topologies that can be constructed employing the H-bridge elements. Note that all these circuit topologies are inverseable (i.e. by swapping the input and the output terminals, a boost converter can be converted into a related buck converter).

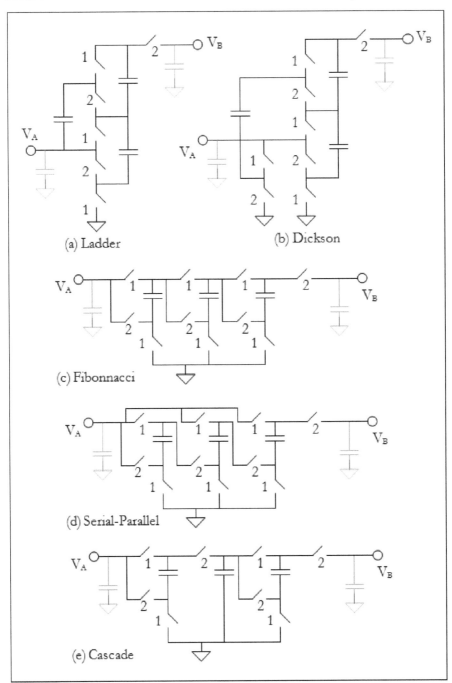

Fig.8.26 Basic switched-capacitor charge pump circuit topologies.

8.2.2 COCKCROFT-WALTON CHARGE PUMPS

The Cockcroft-Walton (CW) voltage multiplier shown in figure 8.27 is a commonly used VLSI passive charge pump dc/dc converter circuit structure. CW multipliers were historically used to develop higher voltages for relatively low current applications such as the bias voltages ranging from tens or hundreds of volts to millions of volts in the electron microscopes, the oscilloscopes, the TV sets and the CRTs applications that use high dc voltages.

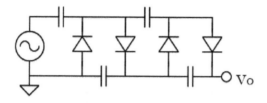

Fig. 8.27 Basic Cockcroft-Walton (CW) voltage multiplier structure

Example 8.8 Shown in figure 8.28 is a Cockcroft-Walton (CW) voltage doubler circuit structure that works with an alternating voltage source (i.e. ac source) in two phases.

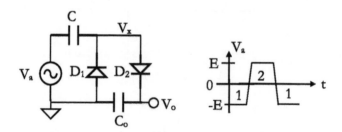

Fig.8.28 Cockcroft-Walton (CW) voltage doubler

This circuit can be analyzed by ignoring the forward diode voltage drops for simplicity as shown in figure 8.29.

- In phase I: D_1 is ON and D_2 is OFF, node Vx of fly capacitor C is charged to 0.

- In phase II: D_1 is OFF and D_2 is ON, node Vx of fly capacitor C is voltage level shifted to 2E to charge Co to the steady state final voltage of 2E.

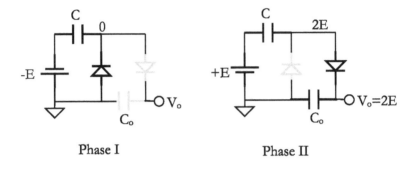

Phase I   Phase II

Fig.8.29 Cockcroft-Walton (CW) voltage doubler circuit operation

Example 8.9 Shown in figure 8.30 is a Cockcroft-Walton (CW) 4x voltage multiplier circuit that works with an alternating voltage source (i.e. ac source) in two phases.

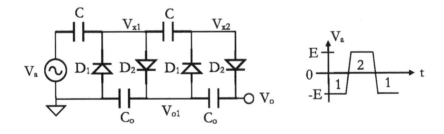

Fig.8.30 Cockcroft-Walton (CW) 4x voltage multiplier

This circuit can be analyzed by ignoring the forward diode voltage drops for simplicity as shown in figure 8.31.

- In phase I: D_1 diodes are ON and D_2 diodes are OFF, node Vx1 of first fly capacitor C is charged to 0.
- In phase II: D_1 diodes are OFF and D_2 diodes are ON, node Vx1 of first fly capacitor C is voltage level shifted to 2E to charge first storage Co to 2E.
- In phase III: D_1 diodes are ON and D_2 diodes are OFF, node Vx2 of second fly capacitor C is charged to 2E.
- In phase IV: D_1 diodes are OFF and D_2 diodes are ON, node Vx2 of second fly capacitor C is voltage level shifted to 4E to charge the second storage capacitor Co to the steady state final output voltage Vo of 4E.

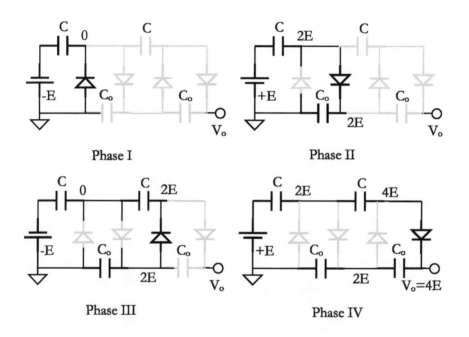

Fig.8.31 Cockcroft-Walton (CW) 4x voltage multiplier circuit operation

There are a few important notes of Cockcroft-Walton voltage multiplier circuit:
- It works with an ac voltage source;
- Phase I and phase III occur at the same time. Phase II and phase IV also occur at the same time.

- The top side of the capacitors C serve as fly capacitors and the node voltages are level-shifted ac source voltage;
- The bottom side of the capacitors Co serve as the storage capacitor that provide the amplified dc voltage of the peak voltage of the source.

The above half-wave CW charge pump circuits suffers from a number of drawbacks, such as the "sag" effect for circuits with large multiplication factor primarily due to the ac impedance of the capacitors in the lower stages. In addition, when supplying an output current, the voltage ripple rapidly increases as the number of stages is increased. For these reasons, CW multipliers with large number of stages are only used for applications with relatively low output current.

Several circuit techniques can be used to partially minimize these non-ideal circuit effects such as using higher capacitance in the lower stages, increasing the frequency of the alternating voltage source and using an ac power source with a square or triangular shaped waveform. By using high frequency source, such as an inverter, or a combination of an inverter and HV transformer, the overall physical size and weight of the CW power supply can be substantially reduced.

Alternatively the full-wave CW charge pump circuit as shown in figure 8.32 can be used to reduce the ripple voltage of the output.

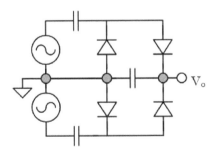

Fig.8.32 Full-wave Cockcroft-Walton (CW) 4x voltage multiplier circuit

Example 8.10 Shown in figure 8.33 is a full-wave Cockcroft-Walton (CW) voltage doubler circuit structure that works with two alternating voltage sources (i.e. ac sources) in two phases.

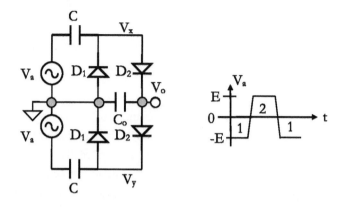

Fig.8.33 full-wave Cockcroft-Walton (CW) voltage doubler

This circuit can be analyzed as shown in figure 8.34. Due to the full-wave circuit structure the output is driven by both phases that improves the circuit's drive capability and minimizes the output ripple.

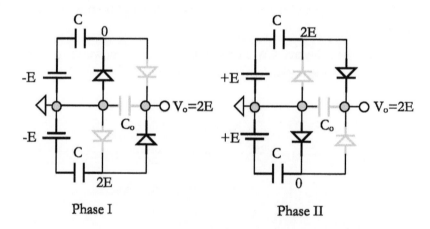

Fig.8.34 full-wave Cockcroft-Walton (CW) voltage doubler operation

Shown in figure 8.35 is an active implementation of this Cockcroft-Walton charge pump circuit that employs MOS switches for synchronous control of the charge pump from source to the load with voltage conversion gain.

VLSI Modulation Circuits

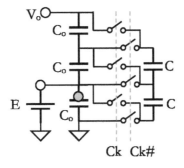

Fig.8.35 Active Cockcroft-Walton charge pump circuit

This steady state response of the unload charge pump circuit works can be analyzed as shown in figure 8.36:

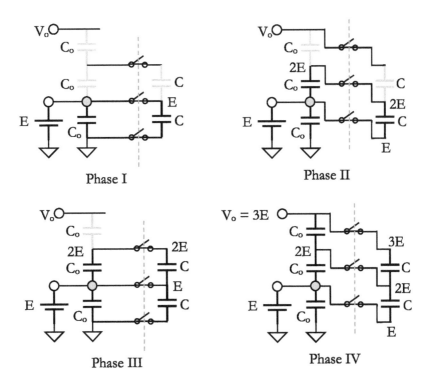

Fig.8.36 Active Cockcroft-Walton charge pump circuit operation

- In phase I (Ck= "0" and Ck# = "1"): The lower side fly capacitor is charged to E.

- In phase II (Ck= "1" and Ck# = "0"): The middle storage capacitor is charged to 2E.

- In phase III (Ck= "0" and Ck# = "1"): The middle fly capacitor is charged to 2E.

- In phase IV (Ck= "1" and Ck# = "0"): The high side storage capacitor is charged to 3E that gives the steady state output voltage of 3E.

Note that in the above analysis, phase I and phase III occur at the same time and phase II and phase IV occur the same time. The charge pump process alternates between the two phases and continuously pump charge from source the output.

8.2.3 VILLARD CHARGE PUMP CIRCUIT

The Villard circuit as shown in figure 8.37 can be viewed as a special implementation of WC switch-capacitor charge pump circuit family.

Villard circuit uses a capacitor and a diode to convert an ac power to a dc power. Such circuit offers the benefit of higher simplicity to shift the dc value of the input ac voltage source. The major drawback of such circuit structure is the degraded ripple characteristics of the output voltage.

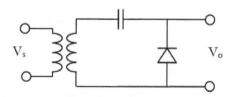

Fig. 8.37 the positive Villard circuit

A Vilard circuit that generates a negative dc voltage shift is shown in figure 8.38.

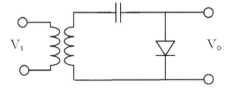

Fig. 8.38 the negative Villard circuit

8.2.4 GREINACHER CHARGE PUMP CIRCUIT

The Greinacher voltage doubler circuit shown in figure 8.39 can also be viewed as a special implementation of the CW circuit family. Greinacher voltage doubler offers performance improvement over the Villard circuit with reduced ripple at the penalty of the added capacitor and diode devices. The output voltage ripple voltage depends on the load resistance and a zero ripples can be achieved under open-circuit load conditions.

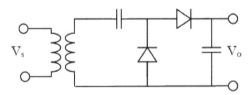

Fig.8.39 Positive Greinacher voltage doubler circuit

The Greinacher circuit can also be used to provide a negative voltage doubling operation as shown in figure 8.40.

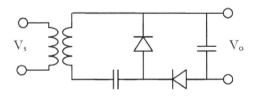

Fig.8.40 Negative Greinacher voltage doubler circuit

The positive and the negative Greinacher voltage doubler circuits can be combined into the voltage quadruple circuit (i.e. full-wave) as shown in figure 8.41.

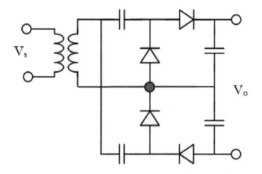

Fig.8.41 Greinacher voltage quadruple circuit

8.2.5 DELON CHARGE PUMP CIRCUIT

Another related circuit structure is the Delon circuit as shown in figure 8.42 is based on a bridge circuit topology for voltage doubling.

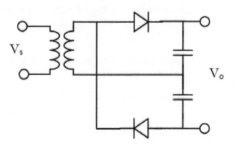

Fig.8.42 Delon half-wave voltage doubler circuit

The ripple performance of such circuit can be improved using a full-wave circuit topology as shown in figure 8.43.

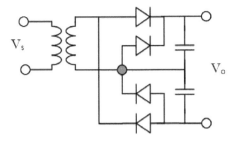

Fig.8.43 Delon full-wave voltage doubler circuit

8.2.6 MARX CHARGE PUMP CIRCUIT

Historically the Marx generator charge pump circuit was developed to generate extremely high-voltage pulse. Shown in figure 8.44 is a conceptual Marx charge pump circuit. Such circuit works as follows. All capacitors are first charged in parallel with the input power supply voltage V_s. Switch S_1 is first closed triggered by voltage breakdown or by external trigger. As the result, capacitor C_1 and C_2 will be in serial to form high voltage to breakdown switch S_2. Short circuit of switch S_2 will cause C_1, C_2, and C_3 capacitors in serial and generate even higher voltage across next switch (S_3).

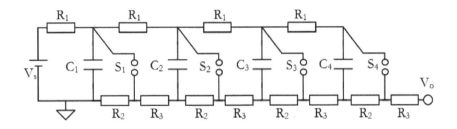

Fig.8.44 Marx generator charge pump circuit

Such breakdown process will continue sequentially until the last switch in the circuit is breakdown and that makes all capacitors in circuit to be connected in series to generate high voltage at the circuit output (i.e. V_O).

8.2.7 DICKSON CHARGE PUMP CIRCUITS

The Dickson charge pump is another commonly used VLSI boost charge pump circuit structure. Shown in figure 8.45 is the active Dickson voltage multiplier circuit structure that uses the diode devices and the alternating two phase input to realize voltage boost (multiplication) operations.

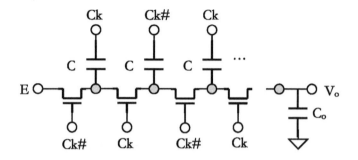

Fig.8.45 Active Dickson charge pump

Example 8.11. Shown in figure 8.46 is active Dickson voltage doubler.

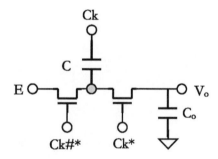

Fig.8.46 Active Dickson voltage doubler

This circuit uses two phase clock to pump charge from source to the load. This ideal circuit operates as shown in figure 3.47:

VLSI Modulation Circuits

- In phase I (Ck# = "1" and Ck = "0"): The fly capacitor is charged by the source.

- In phase II (Ck# = "0" and Ck = "01"): The storage capacitor is charged by the fly capacitor to E + Vc, where Vc is the swing value of the clock signal.

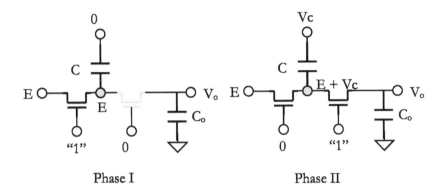

Phase I   Phase II

Fig.8.47 Active Dickson voltage doubler operation

The above active Dickson charge pump analysis has assumed the swing level of the control clock Ck* and Ck#* is high enough such that the ON resistance of the switches are neglect able. In practical design, the generation of such control signal has been a challenge. Alternatively, the passive Dickson charge pump structures can be used to minimize the efforts for generating the charge pump control signal for the circuit.

In the Dickson charge pump circuit shown in figure 8.48, node X has the same phase as clock Ck and it is also a level shifted version of clock Ck. Therefore it may naturally serves as one implementation of Ck* as shown in figure 8.42b. In addition the input NMOS switch will be turn for Ck = "Low" that occurs when X is pulled low by Ck. This implies that by connecting the input NMOS gate to the source E, this may create an equivalent control of the circuit operation as shown in the figure. Such circuit implementation leads to the passive implementation of the Dickson charge pump circuit structures employing the MOS or PN junction diodes.

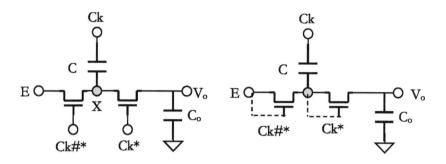

(a) Active implementation (b) Passive implementation

Fig.8.48 Passive Dickson voltage doubler

Shown in figure 8.49 is the VLSI Dickson voltage multiplier circuit structures that use the MOS diode devices and the alternating two phase input to realize voltage boost (multiplication) operations.

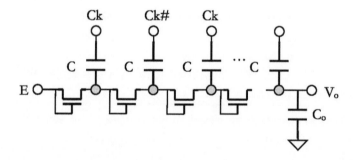

Fig.8.49 MOS Diode-based Dickson charge pump

In the same way the PN junction diode based Dickson charge pump circuit structures can also be generated as shown in figure 8.50.

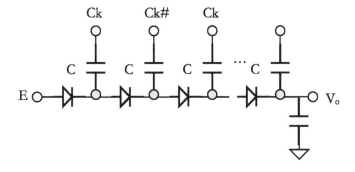

Fig.8.51 Diode-based Dickson charge pump

There are a few circuit enhancements for implementing control signal generation of active Dickson charge pump circuits.

The circuit shown in figure 8.52 is based on the symmetry of the complementary circuit structure to offer self-generated control clock phases with proper voltage level shifting.

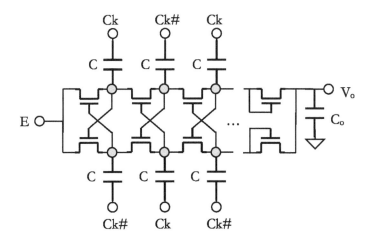

Fig.8.52 Active Dickson charge pump circuit enhancement (I)

Example 8.12. For the Dickson charge pump circuit shown in 8.53, control signal level for the two operation phases can be analyzed as follows.

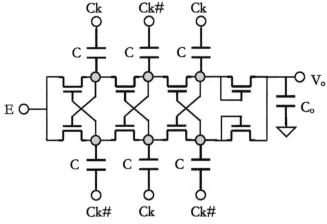

Fig.8.53 Active Dickson charge pump voltage multiplier

- Phase I: In this phase as shown in figure 8.54 the low side input fly capacitor is charged to E. Another terminal of the fly capacitor is pulled to 0 such that the capacitor is charged with charge equals to CE.

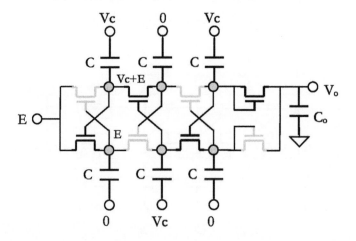

Fig.8.54 Enhanced active Dickson charge pump operation (phase I)

- Phase II: In this phase as shown in figure 8.55 the high side input fly capacitor is charged to E. Another terminal of the fly capacitor is pulled to

0. The lower side middle fly capacitor is charged to (Vc+E). The high side middle and the third fly capacitors are charged to (2Vc+E).

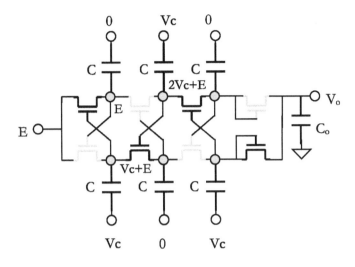

Fig.8.55 Active Dickson charge pump voltage multiplier (phase II)

- Phase III: In this phase as shown in fig.8.56 the low side input fly capacitor is charged to E. The lower side the third fly capacitor is charged to (2Vc+E). The high side the third fly capacitor is charged to (3Vc+E). It results in the output voltage of (3Vc+E-Vth).

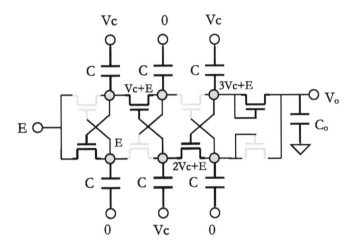

Fig.8.56 Enhanced active Dickson charge pump operation (phase III)

- Phase IV: In this phase as shown in figure 8.57 the high side input fly capacitor is charged to E. Another terminal of the fly capacitor is pulled to 0. The lower side middle fly capacitor is charged to (V_c+E). The high side middle and the third fly capacitors are charged to ($2V_c$+E). The lower side the third fly capacitor is charged to ($3V_c$+E) and the output is ($3V_c$+E-Vth).

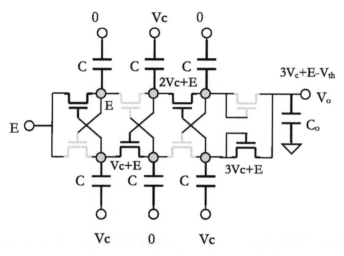

Fig.8.57 Active Dickson charge pump voltage multiplier (phase IV)

Shown in figure 8.58 is another circuit enhancement for the active Dickson charge pump generating level shifted control clock signals.

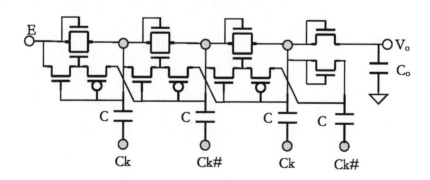

Fig.8.58 Active Dickson charge pump

Such circuit can be analyzed as follows in clock phases.

- Phase I: In this phase as shown in figure 8.59 Ck =0, the input fly capacitor is charged to input source voltage of E.

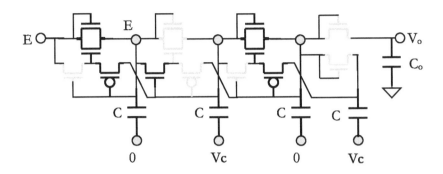

Fig.8.59 Enhance active Dickson charge pump operation (phase I)

- Phase II: In this phase as shown in figure 8.60 Ck = Vc, the second fly capacitor is charged to the voltage of (Vc+E).

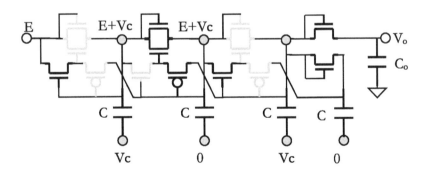

Fig.8.60 Enhance active Dickson charge pump operation (phase II)

- Phase III: In this phase as shown in figure 8.61 Ck =0, the third fly capacitor is charged to voltage of (2Vc+E.)

Signal Processing, Data Conversion, and Power Management

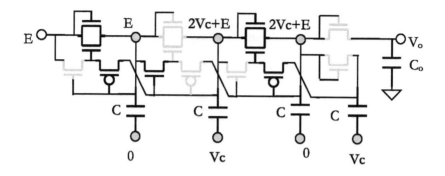

Fig.8.61 Enhance active Dickson charge pump operation (phase III)

- Phase IV: In this phase as shown in figure 8.62 Ck = Vc, the third fly capacitor is charged to the voltage of (3Vc+E).

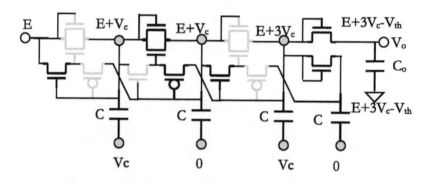

Fig.8.62 Enhance active Dickson charge pump operation (phase IV)

Note that in the above analysis, phase I and III occur at the same time and phase II and phase VI also occur the at the same time. The phases are repeated such that the circuit reach the steady state voltage values after certain time period.

8.2.8 CROSS-COUPLED CHARGE PUMP CIRCUITS

Shown in figure 8.63 is a MOS cross-coupled charge pump circuit structure. This circuit uses two phase clock to pump charge from input source E to the output Vo.

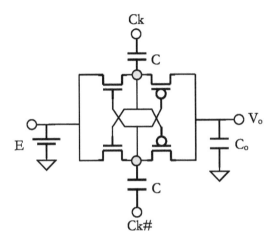

Fig.8.63 Basic cross-coupled charge pump circuit structure

- Phase I (Ck = Vc and Ck# =0): The lower fly capacitor as shown in figure 8.64 is charged to E.

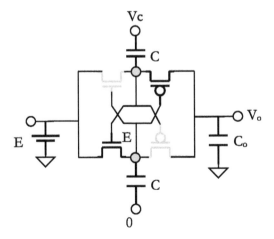

Fig.8.64 Basic cross-coupled charge pump circuit operation (Phase I)

- Phase II (Ck = 0 and Ck# =Vc): The upper fly capacitor as shown in figure 8.65 is charged to E and the storage capacitor is boosted to (E +Vc).

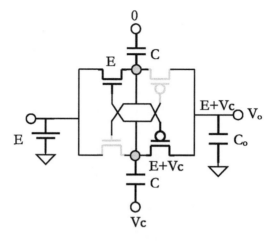

Fig.8.65 Basic cross-coupled charge pump circuit operation (Phase I)

It can be seen that the output voltage is boosted by 2x if E = Vc.

For large storage capacitor, the above charge pump process may take large number of clock cycles to settle to the above steady state voltage value.

Several basic cross-coupled charge pump circuit can be combined to achieve higher conversion gain as shown in figure 8.66.

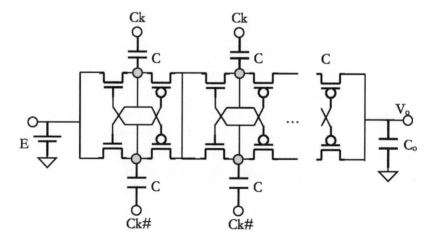

Fig.8.66 Multi-stage cross-coupled charge pump circuit

Such circuit can also be modified by minimizing the number of switched as shown in figure 8.67.

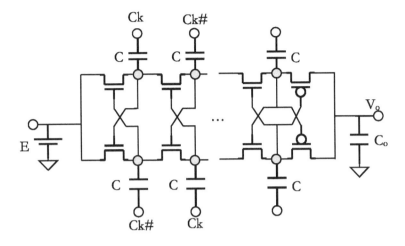

Fig.8.67 Modified multi-stage cross-coupled charge pump circuit

8.3 CHARGE PUMP CIRCUIT CONFIGURATIONS

VLSI switched-capacitor charge pump circuit may have different configurations targeted at various applications.

8.3.1 REGULATED VLSI CHARGE PUMP CIRCUITS

One common problem in the battery-powered electronic system is that the supply voltage may vary significantly in time. For example, the output voltage of a Li+ cell varies from 3.6V to 1.5V during its lifetime before it is recharged. The output voltage of the SC dc/dc converters in general are also sensitive to the load resistance (or output current) and voltage source. In addition, the output voltages are also function of the control clock period (or frequency). This relation can be described using a simple switched-capacitor circuit model shown in figure 8.68 where steady-state output voltage of the converter is given as:

$$V_o = \frac{V_s}{1 + \dfrac{T}{CR_L}} \qquad (8.18)$$

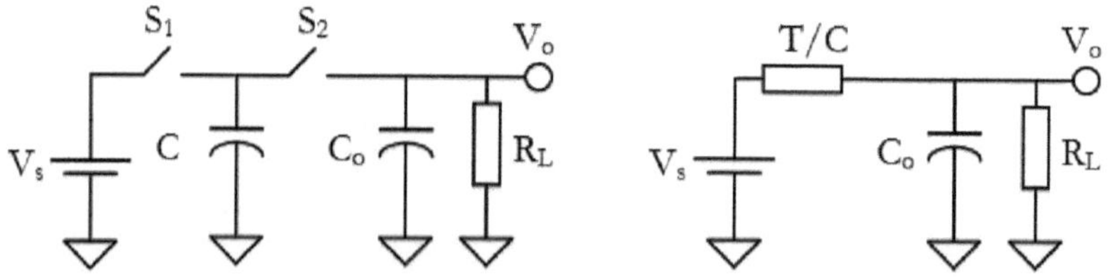

(a) SC dc/dc Converter

(b) Equivalent circuit

Fig. 8.68 Behavioral switched-capacitor DC/DC converter model

It can be seen that the resistive losses decrease at higher the clock frequency and capacitance. In practical applications, the low-ESR capacitors can be used to reduce the internal loss (switching loss). Switching loss is caused by the voltage difference between the flying capacitor and the output capacitor, as well as by on-resistance in the switches. This voltage difference appears across the switches, causing dissipation in the application. Switch's on-resistance can be reduced using sophisticated charge pumps. Since switched-capacitor behaves like a resistance. Thus, we can reduce output resistance and increase the output power by connecting several switched-capacitor devices in parallel.

The above idea can be used to construct integrated charge pumps with regulated output voltage operating in power saving modes, such as the skip mode. When a drop in the output voltage is sensed by the internal comparator, the power-saving skip mode avoids unnecessary switching by activating only the internal oscillator. This helps to reduce the quiescent current and switching dissipation, especially for light loads. The skip mode is preferable for low-power applications, because higher levels of quiescent current reduce overall efficiency.

To minimize output ripple, the circuit can oscillate in a fixed-frequency mode typically range from 50 kHz to 2MHz. A regulation ensures that the flying capacitor is charged through an internal MOS transistor switch with a charging current that depends on the load. A decreasing output voltage, caused by increasing power consumption, charges the capacitor with more energy. As benefits of the fixed-frequency mode, the output ripple is lower and the external components are smaller.

8.3.2 BUCK-BOOST REGULATOR

An improved charge pump circuit implementation for maintaining a constant switching frequency independent of the input voltage is shown in figure 8.69. To generate a constant voltage from one changing input source that might even go below the output voltage, a combined buck/boost converter is typically required. For example, for the Li+ battery power supply with 3.6V initial input and the targeted 3.3V output, the dc/dc converter initially down converts the full battery voltage of 3.6 to 3.3V. When the battery voltage drops below 3.3V, the step-up converter function guarantees the regulated 3.3V output voltage. Such circuit uses a voltage doubler core for boosting the output voltage higher than the input voltage. By adaptively control the turn on duration of the boost converter core, the output voltage of the converter can be regulated to a value independent of the input voltage and the load.

$$V_o = V_{ref} \frac{(R_1 + R_2)}{R_1} \qquad (8.19)$$

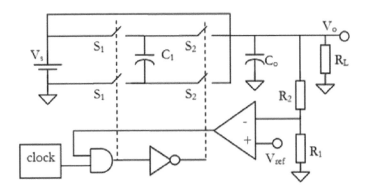

Fig.8.69 Switched-capacitor regulator

In some practical applications, this approach can be implemented with a simple charge pump IC with typical input voltages ranging from 1.6V to 5.5V, and generates an output either fixed (3.3V) or adjustable (2.5V to 5.5V) and delivers output currents up to 100mA.

8.3.3 INVERTER REGULATOR

For applications that need an additional negative voltage output, a regulating charge pump inverter can be used. Shown in figure 8.70 is a regulated SC inverter circuit consisting of a negative voltage doubler and a feedback control circuit. The clock of the negative voltage doubler is on only if the value of the output voltage is higher than the designed output voltage target. Such targeted voltage can be set using the resistor R_1, R_2 and the voltage reference Vref (usually from a bandgap) as:

$$V_o = -V_{ref}\frac{R_2}{R_1} \tag{8.20}$$

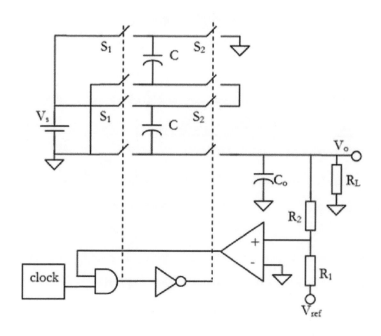

Fig.8.70 Switched-capacitor negative regulator

8.3.4 CONFIGURABLE CHARGE PUMPS

Output voltage levels of SC charge pump circuit can be programmable using the combination of capacitors and their connections. A configurable SC dc/dc converter is shown in figure 8.71. Such circuit can be used to create either 1/2 or 3/2 voltage gains using different connections of capacitors.

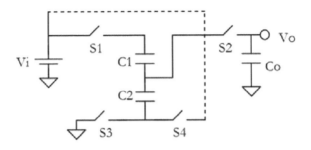

Fig.8.71 Configurable charge pump circuit

Shown in figure 8.72 are the operation phases of the programmable charge pump circuit under the 1/2 gain configuration.

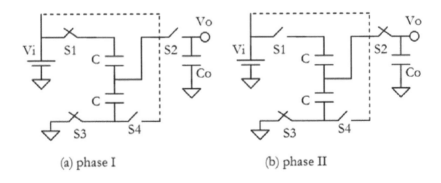

Fig.8.72 Operation phase of 1/2x gain charge pump circuit

In phase I, the switching capacitors form a voltage divider such that the input voltage is divided by two and voltage on each capacitor is half of the input voltage. In phase II, the charge on the bottom switching capacitor is used to charge the output storage capacitor Co. In the steady-state condition with only the capacitive load the output voltage of the charge pump is given as:

$$V_o = \frac{1}{2}V_i \tag{8.21}$$

Shown in figure 8.73 are the operation phases of the programmable charge pump circuit under the 3/2 gain configuration.

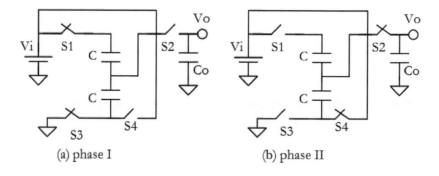

(a) phase I (b) phase II

Fig.8.73 Operation phase of 3/2x gain charge pump circuit

In phase I, the switched-capacitors form a voltage divider such that the input voltage is divided by two and voltage on each capacitor is half of the input voltage. In phase II, the charge on the bottom switching capacitor is added to the input voltage to charge the output storage capacitor Co. In the steady-state condition with only the capacitive load the output voltage of the charge pump is given as:

$$V_o = (1+\frac{1}{2})V_i = \frac{3}{2}V_i \tag{8.22}$$

8.3.5 HYBRID CHARGE PUMP CIRCUIT

In VLSI LDO circuits, the efficiency of the converter circuit is directly related to the difference between the input and output voltages. For application with large input and output voltage difference, a LDO suffers from significant efficiency loss. In such case, hybrid switched-capacitor charge pump circuit shown in figure 8.74 can be used to improve the LDO power efficiency.

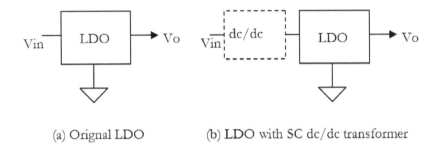

(a) Orignal LDO   (b) LDO with SC dc/dc transformer

Fig.8.74 LDO power efficiency improvement using charge pump circuit

The above hybrid charge pump circuit operates in two phases as shown in figure 8.75 and figure 8.76.
In the first phase, a switched-capacitor is connected in series with the LDO. The input voltage is divided between the capacitor and LDO to reduce the input voltage to the LDO. During this phase, the unused input voltage (and energy) of the supply is stored on a storage capacitor. In the second phase, the charge stored in the capacitor is used to power the LDO. In both phases, the difference between the input and output voltage of the LDO is reduced to improve the conversion efficiency of the charge pump circuit.

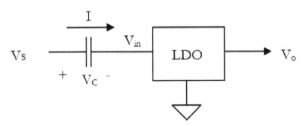

Fig.8.75 Charge storage phase of hybrid charge pump

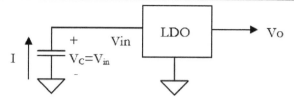

Fig.8.76 Discharge phase of hybrid charge pump

The above hybrid charge pump concept is used in the circuit shown in figure 8.77 to provide a 3.5V voltage from a 5V input.

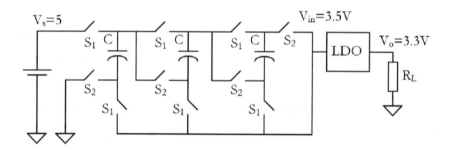

Fig. 8.77 5V/3.3V Hybrid charge pump LDO circuit

This circuit works in two clock phases. In the first phase, the switched-capacitors that are connected in series with the LDO are charged in parallel to 1.5V. The input voltage to the LDO is reduced from the original 5V supply to 3.5V with the capacitor in series. During the second phase, the three switched-capacitors are connected in series to generate 3.5V supply for the LDO. Such as circuit structure can be used to reduce the input voltage of the LDO from 5V to 3.5V, resulting conversion efficiency improvement.

Another hybrid charge pump circuit implementation is shown in figure 8.78 to convert a 5V voltage source to a 1.5V output.

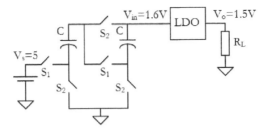

Fig. 8.78 5V/1.5V Hybrid charge pump LDO circuit

In the charging phase, the capacitors in the circuit are charged in series with LDO to 1.7V. The input to the LDO is reduced from the original 5V supply to 1.6V. During the discharge phase, the two switched-capacitors are connected in parallel to supply 1.7V to the LDO.

8.4 CHARGE PUMP CIRCUIT MODELING

VLSI switched-capacitor charge pump circuit preformance can be modeled in variosu ways in pratiical applications based on different circuit aspects, such as conversion gain, ripple voltage, settling time, loading capability and power efficiency. Two types of modeling methods including the the mircoscopic and the macroscopic (or behavrioal) can be used.

8.4.1 CIRCUIT PERFORMANCE PARAMETERS

Some performance parameters can be introduced to model the performance of a specific VLSI charge pump circuit including:

- **Voltage Conversion Gain:** Conversion gain is specified at the steady state output to the input voltage ratio for the unloaded switched-capacitor charge pump circuit.

- **Effective Output Impedance:** It is the impedance of the charge pump circuit seen by the load averaged over one or more control clock periods. Effective output impedance in general specifies the loading capability of the charge pump since any resistive or current load may introduce output voltage variation of the charge pump. Higher impedance means large voltage variation.

- **Effective Input Impedance**: It specifies the loading of the charge pump to the source.

- **Settling Time:** Settling time specifies how long a charge pumps to settle from zero initial voltage to 90% of its steady state output level for a given loading condition. In many charge pump power management circuit. Setting time dedicates how effective a charge pump circuit can be used for micro-scale power management by turning on and off in the circuit operations.

- **Ripple Voltage:** It is the feedthrough voltage noise of the charge pump internal control clock to the steady state output for a given loading condition.

- **Power Efficiency:** Power efficiency is the ratio of the power dissipation at the load to the power measured at the source.

8.4.2 MICRO, MACRO AND SIMULATION MODELING

VLSI switched-capacitor circuit belongs to discrete-time circuit family. However the output voltage of the circuit is usually lowpass filtered before provided to the load. Due to this unique application requirement. A charge pump circuit can be model in different strategies including:

- **Micro-Model:** This is the cycle based model based on the charge redistribution operation of the circuit. Under this model, the output (or input) of the circuit before each control period is expressed by the values before the control cycle with other circuit parameters. Such equation can then be solved for circuit performance parameters such as the setting, ripple, etc. The advantage of the method is the accuracy to specify detail analog performance of the charge pump circuit. The drawback is the high complexity, time and databased requirement to model the circuit.

- **Micro-Model:** In this model the average performance of the circuit over large number of cycles is used by ignoring the carry of the control clock for charge pump operation. This method provides very effective modeling of the charge pump circuit at the system level within a large circuit and system. The drawback is that it does not provide the information of the detail circuit operation and some analog effects at the carry frequency band.

- **Simulation:** In this modeling approach, the circuit is constructed employing the circuit elements and use computer software, such as SPICE and Spectre for circuit performance simulation.

Example 8.13 Shown in figure 8.79 is a charge pump voltage level shifter circuit.

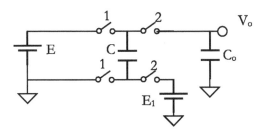

Fig.8.79 Switched-capacitor voltage level shifter circuit

In the micro-scopic modeling method, assume the Vo=0 at t=0. We can analyze the circuit by dividing the operation into two phases:

- Phase I (t = nT): In this phase the equivalent circuit is given as figure 8.80:

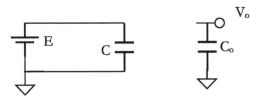

Fig.8.80 Switched-capacitor voltage level shifter circuit (Phae I)

Charge stored on the fly and the storage capacitros are given as:

$$Q(n) = CE \qquad (8.23)$$

$$Q_o(n) = C_o V_o(n) \qquad (8.24)$$

- Phase II (t = (n+1/2)T): In this phase the equivalent circuit is given as figure 8.81:

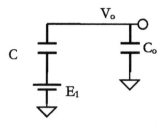

Fig.8.81 Switched-capacitor voltage level shifter circuit (Phae II)

Charge stored on the fly and the storage capacitros are given as:

$$Q(n+1/2) = C[V_o(n+1/2) - E_1] \qquad (8.25)$$

$$Q_o(n+1/2) = C_o V_o(n+1/2) \qquad (8.26)$$

During the phase I to phase II transition, the toatla charge is conserved such that

$$Q(n+1/2) + Q_o(n+1/2) = Q(n) + Q_o(n) \qquad (8.27)$$

We have that

$$V_o(n+1/2) = \frac{C_o}{C_o+C}V_o(n) + \frac{C}{C_o+C}(E+E_1) \quad (8.28)$$

When the circuit is back to phase I again, charge on the storage capacitor remains the same:

$$V_o(n+1) = \frac{C_o}{C_o+C}V_o(n) + \frac{C}{C_o+C}(E+E_1) \quad (8.29)$$

This equation can be solved analytically in various ways by adding the intial condition as:

$$V_o(0) = 0 \quad (8.30)$$

To simplify our analysis we let:

$$\alpha \equiv \frac{C_o}{C_o+C} \quad (8.31)$$

$$\beta \equiv \frac{C}{C_o+C}(E+E_1) \quad (8.32)$$

One easy way is to do it through the iteration steps as:

$$V_o(1) = \alpha V_o(0) + \beta \quad (8.33)$$

$$V_o(2) = \alpha V_o(1) + \beta = \alpha^2 V_o(0) + \alpha\beta + \beta = \alpha^2 V_o(0) + \frac{1-\alpha^2}{1-\alpha}\beta \quad (8.34)$$

$$V_o(3) = \alpha V_o(2) + \beta = \alpha^3 V_o(0) + \frac{1-\alpha^3}{1-\alpha}\beta \quad (8.35)$$

...

This leads to the analystical expression of the output votage versus time as:

$$V_o(n) = \alpha^n V_o(0) + \frac{1-\alpha^n}{1-\alpha} \beta \underset{n\to\infty}{\to} (E+E_1) \qquad (8.36)$$

The macro-model of the circuit can use the following information

A basic VLSI switched-capacitor circuit structure is shown in figure 8.82, where a capacitor is connected with a set of squentially controled signals. Other switched-capacitor circuit can be constructed using this basic circuit structure.

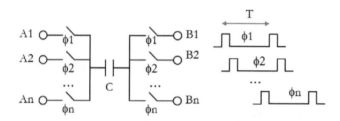

Fig.8.82 multi-phase switched-capacitor circuit structure

A general analysis of such circuit requires the charge redistribution technique and a nonlinear circuit models. A novel SC circuit behavrioal modeling method suitable for signals whose frequency much lower than the control clock frequency (known as the inband signal) was introduced by the author as follows:

- The equivalent continuous-time circuit of the basic SC circuit element looked into the circuit from the k-th terminals $\{A_k, B_k\}$ for in-band signals can be modeled using a voltage source and a linear internal resistor as shown in figure 8.83.

- The equivalent internal resistance of the circuit is proportional to the control clock period and inversely proportional to the capacitance.

- The value of the equivalent voltage source is equal to the terminal voltage difference of the (k-1)-th terminals.

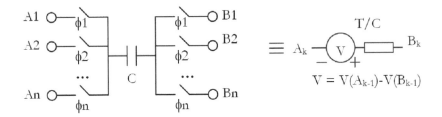

Fig.8.83 B-model of multi-phase switched-capacitor circuit structure

The above behavoral ciruit technique can be used to model the switched-capacitor with two phase non-overlap clock shown in figure 8.84.

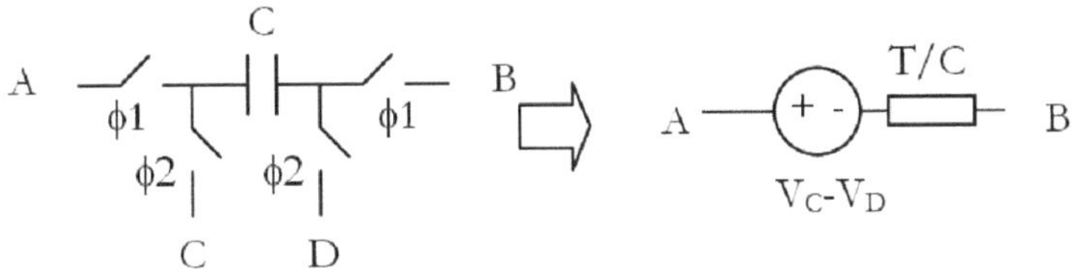

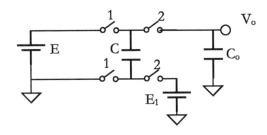

Fig.8.84 B-model of H-bridge switched-capacitor circuit structure

Example 8.14 For the charge pump voltage level shifter circuit shown in figure 8.85, a macro-model can be constructed as figure 8.86.

Fig.8.85 Switched-capacitor voltage level shifter circuit

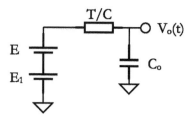

Fig.8.86 Switched-capacitor voltage level shifter macro-model

Such circuit can be solved analytically based on the differential equation as:

$$V_o(t) = V_o(0)e^{-\frac{C}{TC_o}t} + (E+E_1)(1-e^{-\frac{C}{TC_o}t}) \xrightarrow[t \to \infty]{} (E+E_1) \quad (8.37)$$

8.4.3 CHARGE PUMP CIRCUIT OPERATION MODES

Based on the switching frequency, switched impedance and the capacitance of the switched-capacitor charge pump, the circuit can work in various modes and require specific modeling approaches.

- Slow Switching Limit (SSL): Under this operation mode, the switching frequency is slow enough all internal node voltage settle during each witching phase. In such operation condition, the circuit performance for a specific circuit configuration is determined totally by the capacitance and the control clock period T.
- Fast Switch Limit (FSL): Under this operation mode, the switching frequency is too fast for the internal node voltage to settle. In such operation condition, the output impedance of the charge pump is determined by the switch on resistance and not the internal capacitance.
- Other condition: Most practical circuit may operate between the SSL and FSL and the effective output impedance may partially determine by the switch on resistance and the internal fly capacitance.

8.5 VOLTAGE SOURCE AND IMPEDANCE MODELS

The effective voltage source and impedance (known as V-R) based model provides a very effective macro-model for switched-capacitor charge pump circuit analysis in SSL operation mode.

Under the V-R model, a switched-capacitor element as shown in figure 8.87 can be approximated as the voltage source in series as a linear resistor.

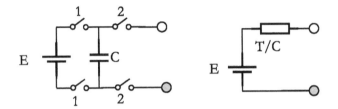

(a) Switched-Capacitor Circuit (b) Phase 2 Equivalent

Fig. 8.87 The V-R macro-model of switched-capacitor circuit

8.5.1 H-BRIDGE CHARGE PUMPS

The V-R macro-model has been very effective for modeling the H-bridge switched-capacitor charge pump circuits.

Example 8.15. For the (1/n) H-bridge voltage divider circuit shown in figure 8.88, the V-R equivalent circuit in phase I is shown in figure 8.89.

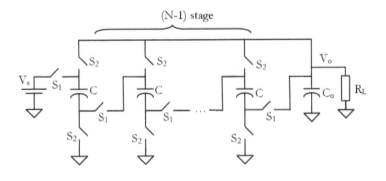

Fig.8.88 H-bridge voltage divider

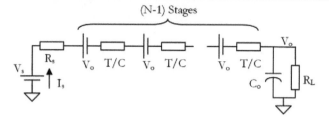

Fig.8.89 V-R macro-model of H-bridge voltage divider

The steady-state current of the source and the output voltage to the load can be derived based on the proposed model as:

$$\begin{cases} I_s[R_s + R_L + (N-1)\dfrac{T}{C}] = V_s - (N-1)V_o \\ V_o = I_s R_L \end{cases} \quad (8.38)$$

The steady-state output voltage and current can be further derived as:

$$\begin{cases} V_o = \dfrac{V_s}{[\dfrac{R_s}{R_L} + N + (N-1)\dfrac{T}{CR_L}]} \\ I_s = \dfrac{V_s / R_L}{[\dfrac{R_s}{R_L} + N + (N-1)\dfrac{T}{CR_L}]} \end{cases} \quad (8.39)$$

We can get the steady-state output voltage Vo as:

$$V_o = V_s / N \quad (8.40)$$

The above circuit can be simplified into circuit shown in figure 8.90, where the equivalent source voltage V_x and the series resistor R_x are given as:

$$\begin{cases} V_X = \dfrac{V_s}{[\dfrac{R_s}{R_L} + N + (N-1)\dfrac{T}{CR_L}]} \\ \\ R_X = \dfrac{R_s + (N-1)\dfrac{T}{C}}{[\dfrac{R_s}{R_L} + N + (N-1)\dfrac{T}{CR_L}]} \end{cases}$$ (8.41)

The transient response of the circuit can be derived based on above parameters as:

$$V_o(t) = V_o(0) + (V_X - V_o(0))(1 - \exp(-\dfrac{t}{R_X C_O}))$$ (8.42)

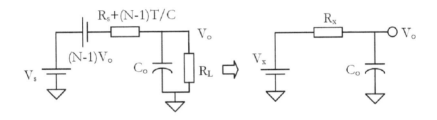

Fig.8.89 V-R macro-model of buck converter

Example 8.16 An H-bridge voltage inversion circuit is shown in figure 8.90.

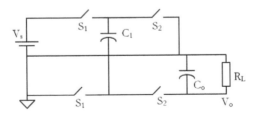

Fig.8.90 H-bridge voltage inversion circuit

The V-R equivalent circuit of the conversion circuit in phase I and phase II are shown in figure 8.91 as.

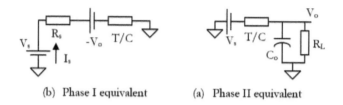

(b) Phase I equivalent   (a) Phase II equivalent

Fig.8.91 V-R macro-model of H-bridge voltage inversion circuit

The steady-state output voltage of the voltage inversion circuit is given as:

$$V_o = \frac{-V_s}{1+\dfrac{T}{CR_L}} \tag{8.43}$$

The transient voltage response can be further derived as:

$$V_o(t) = V_o(0) + (\frac{-V_s}{1+T/(CR_L)} - V_o(0))(1-\exp(-t[\frac{1}{R_L C_o} + \frac{C}{TC_o}])) \tag{8.44}$$

Example 8.17 A H-bridge volatge boost charge pump circuit is shown in figure 8.92, the equivalent circuit of the converter in phase II is shown in figure 8.93.

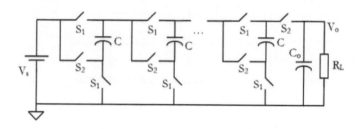

Fig.8.92 H-bridge voltage boost circuit

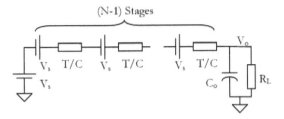

Fig.8.93 V-R macro-model of H-bridge voltage boost circuit

Similarly the equivalent circuit in phase II can be simplified as shown in figure 8.94.

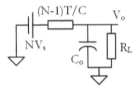

Fig.8.94 Equivalent circuit of H-bridge voltage boost circuit

The steady-state output voltage of the circuit is given as:

$$V_o = \frac{NV_s}{1 + \frac{(N-1)T}{CR_L}} \tag{8.45}$$

The transient voltage response is given as:

$$V_o(t) = V_o(0) + (\frac{NV_s}{1 + \frac{(N-1)T}{CR_L}} - V_o(0))(1 - \exp(-t[\frac{1}{R_L C_O} + \frac{C}{(N-1)TC_o}]))$$

$$\tag{8.46}$$

8.5.2 DICKSON CHARGE PUMPS

An active Dickson charge pump in SSL mode as shown in figure 8.95 can be modeled employing the V-R macro-model as shown in figure 8.95b.

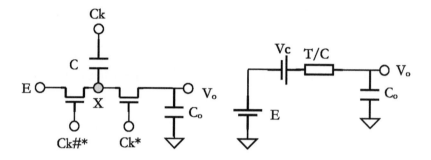

(a) Active Dickson charge

(b) V-R macro-model

Fig.8.95 Dickson voltage doubler V-R macro-model

Similarly the passive Dickson charge pump can also be modeled employing the V-R macro-model by including the diode loss. Shown in figure 8.96 is the charge transfer operation of a passive Dickson charge pump circuit.

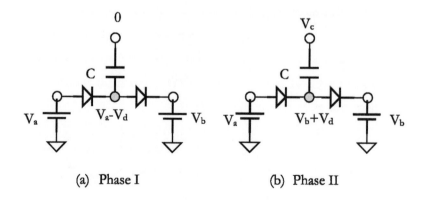

(a) Phase I

(b) Phase II

Fig.8.96 Passive Dickson charge pump charge transfer operation

The total charge transferred from source V_a to the load V_b is given as:

$$Q = C[(V_a - V_d) - (V_b + V_d - V_c)] \tag{8.47}$$

The average current flow from Va to Vb is given as:

$$\bar{I} = \frac{C}{T}[(V_a - V_b) - 2V_d + V_c] \tag{8.48}$$

This leads to the modified V-R macro-model for Dickson charge pump including the diode loss as shown in figure 8.97.

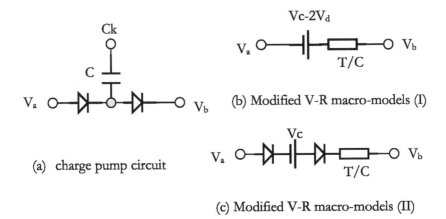

(a) charge pump circuit

(b) Modified V-R macro-models (I)

(c) Modified V-R macro-models (II)

Fig.8.97 Passive Dickson charge pump V-R macro-model

Example 8.18 Shown in figure 8.89 is a 3-stage Cockcroft-Walton (CW) voltage boost circuit structure.

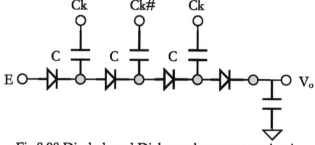

Fig.8.98 Diode-based Dickson charge pump circuit

The modified V-R macro-model can be expressed as figure 8.99.

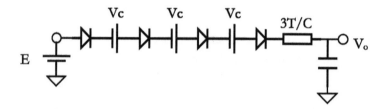

Fig.8.99 Modified passive Dickson charge pump circuit model

8.5.3 COCKCROFT-WALTON CHARGE PUMPS

The V-R macro-model can also be used to model Cockcroft-Walton charge pump circuit family in SSL mode. Shown in figure 8.100 Is the basic CW charge pump circuit structure.

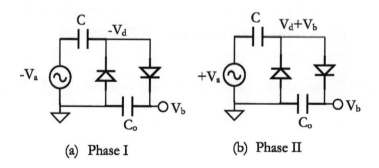

Fig.8.100 Cockcroft-Walton charge pump charge transfer operation

The total charge transferred from source V_a to the load V_b is given as:

$$Q = C[(V_a - V_d) - (V_b + V_d - V_a)] \tag{8.49}$$

The average current flow from Va to Vb is given as:

$$\bar{I} = \frac{C}{T}[(2V_a - V_b) - 2V_d] \tag{8.50}$$

This leads to the modified V-R macro-model for Dickson charge pump including the diode loss as shown in figure 8.101.

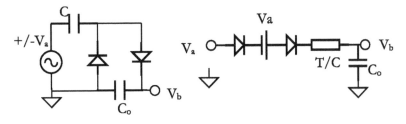

(b) CW charge pump   (b) Modified V-R macro-models

Fig.8.101 Cockcroft-Walton charge pump V-R macro-model

8.5.4 CROSS-COUPLED CHARGE PUMPS

For the cross-couple charge pump circuit shown in shown in figure 8.102 charge transfer from source Va to load Vb can be analyzed to generate the macro-model of the circuit. We can see charge transfer process occurs in both phases.

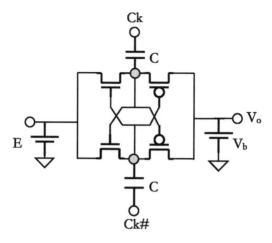

Fig.8.102 Basic cross-coupled charge pump circuit structure

- Phase I (Ck = Vc and Ck# =0): The lower side fly capacitor as shown in figure 8.103a is charged to Va.
- Phase II ((Ck = 0 and Ck# =Vc). The lower side fly capacitor as shown in figure 8.103b is discharged to Vb.

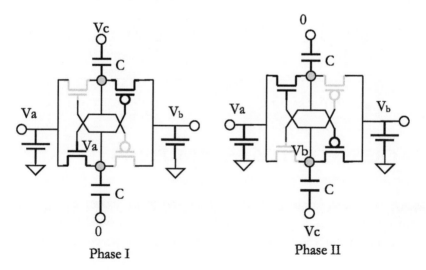

Phase I Phase II

Fig.8.103 Basic cross-coupled charge pump circuit operation

Charge transfer from Va to Vb within one clock period by the lower side fly capacitor is given as:

$$Q = C[(V_a - V_b) + V_c)] \tag{8.51}$$

Due to the symmetry of the circuit structure the average current flow from Va to Vb is twice of the current from the lower side paths:

$$\bar{I} = \frac{2C}{T}[(V_a - V_b) + V_c)] \tag{8.52}$$

This leads to the V-R macro-model of the basic cross-coupled charge pump circuit as shown in figure 8.104

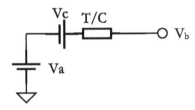

Fig.8.104 Cross-coupled voltage doubler V-R macro-model

Example 8.19 A 3-stage cross-coupled charge pump circuit is shown in figure 8.105. This circuit can be modeled based V-R macro-model as shown in figure 8.106.

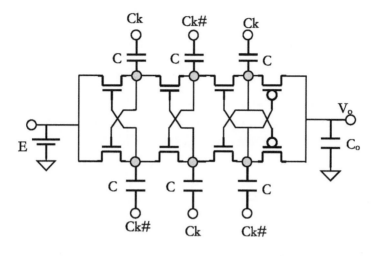

Fig.8.105 3-stage cross-coupled charge pump circuit

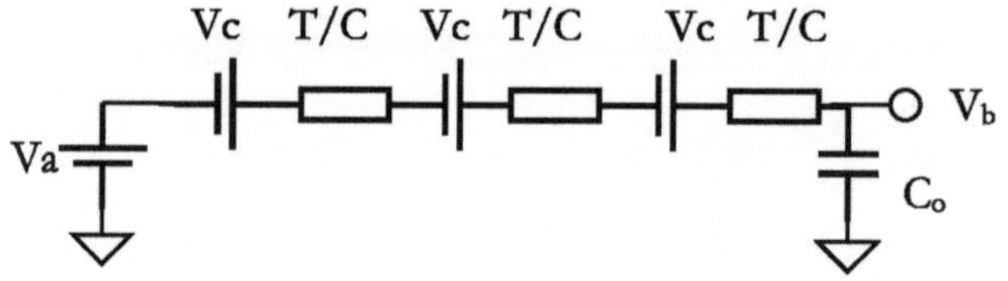

Fig.8.106 V-R 3-stage cross-coupled charge pump circuit model

8.5.5 PARASITIC CAPACITANCE EFFECTS

Practical VLSI charge pump circuit suffers from the parasitic capacitance effects that may degrade the charge pump voltage conversion gain.

For the active Dickson charge pump in SSL mode with parasitic capacitance the two operation phases are shown in figure 8.107.

VLSI Modulation Circuits

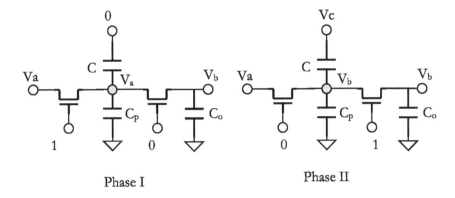

Fig.8.107 Dickson voltage doubler with parasitic capacitance

- Phase I: Both fly and parasitic capacitors are charged to Va.
- Phase II: Both fly and paristic capacitance are charged to Vb.

However since another terminal of the parasitic capacitor is fixed to 0. It does not contribute to the charge pump to the load. As the result, the charge pump to Vb in one clock period is given as:

$$Q = (C+C_p)V_a - (C+C_p)V_b + CV_c \tag{8.53}$$

The average current flow from Va to Vb is given as:

$$\bar{I} = \frac{C+C_p}{T}[V_a - V_b + \frac{C}{C+C_p}V_c] \tag{8.54}$$

This equation can be expressed as a V-R model as shown in figure 8.108 with reduced charge pmup gain and resistance.

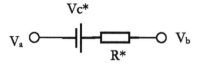

Fig.8.108 Parasitic Dickson charge pump V-R macro-model

Where

$$V_c^* = \frac{C}{C+C_p} V_c \qquad (8.55)$$

$$R^* = \frac{T}{C+C_p} \qquad (8.56)$$

A 2-stage Dickson charge pump with parasitic capacitors is shown in circuit. The phase I equiavlent circuit in SSL mode is shown in figure 8.109.

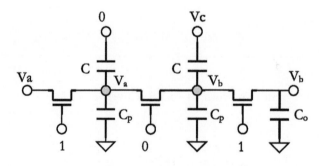

Fig.8.109 Dickson voltage doubler with parasitic capacitance (phase I)

The charge on the fly and parsitic capacitors at the input and the output side are given as:

$$Q_i = (C+C_p)V_a \tag{8.57}$$

$$Q_o = (C+C_p)V_b - CV_c \tag{8.58}$$

The phase II equiavlent circuit in SSL mode is shown in figure 8.110.

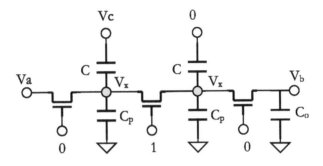

Fig.8.110 Dickson voltage doubler with parasitic capacitance (phase II)

The voltage Vx in this phase can be derived as follows:

$$(C+C_p)V_a + (C+C_p)V_b - CV_c = (2C+2C_p)V_x - CV_c \tag{8.59}$$

$$V_x = \frac{V_a + V_b}{2} \tag{8.60}$$

Charge transferred to Vb during phase I can be expressed as:

$$Q = (C+C_p)V_x - (C+C_p)V_b + CV_c = (C+C_p)\frac{V_a - V_b}{2} + CV_c \tag{8.61}$$

The average current flow from Va to Vb is given as:

$$\bar{I} = \frac{Q}{T} = \frac{C+C_p}{2T}(V_a - V_b + \frac{C}{C+C_p}2V_c) \tag{8.62}$$

Such equation can be expressed as equiavlent circuit shown in figure 8.111.

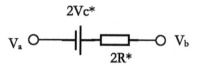

Fig.8.111 Parasitic Dickson charge pump V-R macro-model

Another way to model this parasitic capacitance effect is through the following equiavlent circuit transformation shown in figure 8.112.

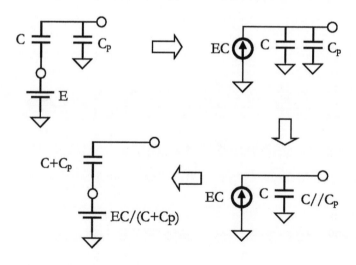

Fig.8.112 Equivalent fly capacitor including the parasitic

VLSI Modulation Circuits

The above circuit tarnsformation implies that

- The equiavlent fly capacitance is increased with a parallel parasitic capacitance as:

$$C \rightarrow C^* = C // C_p = C + C_p \qquad (8.63)$$

-

- The equiavelent charge pump clock swing is reduced by a factor given as:

$$E \rightarrow E^* = \frac{C}{C+C_p} E \qquad (8.64)$$

Example 8.20. For the cross-couple charge pump circuit shown in shown in figure 8.113 With parasitic capacitance Charge transfer from source V_a to load V_b can be analyzed to generate the macro-model of the circuit. We can see charge transfer process occurs in both phases.

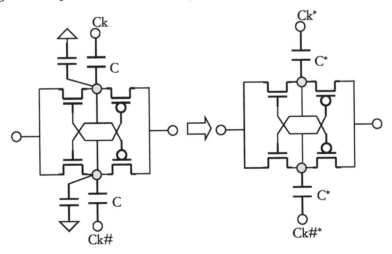

Fig.8.113 Cross-coupled charge pump circuit modeling with parasitic

8.6 TRANSFORMER BASED CHARGE PUMP MODEL

The unidirectional volatge source and impedance (V-R) based micro-model offers a very effective method to model charge pump circuit with fixed voltage source. For many application where volatge source is also a variable that may impacted by the load. A bi-drectional macro-model based on the transformer and impedance (T-R) model can be introduced. Such model offers bi-directional signal and energy path in the circuit simulation.

8.6.1 IDEAL TRANSFORMER CHARACETRISTICS

An ideal transformer is a four terminal circuit element as shown in figure 8.114.

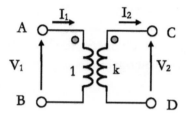

Fig. 8.114 Transformer circuit model

Note that the nots in the circuit symbol indicates the voltage polarities of the transformer. Such ideal transformer element offers few important conversion rules, such as:

- **Rule 1. Voltage Conversion Rule:** Voltage conversion is determined by the conversion ratio k.

$$V_2 = kV_1 \qquad (8.65)$$

- **Rule 2. Current Conversion Rule:** Current conversion can be derived based on the energy conservation as

$$V_1 I_1 = V_2 I_2 \qquad (8.66)$$

Or

$$I_1 = I_2 / k \qquad (8.67)$$

- **Rule 3. Impedance Conversion Rule:** Impedance conversion as shown in figure 8.115 can also be derived based on the energy flow through the circuit as:

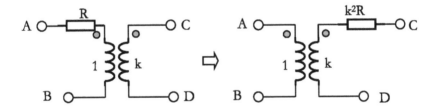

Fig. 8.115 Transformer impedance conversion

- **Rule 4. Transformer Short Anneling Rule:** transformer element can be annealed to short as shown in figure 8.116.

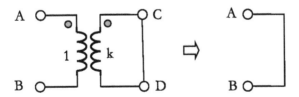

Fig. 8.116 Transformer anneals to short

- **Rule 5. Transformer Open Anneling Rule:** Transformer element can also be annealed to open as shown in figure 8.117.

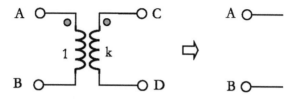

Fig. 8.117 Transformer anneals to open

- **Rule 6. Terminal Impeance Shift Rule:** Impedance on the same side can be moved from one terminal to another one without impact the circuit behavior as shown in figure 8.118.

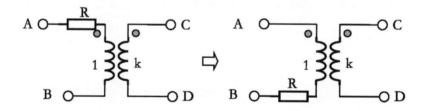

Fig. 8.118 Transformer impedance conversion

- **Rule 7. Transformer Branching Rule:** transformer conversion branches the same way as the impedance path as shown in figure 8.119.

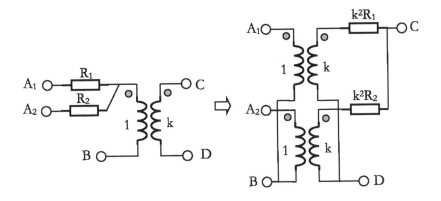

Fig. 8.119 Transformer branching conversion

8.6.2 H-BRIDGE CHARGE PUMPS

A H-bridge in SSL model can be hevarioally modeled employing an ideal transformer in series with a resistor (known as T-R macro-model) as shown in figure 8.120.

The T-R micro-model provides bi-directional path between the input and the output terminals of the switched-capacitor circuits for circuit analysis.

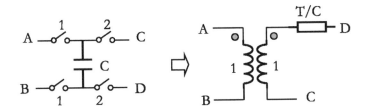

Fig. 8.120 T-R macro-model of H-bridge

Eaxmlpe 8.21. For the H-bridge volatge doubler shown in figure 8.121, the T-R macro-model can be constructed as figure 8.122.

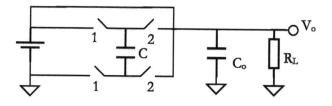

Fig. 8.121 H-bridge doubler circuit

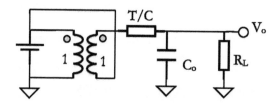

Fig. 8.122 T-R model equivalent of H-bridge voltage doubler (I)

For the sub-circuit structure shown in figure 8.123, we can see that it realizes a lossless voltage doubling.

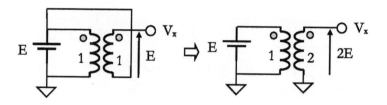

Fig. 8.123 Ideal doubler subcircuit model

The lossy doubler circuit can be modeled based on the circuit shown in figure 8.124.

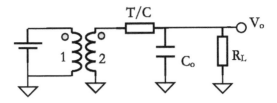

Fig. 8.124 Transformer equivalent of switched-capacitor doubler (II)

Eaxmlpe 8.22. For the H-bridge volatge divider shown in figure 8.125, the T-R macro-model can be constructed as figure 8.126.

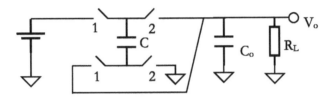

Fig. 8.125 H-bridge volatge divider circuit

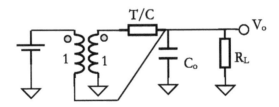

Fig. 8.126 T-R model of H-bridge volatge divider

The above circuit structure can be modified into the circuit structure shown in figure 8.127.

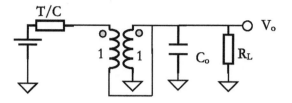

Fig. 8.127 Equivalent T-R model of H-bridge volatge divider

Considering the equivalent of the following sub-circuit structure shown in figure 8.128, we have that

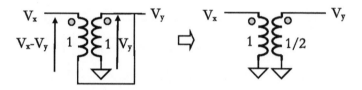

Fig. 8.128 Ideal divider subcircuit model

The voltage lossy voltage divider can be modified as shown in figure 8.129 and figure 8.130.

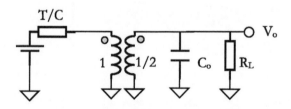

Fig. 8.129 Equivalent T-R model of H-bridge volatge divider

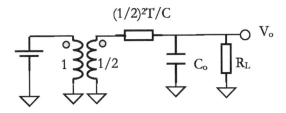

Fig. 8.130 Simplified T-R model of H-bridge volatge divider

Eaxmlpe 8.23. Shown in figure 8.131 is a switched-capacitor 4x voltage boost circuit employing two doubler circuits.

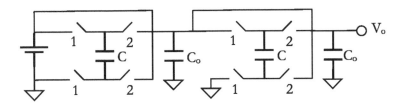

Fig. 8.131 H-bridge 4x voltage boost circuit

The transformer based equivalent circuit model is Shown in figure 8.132 is a switched-capacitor cascaded doubler circuit.

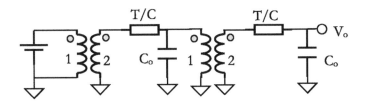

Fig. 8.132 T-R macro-model of 4x voltage boost circuit

By the impeance conversion we can modified this circuit as shown in figure 7.133:

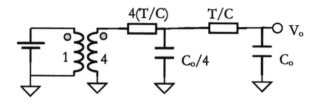

Fig. 8.133 Equivalent T-R macro-model of 4x voltage boost circuit

8.6.3 DICKSON CHARGE PUMPS

The active Dickson charge pump circuit and its T-R model is shown in figure 8.134.

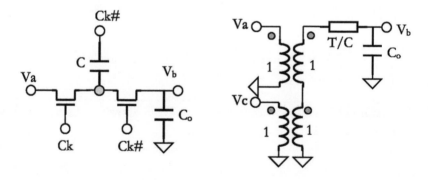

Fig.8.134 Basic Dickson charge pump T-R macro-model (I)

Note that Vc is the peak value of the control clock signal in this model.

If Ck is supplied by an ideal clock source, the above macro-model can also be modified as figure 8.135.

VLSI Modulation Circuits

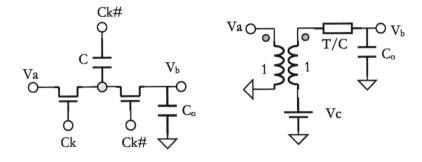

Fig.8.136 Basic Dickson charge pump T-R macro-model (II)

8.6.4 CROSS-COUPLED CHARGE PUMPS

The basic cross-coupled charge pump circuit with ideal control clock can be model using a T-R model is shown in figure 8.137.

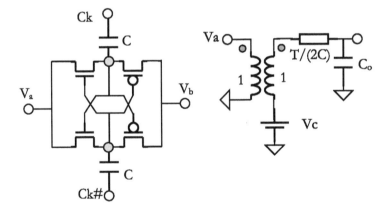

Fig.8.137 Basic Cross-couple charge pump T-R macro-model

Similarly Vc is the peak value of the control clock signal in this model.

8.7 GAIN HOPPING TECHNIQUES

For a charge pump circuit, the current at the source is given as the current is needed to drive the load and the current penalty I_P for the charge pump circuit operation as:

$$I_i = Gain \cdot I_o + I_P \qquad (8.68)$$

The efficiency of the charge pump circuit can be expressed as:

$$\eta = \frac{V_o \cdot I_o}{V_i(Gain \cdot I_o + I_P)} \qquad (8.69)$$

For a fixed gain, Vo and Io, the efficiency of a switched cap regulator is inversely proportional to the input voltage V_i. As V_i goes up, gain can go down to improve efficiency. There are two ways to keep V_o in regulation as V_i goes up including to lower the gain or to increase the effective resistance by lowering switching frequency or changing the linear regulator impedance.

However, the actions to increase the effective impedance will degrade the power efficiency of the circuit. Alternatively gain adjust approach can be used to maintain high efficiency of the circuit. In order to provide high efficiency over a wide range of input voltage, most modern VLSI regulators provide multiple gains and the gains can be switched based on the input voltage.

Reference

[1] B. Arntzen, D. Maksimovic, "Switched-capacitor dc/dc converters with resonant gate drive," IEEE Transactions on Power Electronics, Volume: 13, Issue: 5, Page(s): 892 - 902, 1998.

[2] Lon-Kou Chang, Chih-Huei Hu, "High efficiency MOS charge pumps based on exponential-gain structure with pumping gain increase circuits," IEEE Transactions on Power Electronics, Volume: 21, Issue: 3, Page(s): 826 - 831, 2006.

[3] Lon-Kou Chang, Chih-Huei Hu, "An exponential-folds design of a charge pump styled dc/dc converter," 2004 IEEE 35th Annual Power Electronics Specialists Conference, Volume: 1, Page(s): 516 - 520 Vol.1

[4] G. Thiele, E. Bayer, "Current-Mode LDO with Active Dropout Optimization," IEEE 36th Power Electronics Specialists Conference, Page(s): 1203 - 1208, 2005.

[5] E. Bayer, H. Schmeller, "Charge pump with Active Cycle regulation-closing the gap between linear and skip modes," IEEE 3first Annual Power Electronics Specialists Conference, Volume: 3, Page(s): 1497 - 1502 vol.3, 2000.

[6] M. Bouhamame, J. R. Tourret, L. Lo Coco, S. Toutain, O. Pasquier, "A Fully Integrated dc/dc Converter for Tunable RF Filters," IEEE Custom Integrated Circuits Conference, 2006. Page(s): 817 - 820

[7] Shiming Han, Xiaobo Wu, Xiaolang Yan, "Novel Dual-Output Step Up and Down Switched-capacitor dc/dc Converter," IEEE Conference on Electron Devices and Solid-State Circuits, 2007. Page(s): 871 - 874

[8] Wei-Chung Wu, R. M. Bass, "Analysis of charge pumps using charge balance," IEEE 3first Annual Power Electronics Specialists Conference, 2000. Volume: 3, Page(s): 1491 - 1496 vol.3

[9] T. Van Breussegem, M. Steyaert, "A fully integrated gearbox capacitive dc/dc-converter in 90nm CMOS: Optimization, control and

measurements," IEEE 12th Workshop on Control and Modeling for Power Electronics (COMPEL), Page(s): 1 - 5, 2010.

[10] Y. Beck, S. Singer, "Capacitive Transposed Series-Parallel Topology With Fine Tuning Capabilities," IEEE Transactions on Circuits and Systems I: Regular Papers, Volume: 58, Issue: 1, Page(s): 51 - 61,2011

[11] Hongjiang Song, Yan Song, Tai-hua Chen, "VLSI passive switched-capacitor signal processing circuits: Circuit architecture, closed form modeling and applications," 2008 IEEE International SOC Conference, Page(s): 297 - 300, 2008.

[12] S. Chaisotthee, J. Parnklang, "Low swing CMOS current mode charge pump," 2010 International Conference on Control Automation and Systems (ICCAS), Page(s): 1383 - 1386, 2010.

[13] Xiwen Zhang, Hoi Lee, "An Efficiency-Enhanced Auto-Reconfigurable 2/3 SC Charge Pump for Transcutaneous Power Transmission," IEEE Journal of Solid-State Circuits, Volume: 45, Issue: 9, Page(s): 1906 - 1922, 2010.

[14] R. Perez, "Power conversion techniques for portable EMI sensitive applications," IEEE International Symposium on Electromagnetic Compatibility, 2000. Volume: 1, Page(s): 487 - 491, 2000.

[15] T. Das, P. Mandal, "Switched-Capacitor Based Buck Converter Design Using Current Limiter for Better Efficiency and Output Ripple," 2second International Conference on VLSI Design, Page(s): 181 - 186, 2009.

[16] A. Richelli, L. Colalongo, S Tonoli, Z. Kovacs, "A 0.2V–1.2V converter for power harvesting applications," 34th European Solid-State Circuits Conference, Page(s): 406 - 409, 2008.

[17] Y. P. B. Yeung, K. W. E. Cheng, S. L. Ho, KK. Law, D Sutanto, "Unified analysis of switched-capacitor resonant converters," Industrial Electronics, IEEE Transactions on Volume: 51, Issue: 4, 2004, Page(s): 864-873.

CHAPTER 9
VLSI SWITCHED-CAPACITOR FILTER CIRCUITS

A VLSI switched-capacitor (SC) filter circuit realizes the signal processing operation employing a switched network circuit structure, where the modulation circuit technique is used to enable the charge redistribution circuit operation for the specific signal processing function.

It is interesting to see that a demodulation (in the forms of integration or sample/hold) is usually implemented together with the modulation circuits in the basic switched-capacitor building blocks (such as SC integrator) such that a SC circuit can be effectively modeled similar to a continuous-time circuit structure (such as an active-RC circuit).

The VLSI switched-capacitor circuit techniques offer signal processing circuit implementation fully compactable to CMOS technologies to achieve low power and compact design. SC circuits have been used to simulate resistors with very good accuracy by rapidly charging and discharging the capacitor across two terminal voltages. By using this scheme the RC constants of the signal processing circuits only depend on the clock frequency and the ratio of capacitor values. All of these can be controlled very accurately in VLSI circuit implementations. In addition, the SC circuits are particularly suitable for VLSI circuit realization of high value on-chip resistances, which effectively solves the fundamental problem of applying VLSI analog signal processing circuits in most low frequency signal processing applications.

9.1 VLSI SWITCHED-CAPACITOR CIRCUIT PRINCIPLE

The very basic switched-capacitor structure is shown in figure 9.1. Such circuit operates in two clock phases including a charging phase and a discharging phase.

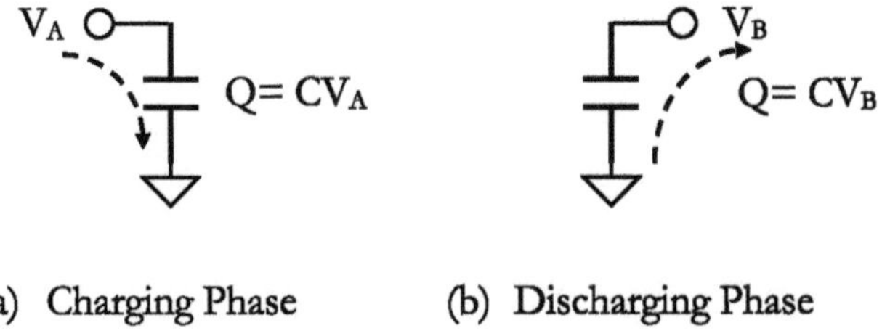

(a) Charging Phase   (b) Discharging Phase

Fig.9.1 Basic switched-capacitor circuit operation

During the charging phase, electrical charge is transferred from the terminal V_a (supply) to the capacitor with the total charge transferred in this phase given as:

$$\Delta q = CV_a - q_o \tag{9.1}$$

q_o is assumed to be the initial charge of the capacitor.

During the discharging phase, the charge is transferred from the capacitor to terminal V_B (load). The total electrical charge transferred from the capacitor to terminal B in this phase is given as

$$\Delta q = C(V_a - V_b) \tag{9.2}$$

If a periodic control clock is used to control this switched-capacitor circuit under the conditions that the frequencies of the signals at the terminal A and B are much lower than that of the clock frequency and the time constant of the system is much greater than the period of the clock, then such a switched-capacitor circuit is equivalent to a linear resistor connected across the two terminals and the equivalent resistance is given as:

$$\Rightarrow R \equiv \frac{(V_a - V_b)}{I} = \frac{(V_a - V_b)}{\frac{\Delta q}{T}} = \frac{(V_a - V_b)}{\frac{C}{T}(V_a - V_b)} = \frac{T}{C} \quad (9.3)$$

The significances the SC simulated resistor shown in Equation (9.3) are as follows:

- Large value resistors can be practically realized using the VLSI processes. Based on Equation (9.3), high value resistor (e.g. 1MΩ) can be easily implemented with capacitor of reasonable value (e.g. 1pF) and clock frequency (e.g. 1MHz) in VLSI form;
- Accuracy of the resistance implemented by SC method is improved compared with the direct VLSI resistor implementation since both clock frequency and capacitance can be controlled much accurately than resistance;
- The RC time constant of on-chip VLSI circuit under SC implementation is highly improved since it only depends on the clock period and the ratio of capacitances, both of them can be controlled very accurately under PVT conditions;
- Since on-chip capacitor typically has better linearity than resistor in VLSI. SC implemented resistor also will improve the linearity of the signal processing circuits.

By replacing the resistor in the active-RC integrator shown in figure 9.2 using a SC resistor, the SC integrator can be constructed.

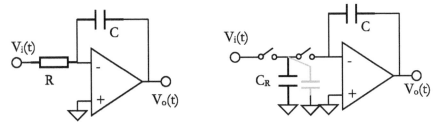

(b) Equivalent RC-circuit   (a) Switched-capacitor circuit

Fig.9.2 VLSI switched-capacitor integrator circuit

The basic SC circuit structure suffers from the fundamental accuracy limitation due to the parasitic capacitance effects of the VLSI implementation, where the parasitic capacitance from the upper plate of the VLSI capacitor contributes to the switched-capacitor value uncertainty. Such a problem can effectively be solved using the so-called parasitic insensitive SC circuit structures as shown in figure 9.3.

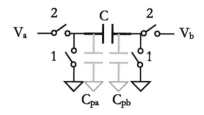

(a) Positive switched-capacitor resistor

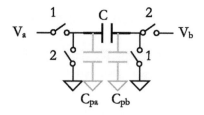

(b) Negative switched-capacitor resistor

Fig.9.3 Parasitic insensitive switched-capacitor resistor structures

In the circuit shown in Fig.9.3a, the capacitor is first discharged with both terminals of the capacitor shorted to ground during phase 1 ($\phi1 = 1$). During phase 2 ($\phi2 = 1$), the capacitor is charged to the input voltage. It can be seen that the parasitic capacitor effect (C_{pa} and C_{pb}) can be eliminated from the resistance Equation (9.3) since the parasitic capacitor is connected to ground and which does not contribute to the current flow from terminal V_a to terminal V_b during $\phi2$ operation. Since here is a direct current path from the input to the output, no delay effect occurs between the input and output signals during this phase:

$$\begin{cases} R = \dfrac{T}{C} \\ \Delta Q(nT + \dfrac{T}{2}) = V_a(nT + \dfrac{T}{2}) \end{cases} \quad (9.4)$$

In the alternative SC circuit shown in Fig.9.3b, the capacitor C and its parasitic capacitor C_{pa} is first charged to the input signal voltage in phase 1 ($\phi 1 = 1$). During phase 2 ($\phi 2 = 1$), the two capacitors are discharged to ground. Similar to the previous circuit, only the capacitor C contributes to the discharged current to terminal V_b, and the parasitic capacitor effect is also eliminated. Since the direction of the discharge current path in such case is opposite to that in the previous circuit, a negative SC resistance with a half clock period delay between the input and output signals is effectively realized:

$$\begin{cases} R = -\dfrac{T}{C} \\ \Delta Q(nT + \dfrac{T}{2}) = V_a(nT + \dfrac{T}{2}) \end{cases} \quad (9.5)$$

9.1.1 CHARGE REDISTRIBUTION ANALYSIS METHOD

VLSI SC circuits are based on the principle of electrical charge redistribution among the capacitors within the SC circuits, where the Opamp or OTA circuits merely serve as the helpers for necessary environment for the charge redistribution operations (isolation, gain, and stabilization of the capacitor terminal voltages.). The charge redistribution method provides an effective way to analyze the SC circuit at all operation frequencies down to the Nyquist rate.

Note that the direct mapping of switched-capacitors to resistors introduced in the previous section is only valid under the oversampling conditions. Under the oversampling condition, the charge redistribution method and the direct mapping method are equivalent to each other.

9.1.2 NON-INVERTED SWITCHED CAPACIOTR INTEGRATOR

For the non-inverted SC integrator circuit shown in figure 9.4, the equivalent circuit during the $\phi 1$ phase ($\phi 1 =1$) is shown in figure 9.5. In this phase, the switched-capacitor C_R is charged to the input signal voltage value and the integration capacitor C holds the charge accumulated in the previous clock cycles.

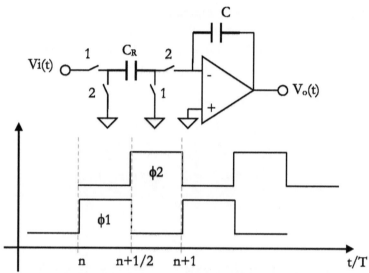

Fig.9.4 Non-inverted switched-capacitor integrator circuit

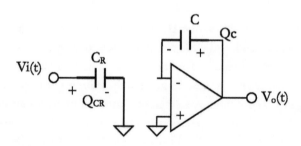

Fig.9.5 Non-inverted switched-capacitor integrator circuit ($\phi 1$ phase)

At the beginning of this phase, the voltages across the two capacitors are given as:

$$\begin{cases} V_o(t) = V_o(nT) \\ V_{CR}(t) = V_i(t) \end{cases} \quad (9.6)$$

At the end of the $\phi 1$ phase, these voltages can be expressed as:

$$\begin{cases} V_o(nT + T/2) = V_o(nT) \\ V_{CR}(nT + T/2) = V_i(nT + T/2) \end{cases} \quad (9.7)$$

The electrical charges stored on the two capacitors can be derived as:

$$\begin{cases} Q_C(nT + T/2) = C \cdot V_o(nT) \\ Q_{CR}(nT + T/2) = C_R \cdot V_i(nT + T/2) \end{cases} \quad (9.8)$$

During the $\phi 2$ phase ($nT + T/2 < t < nT + T$), the equivalent circuit of the SC integrator is shown in figure 9.6.

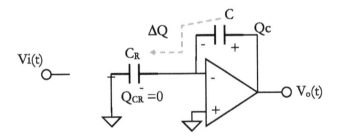

Fig.9.6 Non-inverted switched-capacitor integrator circuit ($\phi 2$ phase)

Note that when the SC circuit configuration is switched from circuit shown figure 9.5 to circuit shown in figure 9.6, electrical charge redistribution occurs.

For example, the switched-capacitor is usually charged to a non-zero voltage at the end of a φ1 phase, while at the end of the φ2 phase it is required to be discharged to the zero voltage based on the fact that the two terminals of C_R will be forced to the same ground voltage during φ2 phase because of the Opamp circuit in negative feedback connection. Since the Opamp circuit has the infinite input impedance, the above discharge process of C_R within φ2 can only be realized through the path across the integration capacitor C. As the result, the integration capacitor C is charged with the same amount of electrical charge when the switched-capacitor is discharged. In the other word, the electrical charge is redistributed between the two capacitors in the circuit.

At the end of φ2 phase, the charges on the two capacitors can be expressed as:

$$\begin{cases} Q_C(t) = Q_C(nT) + Q_{CR}(nT + T/2) \\ Q_{CR}(nT + T) = 0 \end{cases} \qquad (9.9)$$

The output voltage is then given as:

$$V_o(nT + T) = V_o(nT) + \frac{C_R}{C} V_i(nT + T/2) \qquad (9.10)$$

Assuming that the input signal is gated by the opposite phase (φ2 phase), we can further derive the discrete-time input and output transfer function of such SC integrator as:

$$V_o(nT + T) = V_o(nT) + \frac{C_R}{C} V_i(nT + T/2) = V_o(nT) + \frac{C_R}{C} V_i(n) \quad (9.11)$$

The z-domain transfer function can also be derived by applying the z-transform as:

$$\Rightarrow \frac{V_o(z)}{V_i(z)} = \frac{C_R}{C} \frac{z^{-1}}{1-z^{-1}} \qquad (9.12)$$

9.1.3 INVERTED SWITCHED-CAPACITOR INTEGRATOR

The charge redistribution technique can also be used to analyze the operation of the inverted SC integrator circuit as shown in figure 9.7.

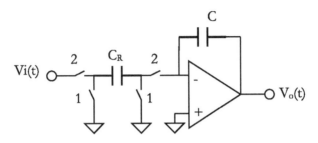

Fig.9.7 Inverted switched-capacitor integrator circuit

For the $\phi 1$ phase ($nT < t < nT+T/2$) this SC integrator is equivalent to the circuit shown in figure 9.8.

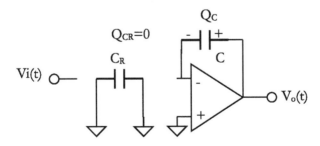

Fig.9.8 Equivalent switched-capacitor integrator circuit ($\phi 1$ phase)

The voltages applied to the two capacitors can be expressed as:

$$\begin{cases} V_o(t) = V_o(nT) \\ V_{CR}(t) = 0 \end{cases} \quad (9.13)$$

The charges on the two capacitors at the end of this phase can be derived as:

$$\begin{cases} Q_C(nT + T/2) = C \cdot V_o(nT) \\ Q_{CR}(nT + T/2) = 0 \end{cases} \quad (9.14)$$

During the ϕ2 phase (nT+T/2 < t < nT+T) the SC integrator is equivalent to the circuit shown in figure 9.9.

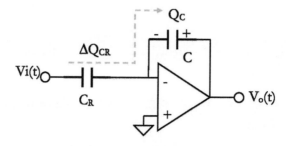

Fig.9.8 Equivalent switched-capacitor integrator circuit (ϕ2 phase)

Similarly charge redistribution occurs when the SC circuit is switched between the ϕ1 and ϕ2 configurations. At the end of ϕ2 phase C_R is charged from zero to the input voltage through the path across the integration capacitor C as:

$$\begin{cases} Q_C(t) = Q_C(nT) - Q_{CR}(nT + T) \\ Q_{CR}(nT + T) = C_R \cdot V_i(nT + T) \end{cases} \quad (9.15)$$

Consequently, the discrete-time input to output voltage transfer function can be derived as

$$V_o(nT+T) = V_o(nT) - \frac{C_R}{C} V_i(nT+T) \qquad (9.16)$$

The z-domain transfer function can further be derived as:

$$\Rightarrow \frac{V_o(z)}{V_i(z)} = -\frac{C_R}{C} \frac{1}{1-z^{-1}} \qquad (9.17)$$

9.1.4 DOUBLY PUMPED SC INTEGRATOR CIRCUIT

Shown in figure 9.10 is a doubly pumped SC integrator circuit structure. The equivalent circuit of such integrator for the $\phi 1$ phase ($nT < t < nT+T/2$) is shown in figure 9.11.

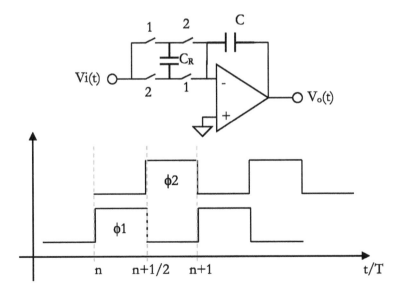

Fig.9.10 Double pumped switched-capacitor integrator circuit

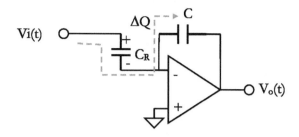

Fig.9.11 Double pumped switched-capacitor integrator circuit (φ1 phase)

The electrical charges stored on the two capacitors at the beginning of φ1 phase are given as:

$$\begin{cases} Q_C(nT) = C \cdot V_o(nT) \\ Q_{CR}(nT) = -C_R \cdot V_i(nT) \end{cases} \quad (9.18)$$

After the charge redistribution process during the φ1 phase, the electrical charges stored on the two capacitors at the end of φ1 phase are given as:

$$\begin{cases} Q_C(nT+T/2) = C \cdot V_o(nT) - C_R \cdot (V_i(nT+T/2) + V_i(nT)) \\ Q_{CR}(nT+T/2) = C_R \cdot V_i(nT+T/2) \end{cases} \quad (9.19)$$

The equivalent circuit of the SC integrator during the φ2 is shown in figure 9.12.

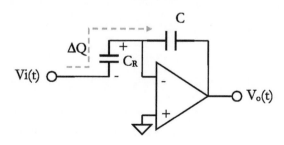

Fig.9.12 Double pumped switched-capacitor integrator circuit (φ2 phase)

The charge redistribution process occurs when the circuit is switched from the φ1 to φ2 circuit configuration. At the end of φ2 phase the electrical charges stored on the two capacitors are given as:

$$\begin{cases} Q_C(nT+T) = C \cdot V_o(nT+T/2) - C_R \cdot (V_i(nT+T) + V_i(nT+T/2)) \\ Q_{CR}(nT+T) = -C_R \cdot V_i(nT+T) \end{cases}$$

(9.20)

Consequently, the discrete-time input to output voltage transfer function of this SC integrator can be derived as:

$$V_o(nT+T) = V_o(nT) - \frac{C_R}{C}(V_i(nT) + 2V_i(nT+T/2) + V_i(nT+T))$$

(9.21)

The z-domain transfer function can be further derived as:

$$\Rightarrow \frac{V_o(z)}{V_i(z)} = -\frac{C_R}{C}\frac{z^{-1} + 2z^{-1/2} + 1}{1-z^{-1}} = -\frac{C_R}{C}\frac{1+z^{-1/2}}{1-z^{-1/2}}$$

(9.22)

9.2 BASIC VLSI SC CIRCUIT ELEMENTS

A typical VLSI SC circuit consists of the following elements including VLSI MOS switch circuits, clock generation circuits, VLSI capacitors, and VLSI Opamp (OTA) circuits.

9.2.1 VLSI ANALOG SWITCH CIRCUIT STRUCTURES

The desired properties of the VLSI SC switch circuits include high OFF resistance and low ON resistance, minimized switching delay between the φ1 and φ2 circuit configurations, minimized influence between switch control and signals, low signal voltage offset. Note that switches based on VLSI MOS transmission gates can meet or approximate the above functions. Practical MOS switches are realized using either NMOS or/and PMOS transistors and they are controlled through the voltage signals on the gates. The basic SC switch in the CMOS process technology uses a single NMOS (or PMOS) transistor as shown in figure 9.13.

Based on the MOS transistor I-V characteristic equations in the linear and cutoff modes, we have that:

$$I = \begin{cases} 0 & V_{gs} - V_T < 0 \\ \beta(V_g - V_T - \dfrac{V_d + V_s}{2})(V_d - V_s) & V_{gs} - V_T \geq 0 \end{cases} \quad (9.23)$$

The effective switch OFF and ON resistances of the basic MOS switch circuit can be derived as:

$$\Rightarrow R = \begin{cases} \infty & V_{gs} - V_T < 0 \\ \dfrac{1}{\beta(V_g - V_T - \dfrac{V_d + V_s}{2})} & V_{gs} - V_T \geq 0 \end{cases} \quad (9.24)$$

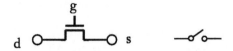

Fig.9.13 Basic MOS switch circuit structure

Note that a typical OFF resistance of the MOS switch is typically in the order of $10^{10} \sim 10^{12}$ Ω and the ON resistance is in the order of 0.5k to 10kΩ. These values are usually PVT dependent.

Note that the MOS pass gate based switch offers zero dc offset in the switch operation. However, the parasitic capacitances between the gate and source and drain may introduce the dc offset and the charge injection (or clock feed through). The total voltage error in a practical circuits caused by the switch charge injection is typically in the mV range. Practically speaking, smaller switch transistor size, slower switching edge rate and/or a small resistance of the voltage source may be used to minimize the charge injection effect. Among various charge injection minimization technique, dummy transistor of half of the switch size as shown in figure 9.14, using inversed switching clock polarity is commonly used in designs to compensate for the charge injection (or clock feedthrough) from the main switching transistor.

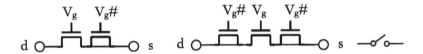

Fig.9.15 NMOS charge-injection compensated switch circuit structures

The ON resistance of VLSI analog switch employing single NMOS (or PMOS) device strongly depends on the signal level. For NMOS switch, high-level signal may significantly degrade the ON resistance that impacts the speed of the switch. On the other hand, a low-level signal will degrade the switch speed in the same way for a PMOS switch. An improved VLSI SC switch circuit structure employing both NMOS and PMOS transistors is shown in figure 9.15.

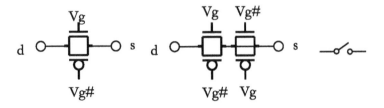

Fig.9.15 CMOS switch circuit structure

Such switch structure offers advantages over the basic N- or PMOS switches by providing reduced ON resistance and charge injections. In practical design, the half-sized dummy transistors as shown in figure 9.15b can also be used in such structure for further charge injection cancellation. Based on the I-V equation in the MOS device in linear mode as:

$$I = \beta_n(V_{gn} - V_{Tn} - \frac{V_d + V_s}{2})(V_d - V_s)$$
$$+ \beta_p(-V_{gp} + |V_{Tp}| + \frac{V_d + V_s}{2})(V_d - V_s) \quad (9.25)$$

The effective ON resistance of the switch can be derived as:

$$\Rightarrow R = \frac{1}{\beta_n(V_{gn} - V_{Tn} - \frac{V_d + V_s}{2}) + \beta_p(-V_{gp} + |V_{Tp}| + \frac{V_d + V_s}{2})} \quad (9.26)$$

9.2.2 NON-OVERLAP CLOCK GENERATION CIRCUIT

A commonly used two-phase non-overlap clock generation circuit structure is shown in figure 9.16.

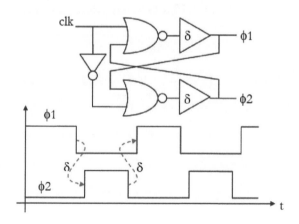

Fig.9.16 VLSI two-phase non-overlap clock generation circuit

Where the non-overlap regions of the two-phase clock are controlled by the delay time of the buffer used.

9.2.3 VLSI OTA CIRCUIT STRUCTURES

Opamp (or OTA) circuits in SC circuits are only required to drive capacitive loads. Consequently Opamp circuit in VLSI SC circuits can typically be optimized for gain, stability and power dissipation with higher output impedance. OTAs are commonly used in many practical SC circuit design applications

Shown in figure 9.17 and figure 9.18 are two commonly used VLSI folded-cascode OTA circuit structures. These circuits offer the features of higher gain and higher output impedance, which are suitable for capacitive load in the SC circuit applications.

In practical SC circuit design, fully differential OTA circuits, such as the one shown in figure 9.19 are typically used for better noise immunity.

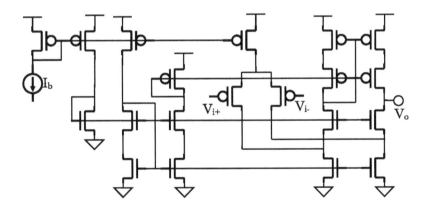

Figure 9.17 VLSI folded-cascode OTA circuit structures

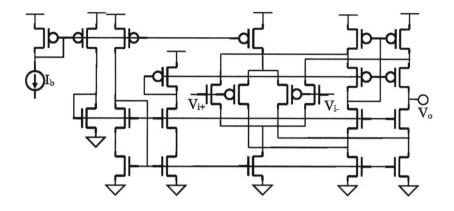

Figure 9.18 VLSI folded-cascode OTA circuit with rail-to-rail input

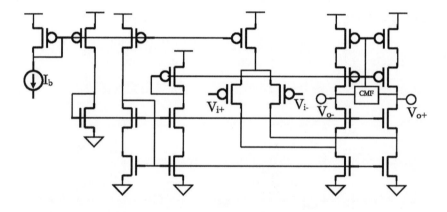

Figure 9.19 VLSI differential folded-cascode OTA circuit structures

9.3 VLSI SAMPLE/HOLD CIRCUIT STRUCTURES

The sample/hold (S/H) circuit as shown in figure 9.20 is an essential building circuit block of DT systems, where the S/H circuit is typically used to avoid frequency response distortion caused by CT signal feed through effect.

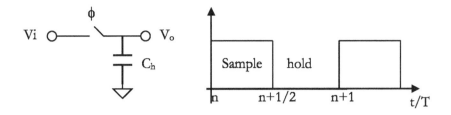

Fig.9.20 Basic sample/hold circuit

The S/H circuit works as follows: the output signal tracks the input signal during the sample phase and the output holds the previous sampled input signal value during the hold phase:

$$\begin{cases} V_o(t) = V_i(t) & nT < t < nT + T_s \\ V_o(t) = V_i(nT + T_s) & nT + T_s < t < nT + T \end{cases} \quad (9.27)$$

Practical S/H circuits suffer from various non-ideal effects as shown in figure 9.21. These non-ideal effects that are characterized by parameters such as the acquisition time, slew rate, aperture uncertainty time, hold-mode droop rate, voltage offset, and clock feed through (CFT).

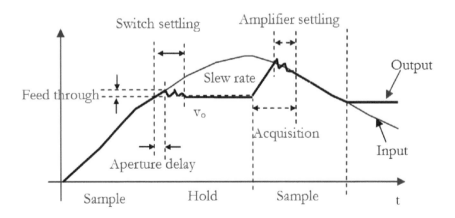

Fig.9.21 Non-ideal effects of practical sample/hold circuits

9.3.1 VLSI BUFFERED SAMPLE/HOLD CIRCUIT

A buffered VLSI S/H circuit employing two Opamp circuits is shown in figure 9.22, where the first Opamp circuit serves as the input buffer and the second one as the output buffer. Such S/H circuit structure offers better isolation at the input and the enhanced driving capability at the output. However, such a circuit suffers from the limitation of large offset error contributed from both Opamp circuits.

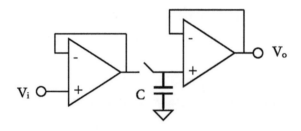

Fig.9.22 Buffered sample/hold circuit structure

An improved VLSI buffered S/H circuit structure is shown in figure 9.23. Such S/H circuit structure effectively eliminates the offset voltage of the output Opamp circuit by connecting the sample capacitor within its feedback loop.

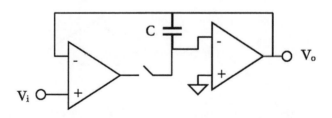

Fig.9.23 Improved sample/hold circuit structure

9.3.2 THE CORRELATED DOUBLE SAMPLING TECHNIQUES

Offset error in VLSI S/H circuits can be minimized using the auto-zero methods based on the correlated double sampling (CDS) techniques. In the CDS method the offset voltage of the sampler circuit is first sampled and stored in the first clock phase and then this stored offset voltage is subtracted from the sampled signal value that occur in the second clock phase. In fact the CDS techniques are also effective in reducing other non-ideal effects, such as the low frequency noise of the devices in the sampler and the power supply noise. Shown in figure 9.24 is a VLSI SC S/H circuit structure that employs a CDS based auto-zero technique.

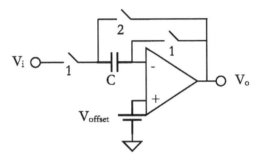

Fig.9.24 CDS sample/hold circuit structure

In the above SC circuit, the non-ideal Opamp circuit with offset and input reference noise are modeled using an effective offset voltage source V_{off} at the offset-free ideal Opamp circuit input. Such circuit works as follows:

- During the sampling phase ($\phi 1$), this offset voltage at the Opamp input sampled and stored in the sampling capacitor C_h. The node voltage V_x and the electrical charge stored on the sampling capacitor C_h at the end of $\phi 1$ phase are given as:

$$\begin{cases} V_x(nT+\frac{T}{2}) = V_{offset}(nT+\frac{T}{2}) \\ Q_h(nT+\frac{T}{2}) = (V_i(nT+\frac{T}{2}) - V_{offset}(n+\frac{T}{2}))C_h \end{cases} \quad (9.28)$$

- During the hold phase (φ2 phase), node voltage on V_x keeps tracking the effective offset voltage because of the Opamp circuit in feedback operation and the charge stored on the sampling capacitor C_h remains the same as the sampling phase. As the result, the node voltages at end of the φ2 are given as:

$$\begin{cases} V_x(nT+T) = V_{offset}(nT+T) \\ V_o(nT+T) = (V_i(nT+\frac{T}{2}) - V_{offset}(n+\frac{T}{2})) + V_{offset}(nT+T) \end{cases} \quad (9.29)$$

Note that for such CDS based S/H circuit:

- The time-independent offset (dc offset), the offset can be totally eliminated from the sampled signal:

$$\begin{cases} V_x(nT+T) = V_{offset}(nT+\frac{T}{2}) \\ V_o(nT+T) = V_i(nT+\frac{T}{2}) \end{cases} \quad (9.30)$$

- The time-dependent offset (ac offset or other input referred noise), CDS circuit provides a high-pass offset shaping similar to a first order sigma-delta modulator:

$$V_o(nT+T) = V_i(nT+\frac{T}{2}) + [V_{offset}(nT+T) - V_{offset}(nT+\frac{T}{2})] \quad (9.31)$$

VLSI Modulation Circuits

$$\begin{cases} H_{offset}(z) = (1-z^{-1}) \\ |H_{offset}(z)|_{z=\exp(j\omega T)} = |2\sin(\frac{\omega T}{2})| \end{cases} \quad (9.32)$$

- Consequently, all low frequency noise sources within the Opamp circuit, such as the $1/f$ noise and power supply noise are attenuated.

An alternative SC CDS circuit technique based on unity-gain buffer circuit configuration is shown in figure 9.25.

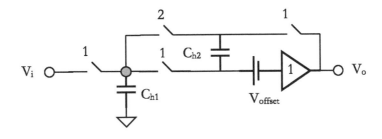

Fig.9.25 CDS sample/hold circuit based on unity gain buffer

Such a circuit works as follows:

- During the $\phi 1$ phase (sampling phase), the input signal and the input referred offset are sampled

$$\begin{cases} Q_{h1}(nT + \frac{T}{2}) = V_i(nT + \frac{T}{2})C_{h1} \\ Q_{h2}(nT + \frac{T}{2}) = (-V_{offset}(nT + \frac{T}{2}))C_{h2} \end{cases} \quad (9.33)$$

- During the $\phi 2$ phase (hold phase), the offset is subtracted from the sampled input signal:

$$V_o(nT+T) = Q_{h1}/C_{h1} - Q_{h2}/C_{h2} - V_{offset}$$
$$= V_i(nT+\frac{T}{2}) + [V_{offset}(nT+\frac{T}{2}) - V_{offset}(nT+T)] \quad (9.34)$$

9.3.3 NON-UNITY GAIN VLSI SC S/H CIRCUIT

Shown in figure 9.26 is a VLSI SC S/H circuit structure that offers non-unity gain circuit operation.

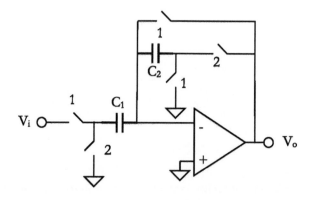

Fig.9.26 non-unity gain CDS sample/hold circuit structure

This circuit works as follows:

- At the end of the sample phase, C_1 is charged and C_2 is discharged. The charge stored on the two capacitor are given as:

$$\begin{cases} Q_{C1} = (V_i(nT+\frac{T}{2}) - V_{offset}(nT+\frac{T}{2}))C_1 \\ Q_{C2} = (-V_{offset}(nT+\frac{T}{2}))C_2 \end{cases} \quad (9.35)$$

- At the end of the hold phase, C_1 is discharged and C_2 is charged and the output voltage can be derived as:

$$V_o(nT+T) = \frac{C_1}{C_2} V_i(nT+\frac{T}{2}) + [V_{offset}(nT+T) - V_{offset}(nT+\frac{T}{2})] \quad (9.36)$$

Note that the input signal can be amplified with desired gain and that the offset of the S/H circuit can be eliminated.

9.3.4 CHARGE INJECTION COMPENSATION TECHNIQUES

Charge injection of the MOS switch in SC circuits may result in the signal processing errors (typically a few mV). A VLSI SC charge injection compensate circuit structure is shown in figure 9.27, where dummy capacitors and switch are added in the positive input terminal of the Opamp. Such a structure matches the total capacitance and switch devices connected to the Opamp positive and negative input terminals. Consequently charge injections on the positive and negative terms of the Opamp circuit are compensated in the output signal.

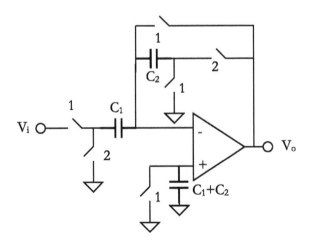

Fig.9.27 Sample/hold circuit with charge injection compensation

An alternative way to reduce the charge injection in the SC circuit employs the so-called bottom plate sampling (BPS) technique. This technique is based on the fact that if the offset or charge injection noise is made time-independent, it can easily be eliminated in the SC circuit based on certain circuit techniques.

The BPS circuit technique focuses on eliminating the signal-dependent charge injection noise.

Shown in figure 9.28 is a VLSI SC sampling circuit structure employing the BPS technique.

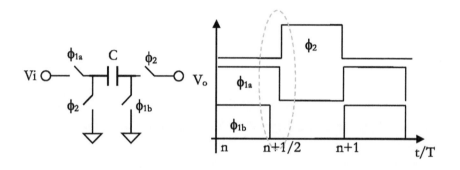

Fig.9.28 BPS based switched-capacitor sample/hold circuit

In this BPS based SC circuit, the switch connected to the bottom plate of the sampling capacitor is turned off earlier than the switch connect to the top (input) plate. As the result, the sampling of the signal in this phase is effectively controlled by the bottom plate switch.

Note that the bottom plat of the sampling capacitor during the sampling phase is always connected to the ground node, independent of the input signal level. As the result, the charge injection resulted from the sampling phase is signal independent.

9.3.5 FULLY DIFFERENTIAL SC S/H CIRCUIT

Shown in figure 9.29 is a VLSI fully differential SC S/H circuit structure. Such a circuit offers the advantages over the single-end S/H circuit structures with better noise immunity, less charge injection and high signal dynamic range and is commonly used in high performances SC circuits. In addition, the CDS and bottom plate sampling technique can also be used in the fully differential VLSI SC S/H circuit structures for further performance enhancements.

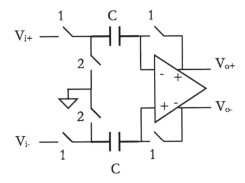

Fig.9.29 VLSI fully differential Sample/hold circuit

Reference:

[1] G. M. Jacobs, D. J. Allstot, R. W. Brodersen, Paul Gary, "Design Techniques for MOS Switched-capacitor Ladder Filters," IEEE *Trans.on Circuits and Systems*, vol. CAS-25, no. 12, pp. 1014-1021, Dec. 1978.

[2] K. Martin, A. S. Sedra, "Exact Design of Switched-Capacitor Bandpass Filter Using Coupled-Biquad Structures," IEEE *Trans. on Circuits and Systems*, vol. CAS-27, no. 6, pp. 469-475, Dec. 1980.

[3] M. S. Lee, G. C. Tempes, C. Chang, M. B. Ghaderi, "Bilinear Switched-Capacitor Ladder Filters," IEEE *Trans. on Circuits and Systems*, vol. CAS-28, no. 8, pp. 881-821, Dec. 1981.

[4] F. Krummenacher, "Micropower Switched-Capacitor Biquadratic Cell," IEEE *J. Solid-State Circuits*, Vol. SC-17, no. 3, pp. 507-512. June. 1982.

[5] R. Gregorian, K. W. Martin, G. C. Tempes, "Switched-Capacitor Circuit Design," *Proc. Of the IEEE*, vol. 71, no. 8, pp.941-966, Aug. 1983.

[6] R. B. Datar, A.S. Sedra, "Exact Design of Stray-Insensitive Switched-Capacitor Ladder Filters," IEEE *Trans. on Circuits and Systems*, vol. CAS-30, no. 12, pp. 888-898, Dec. 1983.

[7] R.K. Henderson, J. I. Sewell, "Matrix Methods for Switched-capacitor Filter Design," ISCAS'88, vol.2, pp.1021-1024, June 1988.

[8] K. Chen, S. Eriksson, "Z-domain Synthesis of Switched-capacitor LDI Ladder Filters," ISCAS'88, vol.2, pp.1013-1016, June 1988.

[9] U-K Moon, "CMOS High-Frequency Switched-Capacitor Filters for Telecommunication Applications," IEEE *J. Solid-State Circuits*, Vol. 35, no. 2, pp. 212-220. Feb. 2000.

[10] E. M. Saad, "Switched-capacitor Circuits and their Applications, an Overview," *Proceedings, Radio Science Conference2002*, pp. 51-59, March 2002.

[11] A. S. Korotkov, "Synthesis of Narrow-band Coupled Resonator SC-filters," MELECON 98, vol.1, pp. 554-557, May 1998.

[12] W. Aloisi, G. Giustolisi, G. Palumbo, "Exploiting the High-Frequency Performance of Low-Voltage Low-Power SC Filters," IEEE Transactions on Circuits and Systems II: Express Briefs, vol. 55, no. 2, pp. 77-84, Feb 2004.

[13] L. Yue, J. I. Sewell, "A comparison Study of SC Biquads in the Realization of SC Filters," IEE Colloquium on Digital and Analogue Filters and Filtering Systems, pp. 101/1 - 101/9, Nov 1993.

[14] J. S. Pereira, A. Petraglia, "Low-Sensitivity Direct-form IIR SC Filters with Improved Phase Linearity," ISCAS'00, vol. 3, pp. 169-172, May 2000.

[15] K. Nakayama, M. Zhiqiang, G. Yamamoto, "A Method to Minimize Total Capacitance in Cascade Realization of SC Filters," Proceedings of the 3second Midwest Symposium on Circuits and Systems, vol.2, pp. 968-972, Aug. 1989.

[16] A. Handkiewicz, "Two-Dimensional SC Filters with Circularly Symmetric Response," European Conference on Circuit Theory and Design, pp. 252-255, Sep 1989.

[17] M. D. Lutovac, D. Tosic, D. Novakovic, "Programmable Low-pass/High-pass SC-filters," MELECON 98, vol. 1, May 1998.

[18] Y. Lu, J. I. Sewell, "A Comparison Study of SC Biquads in the Realization of SC Filters," ISCAS '94, vol.5, pp. 711-714, June 1994.

[19] M. Biey, F. Montecchi, A. Premoli, "Performances of Bandpass Cascaded SC Filters with Reduced Pole-Qs," IEEE International Symposium on Circuits and Systems, vol.2, pp. 1253-1256, June 1988.

[20] A. T. Younis, R. E. Massara, "On the Design of Optimal Switched-capacitor Filters Based on the Use of Lossy Frequency-Dependent Negative Resistance (FDNR) Prototype Structures," IEE 1988 Saraga Colloquium on Electronic Filters, pp. 2/1 - 2/6, May 1988.

[21] E. Sanchez-Sinencio, R. Geiger, J. Silva-Martinez, "Tradeoffs Between Passive Sensitivity, Output Voltage Swing, and total Capacitance in Biquadratic SC Filters," IEEE Transactions on Circuits and Systems, vol. 31, no. 11, pp. 984-987, Nov 1984.

[22] A. S. Korotkov, H. Hauer, R. Unbehauen, "Practical Design of Element Simulation Type SC Filters," IEEE Transactions on Circuits and Systems I: Fundamental Theory and Applications, vol. 47, no. 6, pp. 934-940, June 2000.

[23] H. Baher, "Microelectronic Switched-capacitor Filters," IEEE Circuits and Devices Magazine, vol. 7, no. 1, pp. 33-36, Jan. 1991.

[24] J. J. F. Rijns, H. Wallinga, "Spectral analysis of Double-sampling Switched-capacitor Filters," vol. 38, no. 11, pp. 1269-1279, Nov. 1991.

[25] G. W. Roberts, A.S. Sedra, "Synthesizing Switched-current Filters by Transposing the SFG of Switched-capacitor Filter circuits," IEEE Transactions on Circuits and Systems, vol. 38, no. 3, pp. 337-340, March 1991.

[26] J. Silva-Martinez, E. Sanchez-Sinencio, "Strategic SC Filter Design based on a comparative Study of various S-to-z mappings," IEEE Transactions on Circuits and Systems, vol. 36, no. 11, pp. 1465-1472, Nov. 1989.

[27] M. D. Lutovac, D. Novakovic, I. Markoski, "Selective SC-filters with Lowpassive Sensitivity," Electronics Letters, vol.33, no. 8, pp. 674-675, April 1997.

[28] B. Nowrouzian, "Theory and Design of LDI Lattice Digital and Switched-capacitor Filters," IEE Proceedings Circuits, Devices and Systems, vol. 139, no. 4, pp. 517-526, Aug. 1992.

[29] A. Baschirotto, D. Bijno, R. Castello, F. Montecchi, "A 1 V 1.2 µW 4th Order Bandpass Switched-opamp SC Filter For a cardiac Pacer Sensing Stage," ISCAS'00 Geneva, vol.3, pp. 173-176, May 2000.

[30] J. S. Pereira, A. Petraglia, M. F. Quelhas, "Approximating Linear Phase with IIR SC Filters," ISCAS'02, vol.4, pp. IV-623 - IV-626, May 2002.

[31] J. L. Ausin, G. Torelli, J. F. Duque-Carrillo, J. M. Carrillo, P. Merchan, "Programmable Time-multiplexed SC Filters without Dynamic Range Degradation," IEEE International Conference on Electronics, Circuits and Systems, vol.1, pp. 373-376, Sept. 1998.

[32] P. Filoramo, G. Giustolisi, G. Palmisano, G. Palumbo, "An approach to the Design of Low-voltage SC Filters," ISCAS '98, vol.1, pp. 265-268, June 1998.

[33] J. Goette, A. Kaelin, W. Guggenbuhl, "Criteria and Methods For the Selection of Optimized capacitor values in SC-filters," ISCAS'89, vol.3, pp. 1696 - 1700, May 1989.

[34] J. Silva-Martinez, E. Sanchez-Sinencio, "A Sparing SC Filter Design approach with Reduced Transmission zeros," ISCAS '88, vol.2, pp. 1721 - 1724, June 1988.

[35] B.G. Lofmark, "SC-filter Structures with Reduced Requirements on Op-amp Bandwidth and Settling Time," ISCAS '88, vol.2, pp. 1473-1477, June 1988.

[36] E. S. Sinencio, J. S. Martinez, R. Geiger, "Biquadratic SC Filters with Small GB Effects," IEEE Transactions on Circuits and Systems, vol.31, no. 10, pp. 876-884, Oct 1984.

[37] P. Van Peteghem, W. Sansen, "Power Consumption Versus Filter Topology in SC Filters," IEEE Transactions on Circuits and Systems, vol.33, no.2, pp.150-157, Feb 1986.

[38] A.S. Korotkov, K.H. Feistel, R. Unbehauen, "Synthesis of Switched-capacitor Parametric Bandpass Filters," IEEE Transactions on Circuits and Systems I: Fundamental Theory and Applications, vol. 46, no. 4, pp. 484-490, April 1999.

[39] D. Vazquez, A. Rueda, J. L. Huertas, E. Peralias, "A High-Q Bandpass Fully Differential SC Filter with Enhanced Testability," IEEE Journal of Solid-State Circuits, vol. 33, no. 7, pp. 976-986, July 1998.

[40] A. Baschirotto, R. Castello, "A 1-V 1.8-MHz CMOS Switched-opamp SC Filter with Rail-to-rail Output Swing," IEEE Journal of Solid-State Circuits, vol. 32, no. 12, pp. 1979-1986, Dec. 1997.

[41] B. Maundy, E.I. El-Masry, "A Switched-capacitor Bidirectional Associative Memory," IEEE Transactions on Circuits and Systems, vol. 37, no. 12, pp. 1568-1572, Dec. 1990.

[42] R. Castello, A.G. Grassi, S. Donati, "A 500-nA Sixth-order Bandpass SC Filter," IEEE Journal of Solid-State Circuits, vol. 25, no. 3, pp. 669-676, Jun 1990.

[43] P. M. Van Peteghem, "On the Relationship between PSRR and Clock Feed through in SC Filters," IEEE Journal of Solid-State Circuits," vol. 23, no. 4, pp. 997-1004, Aug. 1988.

[44] P. Filoramo, G. Giustolisi, G. Palmisano, G. Palumbo, "Approach to the Design of Low-voltage SC Filters," IEE Proceedings, Circuits, Devices and Systems, vol. 147, no. 3, pp. 196-200, June 2000.

[45] M. C. Schneider, C. Galup-Montoro, J. C. M. Bermudez, "Explicit Formula for Harmonic Distortion in SC Filters with weakly Nonlinear capacitors," IEE Proceedings Circuits, Devices and Systems, vol. 141, no. 6, pp. 505-509, Dec. 1994.

[46] A. Dabrowski, U. Menzi, G. S. Moschytz, "Design of Switched-capacitor FIR Filters with application to a Low-power MFSK

Receiver," IEE Proceedings, Circuits, Devices and Systems, vol.139, no.4, pp. 450-466, Aug. 1992.

[47] M. F. Quelhas, A. Petraglia, A. P. Baruqui, "Power Consumption Estimation in SC Filters," IEEE International Symposium on Industrial Electronics, vol. 1, pp. 298-300, June 2003.

[48] A. Baschirotto, R. Castello, "A 1V 1.8MHz CMOS Switched-Opamp SC Filter with Rail-to-rail Output Swing, "44th ISSCC, pp. 58-59, Feb. 1997.

[49] A. S. Korotkov, "Predistortion Technique for Element Simulation Type SC-filters," MELECON '96, vol. 3, pp. 1259-1262, May 1996.

[50] J. L. Ausin, R. Perez-Aloe, J.F. Duque-Carrillo, G. Torelli, E. Sanchez-Sinencio, "High-Selectivity SC Filters with Continuous Digital Q-factor Programmability," ISCAS'02, vol.4, pp. IV-631 - IV-634, May 2002.

[51] R. Harjani, B. Vinnakota," Digital Detection of Analog Parametric Faults in SC Filters," Proceedings, 36th Design Automation Conference, pp.772-777, June 1999.

[52] U. Kleine, T. Pasch, "Low-Voltage Medium Q-SC Filters for High Frequency Communication Applications, "1998 IEEE International Conference on Electronics, Circuits and Systems, vol.2, pp. 39-42, Sept. 1998.

[53] U. Marschner, W.-J. Fischer, E.-G. Kranz, "Low Expense Architectures for a Dynamic Spectrum Analyzer based on SC-filters," ISCAS '98, vol.2, pp. 276-279, June 1998.

[54] A. Baschirotto, F. Montecchi, R. Castello, "A 150 Msample/s 20 mW BiCMOS Switched-Capacitor Biquad Using Precise Gain op amps, "4second ISSCC, pp. 212-213, Feb. 1995.

[55] G. H. Wang, K. Watanabe, "Current-Mode Switched-Capacitor Circuit Synthesis by C-invariant Dual Transformation," ISCAS'91, vol.3, pp.1553-1556, June 1991.

[56] X.F. Wania, Johns, D. A. Sedra, A.S.; "Programmable Multiplexed Switched-capacitor Filters," Proceedings of the 33rd Midwest Symposium on Circuits and Systems, vol. 2, pp.973-976, Aug. 1990.

[57] A Baschirotto, R Castello, F. Montecchi, "Finite Gain Compensation Techniques for High-Q bandpass SC Filters," ISCAS'90, vol. 4, pp.23813-2816, May 1990.

[58] R. Geiger, E. Sanchez-Sinencio, "Operational amplifier Gain - bandwidth Product Effects on the Performance of Switched-capacitor Networks," IEEE Transactions on Circuits and Systems, vol. 29, no. 2, pp. 96-106, Feb. 1982.

[59] R. Castello, P. Gray, "Performance Limitations in Switched- Capacitor Filters," IEEE Transactions on Circuits and Systems, vol. 32, no. 9, pp.865-876, Sept. 1985.

[60] D. Bruckmann, U. Kleine, "Novel Voltage Inverter Switches with Minimum Sensitivity Properties," IEEE Transactions on Circuits and Systems, vol. 32, no. 7, pp. 723-726, Jul 1985.

[61] P. Li, J. I. Sewell, "The LUD Approach to Switched-Capacitor Filter Design," IEEE Transactions on Circuits and Systems, vo. 34, no. 12, pp.1611-1614, Dec. 1987.

[62] P. M. Van Peterghem, W. M. C. Sansen, "Power Consumption versus Filter topology in SC Filters," IEEE Journal of Solid-State Circuits, vo. 21, no. 1, pp.40-47, Feb. 1986.

[63] P. Landau, D. Michel, D. Melnik, "A Reduced Capacitor Spread Algorithm for Elliptic Bandpass SC Filters" IEEE Journal of Solid-State Circuits, vol. 22, no. 4, pp.624-626, Aug. 1987.

[64] W. M. C. Sansen, H. Qiuting, K.A.I. Halonen, "Transient Analysis of Charge-Transfer in SC Filters-gain Error and Distortion," IEEE Journal of Solid-State Circuits, vol. 22, no. 2, pp. 268-276, Apr. 1987.

[65] J. L. Ausin, J. F. Duque-Carrillo, G. Torelli, E. Sanchez-Sinencio, "Switched-Capacitor Circuits with Periodical Nonuniform Individual Sampling," IEEE Transactions on Circuits and Systems II: Analog and Digital Signal Processing, vol. 50, no. 8, pp. 404-414, Aug. 2003.

[66] L. Lentola, G. M. Corelazzo, E. Malavasi, A.Baschirotto," Design of SC Filters for Video Applications," IEEE Transactions on Circuits and Systems for Video Technology, vol. 10, no. 1, pp. 14-22, Feb. 2000.

[67] B. Vinnakota, R. Harjani, "DFT for Digital Detection of Analog Parametric Faults in SC Filters," IEEE Transactions on Computer-Aided Design of Integrated Circuits and System, vol. 19, no. 7, pp.789-798, July 2000.

[68] R. P. Martins, J. E. Franca, "Cascade Switched-Capacitor IIR Decimating Filters," IEEE Transactions on Circuits and Systems I: Fundamental Theory and Applications. Vol. 42, no.7, pp.367-376, July 1995.

[69] R. P. Martins, J. E. Franca, F. Maloberti, "An Optimum CMOS Switched-Capacitor Anti-aliasing Decimating Filter," IEEE Journal of Solid-State Circuits, vol. 28, no. 9, pp.962-970, Sept. 1993.

[40] J. L. Huertas, A. Rueda, D. Vazquez, "Testable Switched-Capacitor Filters," IEEE Journal of Solid-State Circuits, vol. 28, no. 7, pp. 719-724, July 1993.

[41] A. Baschirotto, R. Castello, F. Montecchi, "IIR Double-Sampled Switched-Capacitor Decimators for High-Frequency Applications," IEEE Transactions on Circuits and Systems I: Fundamental Theory and Applications, vol. 39, no. 4, pp.300-304, April 1992.

[42] P. Mehta, M. Darwish, T. Thomson, "Switched-Capacitor Filters," IEEE Transactions on Power Electronics, vol. 5, no. 3, pp.331-336, July 1990.

[43] C-Y Wu, Tsai-Chung Yu, Shin-Shi Chang, "New Monolithic Switched-capacitor Differentiators with Good Noise Rejection," IEEE Journal of Solid-State Circuits, vol. 24, no. 1, pp.177-180, Feb. 1989.

[44] J. C. M. Bermudez, M.C. Schneider, C.G. Montoro, "Compatibility of Switched-capacitor Filters with VLSI Processes," IEE Proceedings G, Circuits, Devices and Systems, vol. 139,no. 4, pp. 413-418, Aug. 1992.

[45] E. I. El-Masry, E.G.;Hix, "Novel Switched-capacitor Integrator for High-frequency applications," IEE Proceedings G Circuits, Devices and Systems, vol. 136, no. 5, pp.263-267, Oct. 1989.

[46] D. Novakovic, B. Loncar, P. Osmokrovic, "Efficient High-pass Low-sensitive Selective SC-Filters," 2second International Conference on Microelectronics, 2000, vol.2, pp.739-742, May 2000.

[47] Inchang Seo; R.M. Fox, "Low-Power Switched-Capacitor Filters Using Charge Transfer Integrators," ISCAS '03, vol.1, pp. I-609 - I-612, May 2003.

[48] K. Hajek, J. Sedlacek, "General Multiple LC Prototype Filter Solutions and Optimization," 9th International Conference on Electronics, Circuits and Systems, vol.1, pp.165-168, Sept. 2002.

[49] J.M Canive, A. Petraglia, "On the Testability of SC Filters based on allpass Sections," ISCAS'01, vol. 1, pp. 65-68, May 2001.

[80] A. Baschirotto, "IIR Double-sampled Switched-Capacitor Building Blocks for High-Frequency Decimators," ISCAS'91, vol. 3, pp. 1673-1676, June 1991.

[81] B. Nowrouzian, "A Novel Approach to the Exact Design of LDI Symmetrical Digital and Switched-Capacitor Filters," Proceedings of the 33rd Midwest Symposium on Circuits and Systems, vol.2, pp. 967-972, Aug. 1990.

[82] R. E.Massara, A. T. Younis, "An Efficient Design Method for Optimal MOS Integrated Circuit Switched-Capacitor LDI Ladder Filters," Proceedings of the 33rd Midwest Symposium on Circuits and Systems, 1990, vol.2, pp. 956-959, Aug. 1990.

[83] R. Raut, B. B. Bhattacharrya, S.M. Faruque, "An Application of Systolic Array Design Architecture to Switched-capacitor Filter Circuits," ICASSP-89, vol.4, pp. 2401-2404, May 1989.

[84] M. Nalecz, J. J. Mulawka, "Parasitic-Compensated Building Blocks for Active Switched-Capacitor Filters," ISCAS'88, vol.2, pp.1483-1486, June 1988.

[85] A. M. Davis, H. P. Nguyen, "Exact Synthesis of SC Filters Using Forward Difference Integrators," ISCAS'88, vol.2, pp.1009-1012, June 1988.

[86] S. Setty, C. Toumazou, K. Manetakis, P. Cheung, R. Spence, "Ideas and Concepts for the Automation of Analogue Filters Using Qualitative Reasoning," IEE 15th SARAGA Colloquium on Digital and Analogue Filters and Filtering Systems, pp. 7/1 - 7/9, Nov 1995.

[87] H. Shafeeu, A. Betts, J. Taylor, "Approaches to Ultra-Narrow-Band Analogue IC Filter Design Using Switched-Capacitors," IEE Colloquium on Advances in Analogue VLSI, pp.2/1-2/9, May 1991.

[88] G. Temes, H. Orchard, M. ahanbegloo, "Switched-Capacitor Filter Design Using the Bilinear z-Transform," vol. 25, no. 12, pp.1039-1044, Dec 1978.

[89] M. Lee, C. Chang, "Switched-Capacitor Filters Using the LDI and Bilinear Transformations," IEEE Transactions on Circuits and Systems, vol. 28, no. 4, pp. 265-270, Apr 1981.

[90] E. El-Masry, "Strays-Insensitive State-space Switched-capacitor Filters," IEEE Transactions on Circuits and Systems, vol. 30, no. 7, pp.474-488, Jul 1983.

[91] J. Garcia-Vazquez, E. Sanchez-Sinencio, "Finite Gain-Bandwidth Product Effects on a Pair of Pseudo-N-path SC Filters," IEEE Transactions on Circuits and Systems, vol. 31, no. 6, pp. 583-584, Jun 1984.

[92] C. Campbell, K. Reineck, "A Pole/zero Prewarping Procedure in SC Filter Design," IEEE Transactions on Circuits and Systems, vol. 31, no. 9, pp. 821-825, Sep 1984.

[93] G. Fischer, G. Moschytz, "SC Filters for High-Frequencies with Compensation for Finite-Gain Amplifiers," IEEE Transactions on Circuits and Systems, vol. 32, no. 10, pp.1050-1056, Oct 1985.

[94] T. Inoue, F. Ueno, "On Robustness of the Stability of Multiphase Switched-capacitor Two-ports," IEEE Transactions on Circuits and Systems, vol. 32, no. 6, pp. 522 - 529, Jun 1985.

[95] J. Da Franca, "A Single-path Frequency-translated Switched- Capacitor bandpass Filter System," IEEE Transactions on Circuits and Systems, vol. 32, no. 9, pp. 938-944, Sep 1985.

[96] A. Davis, "Flow Graph Synthesis of Darlington - Cauer Switched-Capacitor Filters," IEEE Transactions on Circuits and Systems, vol. 32, no. 7, pp.727-732, Jul 1985.

[97] M. H. Fino, J. E. Franca; A. Steiger-Garcao, "Automatic Symbolic Analysis of Switched-Capacitor Filtering Networks Using Signal Flow Graphs," IEEE Transactions on Computer-Aided Design of Integrated Circuits and Systems, Volume: 14, Issue: 7, Pages:858-867, July 1995.

[98] J. Crols, M. Steyaert, "Switched-Opamp: An Approach to Realize Full CMOS Switched-Capacitor Circuits at Very Low Power Supply Voltages" IEEE Journal of Solid-State Circuits, Volume: 29, Issue: 8, Pages:936-942, Aug. 1994.

[99] C-Y Wu; P-H Lu; M-K Tsai; "Design Techniques for High-frequency CMOS Switched-Capacitor Filters Using Non-Opamp-Based Unity-Gain Amplifiers" IEEE Journal of Solid-State Circuits, Volume: 26, Issue: 10, Pages:1460-1466, Oct. 1991.

[100] A. Handkiewicz, "Two-dimensional Switched-capacitor Filter Design System for Real-time Image Processing," IEEE Transactions on Circuits and Systems for Video Technology, Volume: 1, Issue: 3, Pages:241-246, Sept. 1991.

[101] K. Nakayama, Y. Takahashi, Y. Sato, Y. Nukada, "An Adaptive SC Line Equalizer System for Four-Wire Full-Duplex and Multirate Digital Transmission," IEEE Transactions on Circuits and Systems, Volume: 35, Issue: 9, Pages: 1073-1081, Sept. 1988.

[102] A.E Salama, "Factored State-space Stray-Insensitive Switched-Capacitor Filters," IEE Proceedings Circuits, Devices and Systems, Volume: 143, Issue: 6, Pages: 325-330, Dec. 1996

[103] T.F. Yong, M.J. Long, "Systematic Method for Design of Switched-capacitor Filters," Electronics Letters, Volume: 28, Issue: 3, Pages: 309-310, Jan. 1992.

[104] Li Qiang; Han Yifeng; Xu Ke; Min Hao; "A Novel Offset Compensation biquad Switched-capacitor Filter Design," 5th International Conference on ASIC, Volume: 1, 21-24 Pages:643-646, Oct. 2003.

[105] J. L. Ausin, J.F. Duque-Carillo, G. Torelli, E. Sanchez-Sinencio, F. Maloberti, "Periodical Nonuniform Individually Sampled Switched-capacitor Circuits," ISCAS'00 Volume: 5, pp. 449 - 452, May 2000 .

[106] H. Matsumoto, Z. Tang, O. Ishizuka, "A Buffer-Based Switched-Capacitor Integrator with Reduced Capacitance Ratio," Proceedings of the 33rd Midwest Symposium on Circuits and Systems, vol.2 Pages:719 - 720 Aug. 1990.

[107] J. A. Hegt, "Signal Flow Graph based Synthesis of Strays-insensitive Switched-capacitor Filters," ISCAS'90, vol.3, pp.2177-2180, May 1990.

[108] I. M. Sahawaneh, A. A. Sakla, E. I. El-Masry, "Comparison between Cascade and Parallel Switched-capacitor Structures," Proceedings of the 3second Midwest Symposium on Circuits and Systems vol.2 Pages: 973 - 976 Aug. 1989.

[109] A.K. Betts, J.T. Taylor, D.G. Haigh, "Synthesis Method for Switched-capacitor FIR Decimators and Interpolators," ISCAS'88, vol.3 Pages: 2463 - 2466 June 1988.

[110] Q. Huang, "A Novel Technique for the Reduction of Capacitance spread in High Q SC Circuits," ISCAS'88, vol.2, Pp. 1249-1252, June 1988.

[111] C. Kurth, G. Moschytz, "Two-port Analysis of Switched-Capacitor Networks Using Four-port Equivalent Circuits in the z-domain," IEEE Transactions on Circuits and Systems Volume: 26, Issue: 3, pp.166-180, Mar 1979.

[112] U. Brugger, D. von Grunigen, G. Moschytz, "A Comprehensive Procedure for the Design of Cascaded Switched-Capacitor Filters," IEEE Transactions on Circuits and Systems, Volume: 28, Issue: 8, Pages: 803-810, Aug 1981.

[113] K. Martin, A. Sedra, "Effects of the Op amp Finite Gain and Bandwidth on the Performance of Switched-Capacitor Filters," IEEE Transactions on Circuits and Systems, Volume: 28, Issue: 8 Pages: 822-829, Aug 1981.

[114] S. Wong, C. Salama, "A Switched Differential Op-amp with Low Offset and Reduced 1/f Noise," IEEE Transactions on Circuits and Systems, Volume: 33, Issue: 11, Pages: 1119-1127, Nov 1986.

[115] F. A. P. Baruqui, A. Petraglia, J.E. Franca, S.K. Mitra, "Efficient Design of Integrated Switched-Capacitor Decimation Filters," IEEE Transactions on Circuits and Systems II: Analog and Digital Signal Processing, Volume: 47, Issue: 11, Pages: 1314-1318, Nov 2000.

[116] B. Raahemi, A. Opal, "Group Delay and Group Delay Sensitivity of Periodically Switched Linear Networks," IEEE Transactions on Circuits and Systems I: Fundamental Theory and Applications," volume: 47, Issue: 1, Pages: 96 - 104 Jan. 2000.

[117] A. Petraglia, "Random Capacitance Ratio Error Effects in the Frequency Response of Switched-capacitor Filters," IEEE-CAS Region 8 Workshop on Analog and Mixed IC Design, pp.112-116, Sept. 1996.

[118] K. van Hartingsveldt, P. Quinn, A. van Roermund, "A 10.7MHz CMOS SC Radio IF Filter with variable Gain and a Q of 55," ISSCC'00, pp.152 - 153, 452, Feb. 2000.

[119] M. M. Amourah, R. L. Geiger, "Gain and Bandwidth Boosting Techniques for High-speed Operational Amplifiers," ISCAS'01, vol. 4, pp.674-677, May 2001.

[120] A. Hossein Nejad-Malayeri, T. Manku, "A 270MHz, 1.8V Fully Differential CMOS Operational Amplifier for Switched-capacitor Channel Select Filters in Wide-band wireless applications," ISCAS 2001. Volume: 1, pp.659-662, May 2001.

[121] B. Manai, P. Loumeau, "A Wide-band CMOS Switched-capacitor Filter for UMTS Direct Conversion Receiver," MWSCAS'01, vol.1, pp.377 - 380 Aug. 2001.

[122] Z. Kaelin, G. S. Moschytz, "Exact Design of arbitrary Parasitic-insensitive Elliptic SC-ladder Filters in the z-domain," ISCAS'88, vol.3, pp.2485-2488, June 1988.

[123] T.S. Fiez, B. Lee, D. J. Allstot, "CMOS Switched-Current Biquadratic Filters," ISCAS'90, vol.3, pp.2300-2303, May 1990.

[124] C-Y Wu; T-C Yu, "New Forward- and backward-mapping SC Differentiator and Their Applications in the Design of Biquad SCF," Proceedings of Technical Papers VLSI Technology, Systems and Applications, pp.169-173, May 1989.

[125] M. J. Svihura, B. Nowrouzian, "A New Approach to the Design of Bilinear-LDI Switched-Capacitor Filters Having Lowpassband Sensitivity," Proceeding. IEEE Pacific Rim Conference on Communications, Computers and Signal Processing, pp.193-197, June 1989.

[126] R. E. Vallee, E. I. El-Masry, "High Performance CMOS Operational-Amplifier," ISCAS'89, vol.2, pp.1475-1478, May 1989.

[127] S. Michael, "Stray Insensitive Programmable Switched-Capacitor Filter," Proceedings of the 3second Midwest Symposium on Circuits and Systems, vol.2, pp.1189-1192, Aug. 1989.

CHAPTER 10
VLSI SWITCHED-CURRENT CIRCUITS

A VLSI switched-current (SI) circuit offers alternative signal processing circuit solution to VLSI switched-capacitor signal processing circuit that avoids the high gain Opamp or OTA circuits and linear capacitors. This attractive feature allows VLSI switched-current circuit to be directly realized using fully digital VLSI process technologies. The basic VLSI SI circuit structures are based on the switched MOS current mirror circuits that use the nonlinear MOS capacitors to store charges. Such a circuit structure allows realizing the modulation and demodulation together in the basic switched-current circuit structure for various signal processing applications.

Similar to the current mirror based VLSI continuous-time current (CTI) circuits, VLSI SI circuits are current mode signal processing circuits that employ the current for signal presentation and processing. The fundamental limitations of SI circuits are the switch charge injection, the finite output impedance, and the mismatch effects of the MOS transistors in the current mirror circuits.

10.1 VLSI SWITCHED-CURRENT CIRCUIT PRINCIPLE

Shown in figure 10.1 is a basic VLSI SI sample/hold circuit structure. Such circuit is based on a switched MOS current mirror employing the MOS switch for the signal sampling and holding operations.

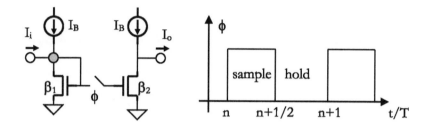

Fig.10.1 Basic VLSI switched-current sample/hold circuit

This basic SI circuit works as follows:

- During the sampling phase ($\phi = 1$), the sampling switch is in the "ON" state and SI circuit works as current mirror. In this phase, the output current signal tracks the input current signal. The input and output current signal relation is fully determined by the size ratio of the two input and output MOS devices as:

$$i_o(t) = -\frac{\beta_2}{\beta_1} i_i(t) \quad (n < \frac{t}{T} < n+\frac{1}{2}) \tag{10.1}$$

- During the hold phase ($\phi = 0$), the sampling switch is in the "OFF" state and the sampled voltage value on the gate of M$_2$ device is held through the entire hold phase. Consequently, the output current in this phase holds the previously sampled input current value:

$$i_o(t) = -\frac{\beta_2}{\beta_1} i_i(t)|_{\frac{t}{T}=n+1/2} \quad (n+\frac{1}{2} \leq \frac{t}{T} < n+1) \tag{10.2}$$

10.2 BASIC SWITCHED-CURRENT SIGNAL PROCESSING ELEMENTS

A set of VLSI signal processing circuits can be constructed based on the SI circuit structures. These circuits can be used to realize the basic VLSI signal processing operations, such as the S/H, the delay, the addition and the scaling operations to support various VLSI analog signal processing applications.

10.2.1 UNIT DELAY CIRCUIT

The SI unity delay circuit is the most basic circuit block in the sampled data signal processing operation, where the output signal and input signal are simply related in the time domain and z-domain by the following expressions as:

$$i_o(nT) = i_i(nT - T) \tag{10.3}$$

And

$$\frac{I_o}{I_i} = z^{-1} \tag{10.4}$$

The VLSI SI unity delay circuit can be constructed from two identical SI S/H circuits, controlled by the non-overlap clocks as shown in figure 10.2.

In practical, it is sometimes convenient to use a half-cycle delay SI circuit in DT signal processing operation, such as the LDI integration. Such a signal processing operation can be expressed by the following equations:

$$i_o(nT) = i_i(nT - T/2) \tag{10.5}$$

And

$$\frac{I_o}{I_i} = z^{-1/2} \tag{10.6}$$

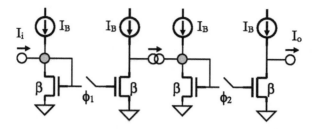

Fig.10.2 Basic VLSI switched-current delay circuit

In this case, the basic VLSI SI S/H circuit of same device size can be directly used. However, one thing should be kept in mind is that this circuit requires the input of the discrete-time signal to be clocked by an opposite clock phase for proper operation due to the transparent nature of the sample/hold circuit at the sample phase.

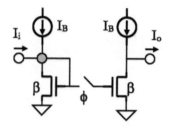

Fig.10.3 Basic VLSI switched-current half-cycle delay circuit

10.2.2 ADDER CIRCUIT

Due to the current nature of the SI signals, addition operation in SI circuit can be directly realized by the wired-addition similar to the CTI circuit implementation.

Note that in practical VLSI SI circuit implementation, it is desired that the addition operation to be realized combined with other signal processing operation (such as delay and scaling operations) to minimize the device count required for overall signal processing operation.

The time domain and the z-domain operations of the combined SI addition with half-cycle delay operation can be expressed as:

$$i_o(nT) = -[i_{i1}(nT - T/2) + i_{i2}(nT - T/2)] \qquad (10.7)$$

$$I_o = -z^{-1/2}(I_{i1} + I_{i2}) \qquad (10.8)$$

It can be seen that the SI circuit shown in figure 10.4 can realize such an operation.

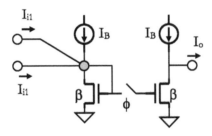

Fig.10.4 VLSI switched-current addition with half-cycle delay circuit

10.2.3 SCALING CIRCUIT

The scaling operation in the VLSI SI circuit as expressed in Equation (10.9) can be directly realized through the proper sizing of the input and output devices in the basic VLSI current mirror circuit. By selecting the output device and the bias current to be α times of that of the input device, we have that:

$$i_o(nT) = -\alpha[i_i(nT)] \qquad (10.9)$$

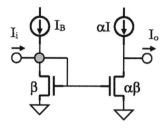

Fig.10.5 VLSI current mode scaler circuit

10.3 VLSI DYNAMIC SI CIRCUIT STRUCTURES

The basic (or the first generation) VLSI SI S/H circuit structures suffer from accuracy limitation due mainly to the non-ideal effects of the current mirror, such as device parameter mismatch, finite output impedance and switching charge injection. In practical VLSI SI circuit design, dynamic S/H (also known as the second generation current mirror) circuit structures can be employed to solve the device mismatch problem. The basic dynamic current S/H circuit is shown in figure 10.6, which consists of a single MOS transistor that is used for both the input and the output operations, controlled by three clock phases.

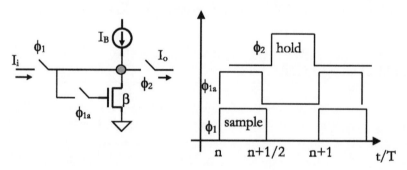

Fig.10.6 VLSI dynamic switched-current sample/hold circuit

Such a dynamic SI S/H circuit works as follows:

VLSI Modulation Circuits

- During the sample phase, the SI S/H circuit is connected as MOS diode circuit as shown in figure 10.7. The gate voltage of the input MOS device tracks input current signal that can be derived from the saturation mode MOS device V-I characteristic equation as

$$V_g(t) = V_T + \sqrt{\frac{2(I_i(t)+I_B)}{\beta}} \quad (nT < t < nT+T/2) \quad (10.10)$$

This voltage is stored on the parasitic gate capacitor of the input MOS device. Note that during this phase, the output current is always zero that is not available to external.

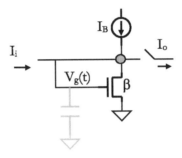

Fig.10.7 Equivalent dynamic SI sample/hold circuit in sample phase

- During the hold phase, the sampled charge trapped on the parasitic gate capacitor of the input MOS device will continue being trapped during this phase as shown in figure 10.8. As the result, the drain current of the input MOS device will maintain the same value just before the sampling switch to the gate of the input MOS device is turned off.

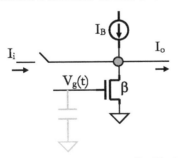

Fig.10.8 Equivalent dynamic SI sample/hold circuit in hold phase

$$i_o(t) = -i_i(t)\big|_{\frac{t}{T}=n+1/2} \quad (nT + \frac{T}{2} \le t < nT + T) \tag{10.11}$$

Note that since both sample and hold circuits use the same MOS device in a time-sharing fashion, the size mismatch between the sample and hold devices can be effectively eliminated.

10.4 SWITCHED-CURRENT CHARGE INJECTION COMPENSATION TECHNIQUES

Similar to the VLSI SC circuits, the VLSI SI circuits suffer from the charge injection effects, resulted from the switching device in the SI circuits. Charge injection has been one of the fundamental limitations in practical VLSI SI circuit structures and a large number of charge injection compensation techniques have been developed in past years.

10.4.1 CHARGE INJECTION ATTENUATION CIRCUITS

A VLSI SI circuit charge injection attenuation technique is shown in figure 10.9. This technique relies on adaptive controlling the capacitance of sampling charge storage node such that a larger capacitance occurs at the end of the sample phase based on the Miller capacitor enhancement techniques.

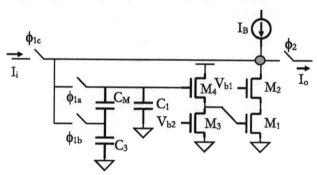

Fig.10.9 Dynamic SI sample/hold circuit with charge injection cancellation

This circuit operates as follows.

- During the sample phase, switch S1a, S1b, and S1c are turned off in sequential. Charge injection occurs when switch S1a is turned off. However, since the other side of added miller capacitor C_M is connected to the drain of M2 through switch S1b. Such a connection results in significantly higher the effective sampling gate capacitance because of Miller effect.
- The node capacitance looked from S1a is enlarged in this case by the loop gain A and it is approximately given as

$$C_{eff} = C_1 + (1+A)C_M \qquad (10.12)$$

- Due to large capacitance presented at the sampling node, the charge injection due to switch S1a is highly attenuated.
- When S1b is turned off, charge may still inject to the storage node through C_M. Such a charge injection effect can be minimized using smaller C_M value.

An alternative VLSI SI charge attenuation circuit structure based on dynamic Miller effect compensation method is shown in figure 10.10. Such a circuit has a controllable loop gain. The loop-gain of the circuit is set to the low gain state during sample phase and the hold phases. The loop is set to high gain state during the sample to hold transitions. The Miller effect enhanced by the high gain feedback loop helps to minimize the charge injection effect.

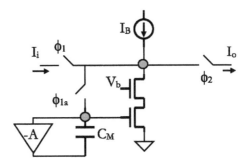

Fig.10.10 Alternative dynamic SI sample/hold with charge injection attenuation

10.4.2 CHARGE INJECTION CANCELLATION CIRCUITS

The SI charge injection cancellation circuit technique based on the replica charge injection cancellation effect is shown in figure 10.11. As the dummy path is selected to be identical to the main signal path, the charge injection from the main signal path and the dummy path is expected to be equal and will cancel each other in the main signal path. The above replica circuit technique can also be implemented using the dynamic SI circuit as shown in figure 10.12.

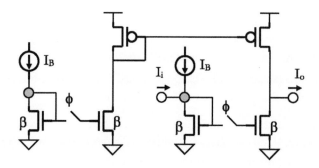

Fig.10.11 VLSI replica switched-current charge injection cancellation circuit

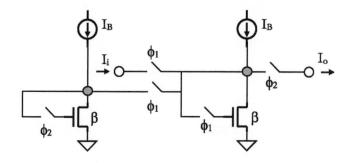

Fig.10.12 Replica dynamic switched-current charge injection cancellation circuit

An alternative SI circuit is shown in figure 10.13. In this circuit, the two switching transistors are designed to be identical, and one of the output branches is double sized. The output current is taken as the difference of the two output ports.

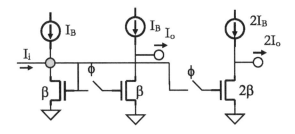

Fig.10.13 Alternative replica charge injection cancellation circuit

Since the two output currents are scaled by a factor of 2. On the other hand, the charge injection induced current errors are scaled by a factor of 1. Consequently the difference of the current from the two output ports eliminates the charge injection induced error.

10.4.3 ALGORITHMIC CHARGE CANCELLATION

The algorithmic charge injection cancellation techniques based on iteration sampling method for charge injections compensated in sequential steps is shown in figure 10.14. Such circuit employs a two-step VLSI SI algorithmic charge injection cancellation technique.

This circuit consists of two current sample/hold elements, including the coarse current sample/hold element controlled by clock $\phi 1a$ and the fine current sample/hold circuit elements controlled clock $\phi 1b$ and $\phi 1c$. The circuit works as follows:

- During the operation, the input signal is first sampled and stored in a coarse current mirror, which has relaxed current resolution and is tolerant to high charge injection error;

- Then the current difference between input current and the sampled coarse current is sampled and held in the fine current mirror;
- Since the fine current mirror only needs to handle small current signal range, the error introduced by fine current mirror sampling charge injection weakly depends on the signal level and may be cancelled relatively easily.

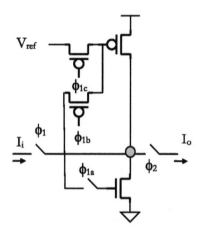

Fig.10.14 Two-step switched-current mirror circuit structure

The algorithmic charge injection schemes can also be extended to multi-step approaches as shown in figure 10.15.

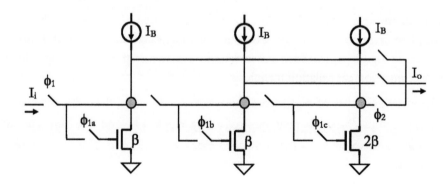

Fig.10.15 Multi-step switched-current mirror circuit structure

10.4.4 FULLY DIFFERENTIAL VLSI SI CIRCUITS

A differential SI circuit structure is shown figure 10.16. Such SI circuit structure can also be used to minimize the signal independent charge injection. Such circuit technique is based on the fact that the charge injection mainly appears as common-mode error, and that can be minimized by the common-mode-rejection ratio (CMRR).

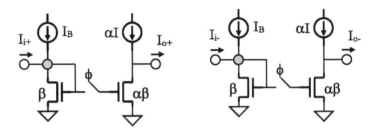

Fig.10.16 Differential switched-current mirror circuit structure

However the signal dependent charge injection, which appears as differential mode error cannot be effectively eliminated by this method.

10.4.5 CONTROLLED CLOCKING AND INPUT TECHNIQUES

The charge injection effect may also be minimized by employing either adaptive controlling of the clocking voltage level with respect to the input signal level or through keeping the input signal voltage (not the current) level constant with the operation in negative feedback. Under such conditions, the charge injection can be made signal independent and therefore can be effectively minimized or eliminated using other circuit techniques. A "zero-switching" SI circuit implementation based on such design concept is shown in figure 10.17.

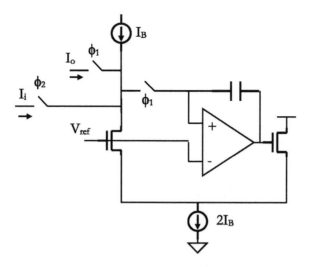

Fig.10.17 Zero-switching current mirror circuit structure

Reference:

[1] A. Ng, J. I. Sewell, "Switched-Current Techniques under Pressure: High Order and Wideband Filter Design," IEE Colloquium on Analog Signal Processing. pp. 4/1-411, Oct. 1998.

[2] Y. Lu, J. I. Sewell, "First or second Generation SI Cells. A Comparison of Sensitivity from an SI Filter System Viewpoint," IEE Colloquium on Digital and Analogue Filters and Filtering Systems, pp. 11/1 - 11/5, Nov 1994.

[3] J. A. Barby, "Switched-current Filter Models for Frequency Analysis in the Continuous-Time Domain," ISCAS '93, vol.2, pp. 1427-1430, May 1993.

[4] Y. Lu, J. I. Sewell, "Multirate SC and SI Filter System Design by XFILT," ISCAS '95, vol. 3, pp. 2257-2260, May 1995.

[5] Z. Q. Shang, J. I. Sewell, "Efficient Analysis of Some Nonlinearities in SC and SI Filter Networks," IEE Colloquium on Digital and Analogue Filters and Filtering Systems, pp. 10/1 - 10/5, Nov 1994.

[6] G. W. Roberts, A. S. Sedra, "Synthesizing Switched-Current Filters by Transposing the SFG of Switched-Capacitor Filter Circuits," IEEE Transactions on Circuits and Systems, vol.38, no.3, pp. 337-340, March 1991.

[7] B. Jonsson, S. Eriksson, "A Low Voltage Wave SI Filter Implementation using Improved Delay Elements," ISCAS '94, vol.5, pp. 305-308, June 1994.

[8] M. Goldenberg, R. Croman, T.S. Fiez, "Accurate SI Filters using RGC Integrators," IEEE Journal of Solid-State Circuits, vol. 29, no. 11, pp. 1388-1395, Nov. 1994.

[9] T. S. Fiez, D. J. Allstot, "CMOS Switched-Current Ladder Filters," IEEE Journal of Solid-State Circuits, vol. 25, no. 6, pp. 1360-1367, Dec. 1990.

[10] B. Jonsson, S. Eriksson, "Current-Mode N-port Adaptors for Wave SI Filters," Electronics Letters, vol. 29, no. 10, pp. 925-926, May 1993.

[11] M. H. Fino, L. J. Mourao, "SymbSI-A Program for the Symbolic Signal flow Graph Generation of Switched-current Circuits," 1998 IEEE

International Conference on Electronics, Circuits and Systems, vol.3, pp. 211-214, Sept. 1998.

[12] M. Helfenstein, A. Muralt, G. S. Moschytz, "Direct Analysis of Multiphase Switched-Current Networks using Signal-flow Graphs," ISCAS '95, vol.2, pp. 1476-1479, May 1995.

[13] M. Helfenstein, A. Muralt, G. Fischer, P. Zbinden, D. Pfaff, F. Frey, G.S. Moschytzl, "SC and SI Filters in Baseband Applications: A Comparison," ISCAS '97, vol.1, pp. 297-300, June 1997.

[13] A. Handkiewicz, P. Sniatala, M. Lukowiak, "Low-voltage High-performance Switched-current Memory Cell," Proceedings, Tenth Annual IEEE International ASIC Conference and Exhibit, pp. 12-16, Sept. 1997.

[14] J.L. Chan, S.S. Chung, "Universal Switched-Current Integrator Blocks for SI Filter Design," Proceedings of the ASP-DAC '99, vol.1, Jan. pp. 261-264, 1999.

[15] L.J. Mourao, M.H. Fino, "Automatic Symbolic Evaluation of Nonideal Effects in SI Circuit Behavior," Third International Workshop on Design of Mixed-Mode Integrated Circuits and Applications, pp. 40-43, July 1999.

[16] J. Schechtman, A.C.M. De Queiroz, L.P. Caloba, "A Practical Implementation Scheme for Component Simulation SI Filters," Proceedings., Proceedings of the 38th Midwest Circuits and Systems, vol.1, pp. 174-177, Aug. 1995.

[17] J.B. Hughes, K.W. Moulding, "A Switched-Current Double Sampling Bilinear z-transform Filter Technique," ISCAS '94, vol.5, pp. 293-296, June 1994.

[18] N.C. Battersby, C. Toumazou, "Towards High Frequency Switched-current Filters in CMOS and GaAs Technology," ISCAS '93, vol.2, pp. 1239-1242, May 1993.

[19] T. Fiez, D. Allstot, "A CMOS Switched-Current Filter Technique," Digest of Technical Papers, 37th ISSCC, pp. 206-207, Feb. 1990.

[20] G.K. Balachandran, P.E. Allen, "Switched-Current Circuits in Digital CMOS Technology with Low Charge-injection Errors," IEEE Journal of Solid-State Circuits, vol. 37, no.10, pp. 1271-1281, Oct. 2002.

[21] B. Raahemi, A. Opal, "Group Delay and Group Delay Sensitivity of Periodically Switched Linear Networks," IEEE Transactions on Circuits and Systems I: Fundamental Theory and Applications, vol. 47, no. 1, pp. 96-104, Jan. 2000.

[22] T-H Kuo, S-Y Lee, K-D Chen, "Offset Current Cancellation Based on a Multiple-Path feedback Compensation (MPFC) Technique for Switched-current Circuits and Systems," IEEE Transactions on Circuits and Systems II: Analog and Digital Signal Processing, vol. 44, no. 4, pp. 299-309, April 1997.

[23] A. Yufera, A. Rueda, "Studying The Effects of Mismatching and Clock-feed Through in Switched-Current Filters using Behavioral Simulation," IEEE Transactions on Circuits and Systems II: Analog and Digital Signal Processing, vol. 44, no. 12, pp. 1058-1067, Dec. 1997.

[24] A. Worapishet, R. Sitdhikorn, .B. J. Hughes, "Architecture for Low-voltage Switched-current Complex Bandpass Filters," Electronics Letters, vol. 38, no. 12, pp. 535-536, June 2002.

[25] A.E.J. Ng, J.I. Sewell, "Ladder Decompositions for wideband SI Filter Applications," IEE Proceedings, Circuits, Devices and Systems, vol. 145, no. 5, pp. 306-313, Oct. 1998.

[26] T.-H. Kuo, S.-Y. Lee D.-J. Lu, T. Jih, J.-J. Tsaur, "3.3 V Mixed-mode IC Design using Switched-current Techniques for Speech Applications," IEE Proceedings Circuits, Devices and Systems, vol. 144, no. 6, pp. 367-374, Dec. 1997.

[27] Kuan-Dar Chen, Tai-Haur Kuo, Shuenn-Yuh Lee, "A Novel Offset Current Cancellation Technique for Switched-current Circuits and Systems," ISCAS '96, vol. 1, pp. 417-420, May 1996.

[28] A. Handkiewicz, M. Kropidlowski M. Lukowiak, "Computer Tools for Switched-Current Filter Design, "MWSCAS-2002, vol. 2, pp. II-501 - II-504, Aug. 2002.

[29] Qingyun Gao, Shicai Qin, Xiangluan Jia, Yonggang Song, "Computer Aided Design of Switched-current Filters, "4th International Conference on ASIC, pp. 94-97, Oct. 2001.

[30] D.L. Shlemin, "A Design Technique for High Q-factor SI Filters, "APEIE-98, vol. 1, pp. 142-152, Sept. 1998.

[31] I. Yusim, G. Ionis, K. Suyama, "Simulator for Switched-current Integrated Circuits," proceedings of the IEEE 1998 Custom Integrated Circuits Conference, pp. 459-462, May 1998.

[32] A. Yufera, A. Rueda, J.L Huertas, "Switched-Current Wave Analog Filters," ISCAS '92, vol. 2, pp. 859-862, May 1992.

[33] T.S. Fiez, B. Lee, D.J. Allstot, CMOS Switched-Current Biquadratic Filters," IEEE International Symposium on Circuits and Systems, vol.3, pp. 2300-2303, May 1990.

[34] J.B. Hughes, K.W. Moulding, J. Richardson, J. Bennett, W. Redman-White, M. Bracey, R.S. Soin, "Automated Design of Switched-current Filters," IEEE Journal of Solid-State Circuits. vol. 31, no. 7, pp. 898-907, July 1996.

[35] T.S. Fiez, G. Liang, D.J. Allstot, "Switched-current Circuit Design Issues," IEEE Journal of Solid-State Circuits, vol. 26, no. 3, pp. 192-202, Mar 1991.

[36] J.B. Hughes, "Top-down Design of A Switched-Current Video Filter," IEE Proceedings, Circuits, Devices and Systems, vol. 147, no. 1, pp. 73-81, Feb. 2000.

[37] A. Handkiewicz, M. Kropidlowski, M. Lukowiak, M. Bartkowiak, "Switched-Current Filter Design for Image Processing Systems," 13th Annual IEEE International ASIC/SOC Conference, pp. 8-12, Sept. 2000.

[38] R.T. Goncalves, S. Noceti Filho, M.C. Schneider, C. Galup-Montoro, "Digitally Programmable Switched-current Filters," ISCAS '96, vol.1, pp. 258-261, May 1996.

[39] A.D. Remenyuk, E.V. Astrova, L.I. Korovin, I.G. Lang, V.B. Shuman, "Diffraction Effects in Macroporous Silicon with Through Pores," Proceedings of 2003 5th International Conference on Transparent Optical Networks, vol.1, pp. 273-275, July 2003.

[40] Xin Li, Xuan Zeng, Dian Zhou, Xieting Ling, "Behavioral Modeling of Analog Circuits by Wavelet Collocation Method," ACM IEEE International Conference on Computer Aided Design, pp. 65-69, Nov. 2001.

[41] Yue Lu; J.I. Sewell, "A Systematic Approach for Ladder Based Switched-current Filter Design," ISCAS '95, vol.3, pp. 2253-2256, May 1995.

[42] R.T. Goncalves, S.N. Filho, M.C. Schneider, C. Galup-Montoro, "Programmable Switched-current Filters using MOSFET-only Current Dividers," Proceedings of the 38th Midwest Circuits and Systems, vol.2, pp. 1046-1049, Aug. 1995.

[43] A. Yufera, A. Rueda, J.L. Huertas, "A Study of the Sensitivity of Switched-current Wave Analog Filters to Mismatching and Clock-Feedthrough Errors," ISCAS '94, vol.5, pp. 317-320, June 1994.

[44] Hpngjiang Song, "Current Mode Approaches to Low Voltage/Power VLSI Design for Portable Mixed Signal System," Ph.D. Thesis, 1996.

CHAPTER 11
VLSI SWITCHED-INDUCTOR DC/DC CONVERTER CIRCUITS

The inductor-based (or switched-inductor) dc/dc converters offer several advantages compared to linear regulators such as better power conversion efficiency, smaller form factor, and fewer thermal management issues. The switched-inductor dc/dc converter avoids the power loss in linear regulator that uses a resistive voltage drop to regulate the voltage. A switched-inductor regulator has a voltage drop 90 degrees out of phase with the associated current such that the stored energy can be recovered in the discharge phase of the switching cycle.

Because less energy is lost in the conversion, smaller components and less thermal management are required in the switched-inductor dc/dc converter. In addition, the energy stored by the inductors in a switched-inductor dc/dc converter can be transformed to output voltages that are higher than the input (boost), negative (inverter), or even is transferred through a transformer to provide electrical isolation with respect to the input.

11.1 INDUTOR-BASED DC/DC CONVERTER BASIS

The operation of a VLSI switched-inductor dc/dc converter can be analyzed based on various circuit models.

11.1.1 VOLTAGE DIVIDER BASED MODEL

Shown in figure 11.1 is a conceptual linear dc/dc converter circuit (such as LDO) where a lower supply voltage level is generated from a higher voltage source using a resistive divider (buck conversion).

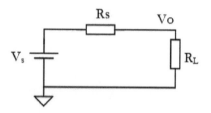

Fig.11.1 Buck dc/dc converter model

This linear dc/dc voltage converter offers a voltage conversion operation that can be expressed as:

$$V_o = \frac{1}{1 + R_s / R_L} V_s \qquad (11.1)$$

A specific output voltage can be generated by selecting a proper R_s/R_L ratio. However, such a resistive circuit has a significant power conversion loss to support the resistive division operation. This is very true when the R_S/R_L is large (i.e. large difference between V_o and V_s).

The power conversion efficiency can be improved if the power loss in R_s is reduced by using a lossless switched-inductor equivalent resistance R_s in the voltage divider as shown in figure 11.2. When the control clock period is significantly short such that the variations in terminal voltages (V_A and V_B) are small enough, we have the voltage and current relations of the converter circuit in the main switch S and the synchronous switch S# phase respectively as:

$$V_A - V_B = L\frac{(I_2 - I_1)}{kT} \qquad (11.2)$$

$$V_B = L\frac{(I_2 - I_1)}{(1-k)T} = I_{av}R_L \qquad (11.3)$$

Where I_{av} is the average inductor current flowing to terminal B within one clock period.

$$I_{av} = \frac{(I_2 + I_1)}{2} \qquad (11.4)$$

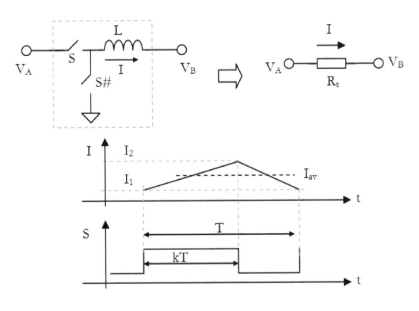

Fig.11.2 Switched-Inductor resistor model

We can derive the equivalent switched-inductor resistance R_s of the converter as:

$$R_s = \frac{V_A - V_B}{\overline{I}} = \frac{(1-k)}{k} R_L \qquad (11.5)$$

Consequently the output voltage of the dc/dc converter employing the lossless switched-inductor voltage divider shown in figure 11.1 is given as:

$$V_o = \frac{1}{1 + R_s / R_L} V_s = \frac{1}{1 + (1-k)/k} V_s = k V_s \qquad (11.6)$$

For the dc/dc converter shown in figure 11.3, a higher output voltage than the source voltage can also be achieved using a negative equivalent source resistance R_s.

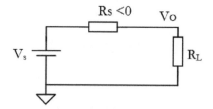

Fig.11.3 Boost dc/dc converter model

Such negative resistance can be realized using a switched-inductor circuit structure shown in figure 11.4.

For this switched-inductor circuit structure we may also assume that the terminal voltage within one clock period is approximately constant when a sufficiently high switching frequency is used. The relationships between the terminal voltages and the inductor current in the two control clock phases can be derived as:

$$V_A = L \frac{(I_2 - I_1)}{kT} \qquad (11.7)$$

$$V_B - V_A = L\frac{(I_2 - I_1)}{(1-k)T} \qquad (11.8)$$

We can further get that

$$V_B - V_A = kV_B \qquad (11.9)$$

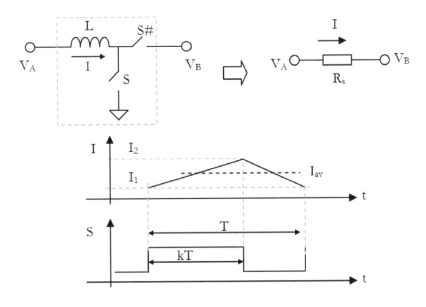

Fig.11.4 Switched-Inductor resistor model

The equivalent source resistance can then be derived as:

$$R_s = \frac{V_A - V_B}{I_{av}} = -\frac{kV_B}{I_{av}} = -kR_L \qquad (11.10)$$

We can see that such a dc/dc converter offers a boost voltage conversion given as:

$$V_o = \frac{1}{1+R_s/R_L}V_s = \frac{1}{1-k}V_s \qquad (11.11)$$

It is important to note that synchronous switch S# is open when the main switch S is closed, and the same is true conversely. To prevent the dc-path problem (both top and bottom switches are on simultaneously), the switching scheme must be break-before-make.

11.1.2 LC FILTER BASED MODEL

Shown in figure 11.5 is a simple switching-mode power supply that can be used to control the power to a load R_L.

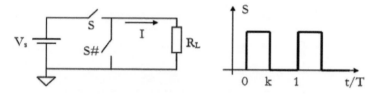

Fig.11.5 Switching mode power transfer circuit

In this circuit, turning on and off the switch can be used to control the average current and power dissipation of the load R_L by using controlling the duty-cycle k as:

$$I_{av} = k\frac{V_s}{R_L} \qquad (11.12)$$

$$P_{av} = k\frac{V_s^2}{R_L} \qquad (11.13)$$

In another word, the input voltage and the duty-cycle of the switch control voltage determine the equivalent voltage observed by the load:

$$V_{eq} = kV_s \qquad (11.14)$$

The current discontinuity of the basic switching-mode power supply circuit can be minimized using a LC lowpass filter as shown in figure 11.6 to eliminate the high frequency components in the load voltage and current.

Under this approach, the dc/dc converter can be modeled using an equivalent switching voltage source Vi and a LC filter with the transfer function given respectively as

$$V_i(t) = \begin{cases} V_s & 0 \le t < kT \\ 0 & kT \le t < T \end{cases} \qquad (11.15)$$

$$\frac{V_o(s)}{V_i(s)} = \frac{k}{LCs^2 + \frac{L}{R}s + 1} \qquad (11.16)$$

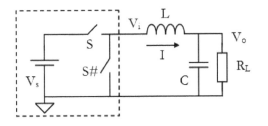

Fig.11.6 Switched-Inductor Buck Converter

Where L, C and R_L are the switched-inductor, the filter capacitor and the load resistor respectively. The control clock frequency is 1/T and the duty-cycle is k.

This transfer function model can be used to characterize the dc/dc converter for the settling time, voltage ripple and other converter performance parameters.

11.1.3 SYNC- AND ASYNCHRONOUS MODES

It is interesting to note that in the dc/dc converter shown in figure 11.7 the synchronous switch S# in the synchronous converter can be replaced by a diode to provide necessary path for the inductor current when the switch S is off. Without this current path an extremely high voltage would build up. This may result in some undesired phenomenon such as circuit breakdown.

Similar to the buck converter, the synchronous switch S# in boost dc/dc converter circuit synchronous converter can also be replaced by a diode as shown in figure 11.8.

A synchronous dc/dc converter may offer improved conversion efficiency than the asynchronous dc/dc converter. When the synchronous dc/dc switch closes, a current flows through the MOS channel. Because of the very low-channel resistance for the power MOS transistors, the forward voltage drop of the rectifying diode can be reduced to a few millivolts therefore it can provide conversion efficiencies well above 90%. However the diode rectify dc/dc converter show large power-loss factors during dc/dc conversion due to the forward voltage drop loss of the rectifying diode where the power dissipated is simply determined by the forward voltage drop multiplied by the current going through it. In addition, the reverse recovery for silicon diodes also contributes to the loss. These power losses reduce overall efficiency and require thermal management in the form of a heat sink or fan. To minimize this loss, dc/dc converter may use Schottky diodes to reduce the forward-voltage drop with good reverse recovery.

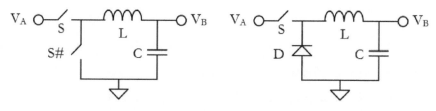

a) Synchronous converter   b) Diode based buck converter

Fig.11.7 Buck dc/dc converter structures

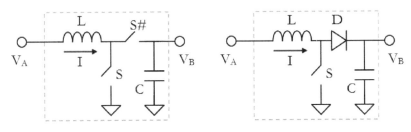

a) Synchronous converter   b) Diode based boost converter

Fig.11.8 Boost dc/dc converter structures

11.2 BASIC DC/DC CONVERTER ARCHITECTURES

The switched-inductor dc/dc converters use switched network to transfer energy from input to output. There are several basic dc/dc converter circuit topologies including the step-down (buck), the step-up (boost), the inverter, the flyback and their combined configurations.

11.2.1 STEP DOWN CONVERTER

For the synchronous buck converter circuit shown in figure 11.9, the switch S and S# are turned ON alternatively. In the following analysis we assume that the output voltage is kept approximately consistent using a large capacitance C and that the duty-cycle of the control clock is k.

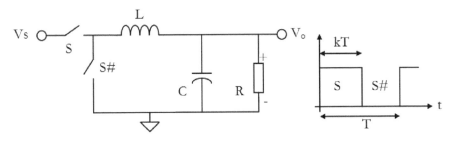

Fig.11.9 Synchronous buck converter

When the switch S is ON, the inductor and the output load get energy from the input Vs as shown in figure 11.10.

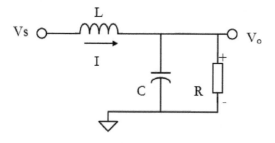

Fig.11.10 Equivalent circuit switch S is ON

The ramping up of the current in the inductor can be expressed as:

$$L\frac{\Delta I}{kT} = V_s - V_o \qquad (11.17)$$

When the converter is in the steady-state condition the ramping down current of the inductor in the circuit shown in figure 11.11 is equal to the ramping up current of the inductor when switch S# is ON.

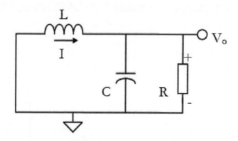

Fig.11.11 Equivalent circuit switch S# is ON

We have that

VLSI Modulation Circuits

$$L\frac{\Delta I}{(1-k)T} = V_o \qquad (11.18)$$

We can further derive the steady-state voltage of the circuit as:

$$k(V_s - V_o) = (1-k)V_o \qquad (11.19)$$

Or

$$V_o = kV_s \qquad (11.20)$$

It is important to see that (1) since the dc value of the inductor current is the same as the output current, the output voltage is equal to the duty-cycle times the input voltage; (2) since the average current through the capacitor must be zero, the average current flowing through the inductor is equal to the load current. This implies that the inductor saturation current must be selected to be higher than the load current, since the average voltage across the inductor must be zero, the output voltage of the converter is proportional to kVs.

Recalling from the analysis of switch S in ON state, the inductor charging equation is given as:

$$\frac{dI}{dt} = \frac{V_s - V_o}{L} \qquad (11.21)$$

The ramping-up of the inductor current can be expressed as:

$$2\Delta i_L = \frac{V_s - V_o}{L}kT = \frac{V_s - V_o}{Lf}k \qquad (11.22)$$

Similarly the ramping-down of the inductor current can be expressed as:

$$2\Delta i_L = \frac{-V_o}{L}(1-k)T = \frac{V_s - V_o}{Lf}(1-k) \qquad (11.23)$$

Both equations represent the same value since it is in the steady-state condition. After substituting, Vo = kVs, and recognizing that in order to minimize ripple, large values of f and Vs must be used. A criteria of inductor selection in the buck dc/dc converter can be given as:

$$L = \frac{V_o}{2f_{\min}\Delta i_L}(1 - \frac{V_o}{V_s}) \qquad (11.24)$$

In the dc/dc converter inductor selection, we should also check the physical limitation, the maximum current limitation, the EMI interference tolerated and the cost.

Shown in figure 11.12 is asynchronous switched-inductor step down dc/dc converter circuit structure. In the circuit operation, the input terminal of the inductor will be put to voltage Vi when the switch is ON. This voltage causes the inductor current to ramp-up. When the switch is OFF, the current will continue flowing through the inductor but now flowing through the diode. We can initially assume that the current through the inductor does not reach zero, thus the voltage at Vx will now be only the voltage across the conducting diode during the full OFF time. The average voltage at Vx will depend on the average ON time of the transistor making the inductor current continuous. Under such operation the analysis of this dc/dc converter is the same as the synchronous buck converter discussed earlier.

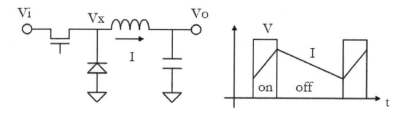

Fig.11.12. Buck dc/dc converter operation

11.2.2 STEP UP CONVERTER

The boost switched-inductor dc/dc converter can be used when a higher output voltage than input is required. For the synchronous switched-inductor step-up dc/dc converter shown in figure 11.13, the switch S and S# are turned ON and OFF alternatively. Assume the output voltage is approximately constant (i.e. a large storage capacitance C is used) and the duty-cycle of the control clock is k.

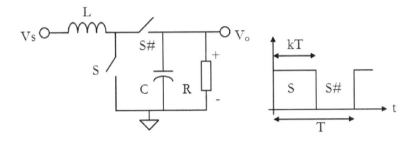

Fig.11.13 Synchronous boost converter

When the switch S is ON, the inductor gets energy from the input Vs as shown in figure 11.14.

The ramping up of the current in the inductor can be expressed as:

$$L\frac{\Delta I}{kT}=V_s \qquad (11.25)$$

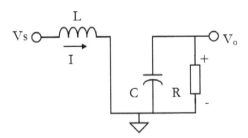

Fig.11.14 Equivalent circuit switch S is ON

For the steady-state operation the ramping-down of the current in the circuit shown in figure 11.15 is equal to the ramping-up amount of the current when switch S# is ON.

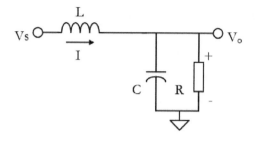

Fig.11.15 Equivalent circuit switch S# is ON

We have that

$$L\frac{\Delta I}{(1-k)T} = V_o - V_s \qquad (11.26)$$

We may further derive the steady-state voltage of the circuit by combining the two equations as:

$$V_o = \frac{V_s}{(1-k)} \qquad (11.27)$$

We can see that such circuit offers a conversion voltage gain that is larger than 1.

Alternatively the boost dc/dc converter can be realized in the asynchronous mode as shown in figure 11.16. If the inductor current always remains continuous conduction, the steady-state output voltage calculation is the same as the synchronous boost converter.

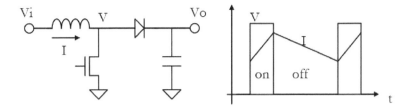

Fig. 11.16. Boost Converter Circuit

11.2.3 STEP UP/DOWN CONVERTERS

Shown in figure 11.17 is a synchronous step up/down dc/dc converter circuit.

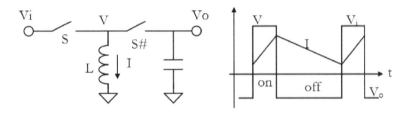

Fig. 11.17. Buck/Boost Converter Circuit

In this buck-boost dc/dc converter, $V = V_i$ when the transistor is ON and $V = V_o$ when the transistor is OFF. For zero net current change over a period the average voltage across the inductor is zero

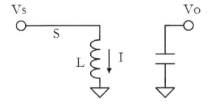

Fig. 11.18. Buck/Boost Converter Circuit in S phase

The steady-state ripple current for unloaded operation condition can be analyzed as follows.

The ramping-up current in the first phase can be expressed as:

$$V_s = L\frac{\Delta I}{kT} \quad (11.28)$$

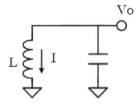

Fig. 11.19. Buck/Boost Converter Circuit is S# phase

Similarly the ramping-down current in the second phase can be expressed as

$$V_o = -L\frac{\Delta I}{(1-k)T} \quad (11.29)$$

Under the steady-state operation condition, the ramping-up and ramping-down current are equal and the conversion gain can be expressed as:

$$\frac{V_o}{V_s} = -\frac{k}{1-k} \quad (11.30)$$

We may also obtain the output to the input current ratio of the circuit as:

$$\frac{I_o}{I_i} = -\frac{1-k}{k} \quad (11.31)$$

Since the duty-cycle k is between 0 and 1 the output voltage can vary between lower or higher than the input voltage. The negative sign indicates a reverse sign of the output voltage.

The achievable voltage conversion gains by the basic dc/dc converters are summarized in figure 11.20. Note that only the buck converter shows a linear relationship between the control duty-cycle k and output voltage. The buck-boost converter offers unit gain at 50% duty-cycle.

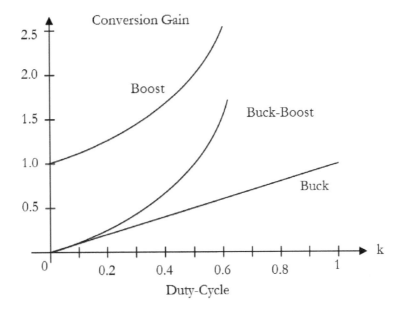

Fig. 11.20 Comparison of dc/dc converter voltage ratio

11.2.4 CUK DC/DC CONVERTER

Shown in figure 11.21 is the Cuk converter circuit structure that can be used to realize the buck, the boost and the buck-boost conversions.

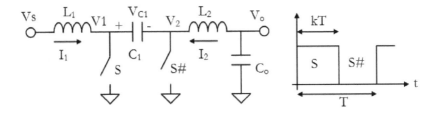

Fig. 11.21 Synchronous Cuk dc/dc converter

The Cuk converter employs the capacitor for energy transfer. The Cuk circuit can be analyzed using the duality principle on the buck-boost converter.
This circuit works as follows. When switch S is ON, the equivalent circuit under the steady-state condition is shown in figure 11.22.

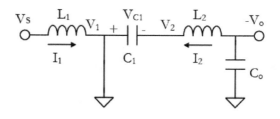

Fig. 11.22 Cuk dc/dc Converter in S phase

The current changes in this state can be expressed as:

$$\Delta I_1 = \frac{V_s}{L_1} kT \qquad (11.32)$$

$$\Delta I_2 = \frac{V_{C1} - V_o}{L_2} kT \qquad (11.33)$$

When switch S is OFF, the equivalent circuit is shown in figure 11.23. The current changes in this state can be expressed as:

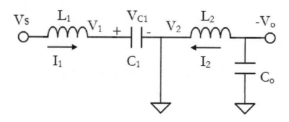

Fig. 11.23 Cuk dc/dc Converter in S# phase

$$\Delta I_1 = \frac{V_{C1} - V_s}{L_1}(1-k)T \tag{11.34}$$

$$\Delta I_2 = \frac{V_o}{L_2}(1-k)T \tag{11.35}$$

Solving for above equations we have the steady-state expression of the input and the output voltages as:

$$\frac{V_o}{V_s} = -\frac{k}{1-k} \tag{11.36}$$

$$\frac{V_{C1}}{V_s} = \frac{1}{1-k} \tag{11.37}$$

It can be seen that a Cuk converter has a voltage gain same to the buck-boost converter. However the Cuk converter offers the advantage that the input and output inductors create a smooth current at both sides of the converter while the buck, boost and buck-boost have at least one side with pulsed current.

Similarly, the diodes can replace the synchronous switches as shown in figure 11.24 to provide the asynchronous dc/dc conversion.

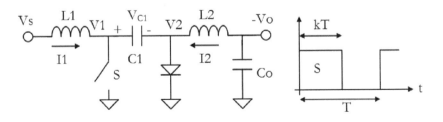

Fig. 11.24 CUK dc/dc converter

In such case, the analysis of the circuit is similar to the synchronous convert as soon as the diode is forward biased in the S# phase. However such implementation has the penalty of higher power loss due to the diode loss.

11.2.5 SEPIC DC/DC CONVERTER

Shown in figure 11.25 is the Sepic converter circuit structure that is similar to the Cuk converter, which also uses the capacitors for energy transfer.

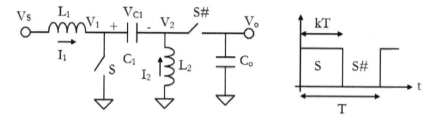

Fig. 11.25 Sepic dc/dc Converter

During the first phase (the S phase), the equivalent circuit under the steady-state condition is shown in figure 11.26. The changes of the currents in this phase can be expressed as

$$\Delta I_1 = \frac{V_s}{L_1} kT \tag{11.38}$$

$$\Delta I_2 = \frac{V_{C1}}{L_2} kT \tag{11.39}$$

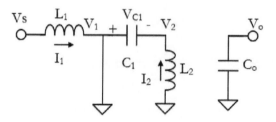

Fig. 11.26 S phase equivalent of Sepic dc/dc Converter

During the second phase (the S# phase), the equivalent circuit is shown in figure 11.27.

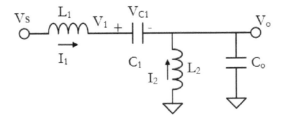

Fig. 11.27 S# phase equivalent of Sepic dc/dc Converter

The current changes in this phase are given as:

$$\Delta I_1 = \frac{V_o + V_{C1} - V_s}{L_1}(1-k)T \qquad (11.40)$$

$$\Delta I_2 = \frac{V_o}{L_2}(1-k)T \qquad (11.41)$$

The steady-state relation between the input and the output is given as:

$$V_o = \frac{k}{1-k}V_s \qquad (11.42)$$

$$V_{C1} = V_s \qquad (11.43)$$

It can be seen that the Sepic converter offers a positive buck/boost conversion operation.

Similarly the synchronous switch S# can be replaced by a diode in the asynchronous Sepic converter as shown in figure 11.28. The analysis of the circuit performance is similar to the synchronous Sepic converter circuit if the current is continuous.

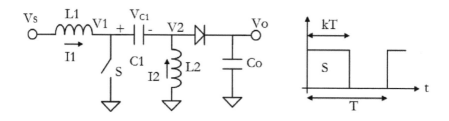

Fig. 11.28 Sepic dc/dc Converter

11.2.6 FLYBACK CONVERTER

In many switched-inductor dc/dc converter applications, the isolated output and input may be required to meet safety standards and to provide impedance matching. In these cases, the basic switched-inductor dc/dc converter topologies can be modified to provide the isolation among the ports.

Shown in figure 11.29 and figure 11.30 are two flyback converter structures developed from the buck-boost converters where the inductors are replaced by a transformer. This buck-boost converter works by storing energy in the inductor during the ON phase and releasing it to the output during the OFF phase.

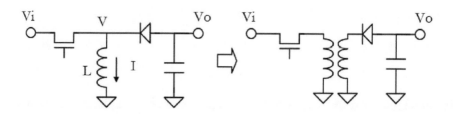

Fig.11.29. Isolated buck dc/dc converter

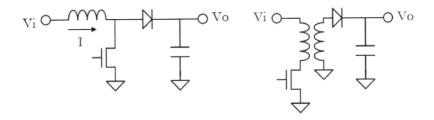

Fig. 11.30. Flyback converter re-configured

11.2.7 FORWARD CONVERTER

The forward converter uses an ideal transformer converting the input ac voltage to an isolated secondary output voltage. For the circuit shown in figure 11.31, when the transistor is ON, Vs appears across the primary and then generates a voltage at the other side of the transformer as:

$$V_x = \frac{N_1}{N_2} V_i \qquad (11.44)$$

The diode at the secondary output ensures that only positive voltages are applied to the output circuit while the second diode provides a circulating path for inductor current if the transformer voltage is zero or negative.

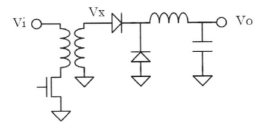

Fig.11.31. Forward dc/dc Converter

The problem associated with the above circuit is that only the positive voltage is applied across the core, thus flux can only increase with the application of the supply. The flux will increase until the core saturates when the magnetizing current increases significantly and circuit failure occurs. The transformer can only sustain operation when there is no significant dc component to the input voltage. While the switch is ON there is a positive voltage across the core and the flux increases. When the switch turns OFF we need to supply negative voltage to reset the core flux. The circuit in figure 11.32 shows a tertiary winding with a diode connection to permit reverse current flow to resolve above issue.

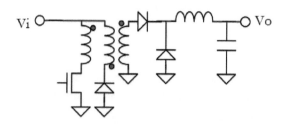

Fig.11.32. Forward converter with tertiary winding

11.2.8 RESONANT CONVERTER

For the DC/DC converter circuits shown in figure 11.33 and figure 11.34, when the clock frequency is equal to the resonant frequency of the LC tank, they form the resonant DC/DC converters.

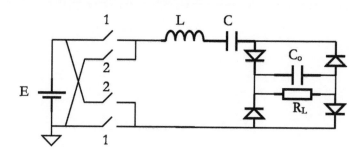

Fig. 11.33 Resonant DC/DC converter circuit (I)

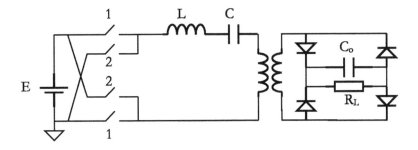

Fig. 11.34 Resonant DC/DC converter circuit (II)

11.3 DC/DC REGULATORS

Open loop DC/DC converters suffer from high sensitivities of the PVT and loading conditions. As the result, closed-loop DC/DC converters are commonly used in DC/DC regulator circuits. The objective of closed-loop DC/DC converter regulators are to make output of regulator follow precisely the reference input. The objective of the closed-loop are that

- Minimize the steady state voltage error.
- Minimize the sensitivity of the parameter variation.
- Compensate for the gain and phase responses of the circuit over the application frequency range.
- Stabilize the circuit operation.
- Minimize the output voltage variation due to the load condition change.
- Linearize the circuit operation.

There are a few commonly used DC/DC regulator families based on the feedback loop configurations used, such as the hysteretic, voltage and current mode regulator circuits.

11.3.1 HYSTERETIC REGULATOR

Hysteretic regulators employ bang-bang control schemes. This regulator family can be implemented in various ways, such as the constant on-time control method. Some general characteristics of the hysteretic regulator circuits include:

- The simplest control of the regulator feedback loop without the need for loop compensation
- Fast loop response time.
- Few external components.
- Fast load response, and
- High power efficiency over wide load conditions.

However, hysteretic regulators generally suffer from some common limitations, such as:

- Inherent varying switching frequency of the hysteretic regulator may make the filtering of the switching noise difficult and create the EMI issue in some electronic systems.
- Higher voltage ripple since ripple is part of the circuit operation.
- Higher sensitive to the output noise since they directly use the output voltage for controlling the switch.
- Requirement of additional overcurrent protection since at certain load condition the current of the switch can be significantly large.

Shown in figure 11.35 is a VLSI hysteretic regulator circuit implementation.

- When the feedback voltage V_{fb} drops below than the low reference voltage, the switch will be ON and the input is used to charge the load through the inductor. This cause the output voltage increase. The current in the inductor also increases.

- When the feedback V_{fb} is above than the high voltage reference V_{ref+}, the switch will be OFF and the output voltage will slowly drop. This also causes the current in the inductor decreases.

·

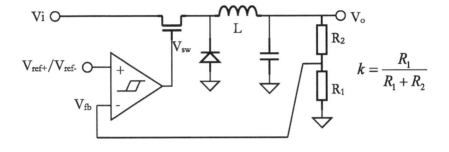

Fig 11.35 Hysteretic buck regulator circuit

Such circuit works as follows as shown in figure 11.36:

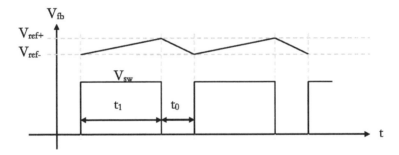

Fig 11.36 Hysteretic regulator circuit operation

- As the result, the output voltage is controlled to with the two reference as:

$$\frac{V_{ref-}}{k} \leq V_o \leq \frac{V_{ref+}}{k} \tag{11.45}$$

Shown in figure 11.37 and figure 11.38 are alternative hysteretic regulator circuit implementation and operation that is based on the constant-on time control schemes.

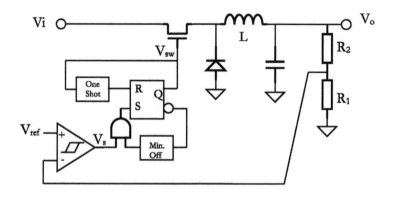

Fig 11.37 Hysteretic buck regulator circuit with constant-on time

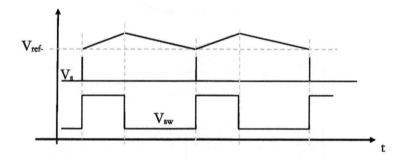

Fig 11.38 Constant-on time regulator circuit operation

In this circuit implementation, the switch will be turn on by a control clock pulse with a constant on time. The off time is determined by the output voltage that drops to the reference level.

11.3.2 VOLTAGE MODE REGULATORS

The voltage mode DC/DC regulators are based on the voltage mode feedback control that offer the features of

- Single loop and offers good noise margin and
- Low output impedance.

The major drawbacks of voltage mode regulator include:

- Slow voltage feedback loop.
- Requirement of double-pole loop compensation.
- Senstivity of loop compensation to the inductor, and
- Senstivity of loop compensation gain to Vin.

Shown in figure 11.39 is a VLSI realization of the voltage mode buck regulator.

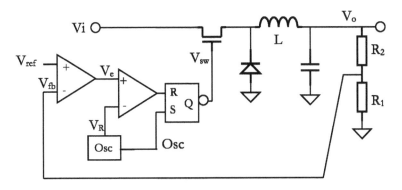

Fig 11.39 Voltage mode buck regulator circuit

This circuit employs an oscillator to generate the set pulse for the RS latch. It also generates a ramp signal for the comparator. The scaled output voltage is compared with the reference to provide an error signal Ve. This error voltage is compared with the ramp signal from the oscillator to generate the reset control signal. Shown in figure 11.40 is an operation of the circuit.

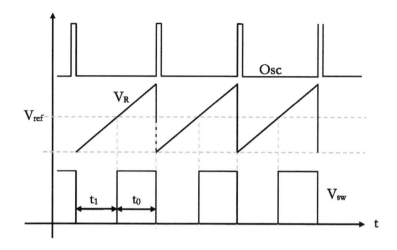

Fig 11.40 Voltage mode regulator circuit operation

In the voltage mode boost regulator circuit implementation shown in figure 11.41 the circuit samples the output voltage and subtracts this from a reference voltage to establish a small error signal (V_E).

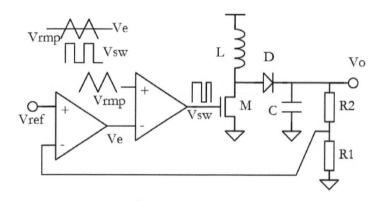

Fig 11.41 Voltage mode regulator circuit

This error signal is then compared to an oscillator ramp signal. The comparator output provides a varying duty-cycle (i.e. PWM) output to control the power switch. When the circuit output voltage changes, V_{ERROR} also changes and thus causes the comparator threshold to change. Consequently, the output pulse width (PWM) also changes. This duty-cycle change is then used to adjust the output voltage to minimize the error signal. This type of switched-inductor regulator topology is typically classified as a voltage-mode controller (VMC) because the feedback regulates the output voltage.

When the regulator control loop reaches its steady-state, the output of the dc/dc converter is simply a ratio of the reference voltage independent of the load current and the input voltage supply:

$$V_o = \frac{R_1 + R_2}{R_1} V_{ref} \qquad (11.46)$$

11.3.3 CURRENT MODE REGULATORS

Current mode controll loops are commonly used in DC/DC regulators to improve loop stability and dynamic responses of the DC/DC converters. Such regulators offer a few advantages include:

- Fast response to output current changes.
- Inherent cycle based current protection of the regulator.
- Simplified loop compensation by elimiating the inductor in the voltage feedback loop.

Major drawbacks of current mode regulators include:

- Higher noise sensitivity to the current spikes,
- More complex circuit design with two feedback loops.
- Requirement of slope compensation in design for DC>0.5 due to the instability that may cause oscillations at subharmonics of the switching frequency.

With the fast current loop significantly faster than the voltage control loop, the feedback loop in the regulator can be reduced to a first order loop in the circuit implementation, where the inductor does not affect the loop compensation. Such circuit only require single pole loop compensation in the circuit implementation and they offer the parallelability with load sharing applications.

The current loop regulates the output current and, with infinite loop gain, the output is a high-impedance source. In such circuit, the fast current loop is nested with a slower voltage loop. A ramp is generated by the slope of the inductor current and compared with the error signal. So, when the output voltage sags, the current mode loop supplies more current to the load. The advantage of such circuit is its ability to control the inductor current. Different from the voltage mode regulator where the inductor current is not directly controlled. This can become a problem because the inductor, in conjunction with the output filter capacitor, forms a resonant tank that can ring and even cause oscillations.

Shown in figure 11.42 is a circuit implementation of current mode buck DC/DC regulator. This circuit sense the inductor current through the capacitor C and the resistor R_{cs} path. As shown in figure11.43, the latch for the switch is set each cycle by an oscillator clock and the reset is generated by the current loop then the sensed the current (times the sensing resistor) and compared with the error voltage from the reference and the voltage feedback.

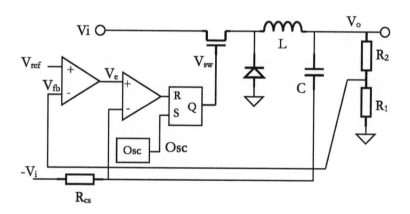

Fig 11.42 Current mode buck regulator circuit

VLSI Modulation Circuits

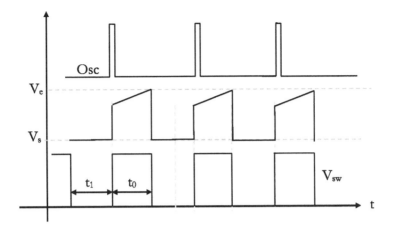

Fig 11.43 Current mode regulator circuit operation

Shown in figure 11.44 and figure 11.45 is an alternative circuit implementation of the current mode DC/DC regulator. In this circuit implementation, current sensing is realized through a replica current path through a sensing resistor Rcs.

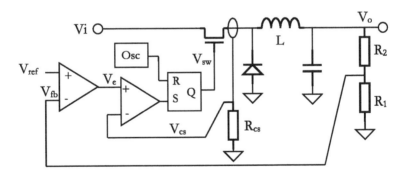

Fig 11.44 Current mode buck regulator circuit

The operatio of such circuit is shown in figure 11.45. The osciilator clock initializes the switch ON transition in each cycle. The OFF time is determined by the current sensing circuit that is controlled by an error amplifier. As the result, bith the switching current and the output voltage are regulated.

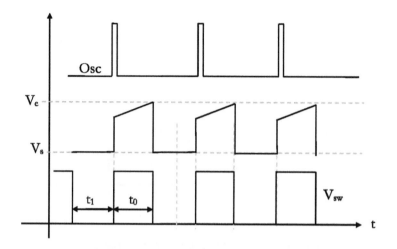

Fig 11.45 Current mode regulator circuit operation

In the similar way, the current mode DC/DC regulator can also be implemented in the boost converter as shown in figure 11.46.

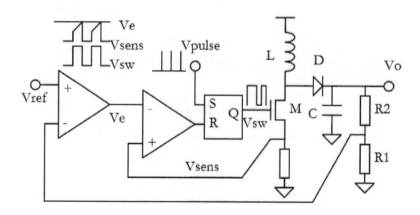

Fig 11.46 Current mode boost regulator circuit

When the control loop reaches its steady-state control in this circuit, the output of the DC/DC converter is a simply a ratio of the reference voltage independent of the load current and the input voltage supply:

$$V_o = \frac{R_1 + R_2}{R_1} V_{ref} \tag{11.47}$$

Shown in figure 11.47 is a method for sensing the inductor current in the current mode regulator employing the inductor ESR resistor and the a RC network. Assuming the RC network impedance is significantly high than the inductor path impedance. We have that

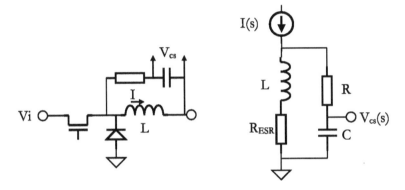

Fig.11.47 Current sensing circuit

$$V_{cs}(s) = I(s)(Ls + R_{ESR}) \frac{\frac{1}{sC}}{R + \frac{1}{sC}} \tag{11.48}$$

If we let

$$\frac{L}{R_{ESR}} = RC \tag{11.49}$$

We have that

$$I(s) = \frac{V_{cs}(s)}{R_{ESR}} \tag{11.50}$$

It offers a method to sense the inductor current through the voltage on the parallel capacitor terminals.

11.4 DC/DC CONVERTER OPERATION MODES

To avoid efficiency losses, there are three commonly used ways in the switched-inductor dc/dc converter regulators to handle light load situation including the ZCD (zero crossing detect), PFM (pulse frequency modulation), and LDO (no switching). The first two are operated on the discontinuous mode (DCM), where neither switch is conducting in a time period in the dc/dc converter.

11.4.1 DISCONTINOUS MODE

When the current in the inductor L remains positive then either the transistor S or the diode D in figure 11.48 must be conducting. For continuous conduction the voltage Vx is either V_{in} or 0.

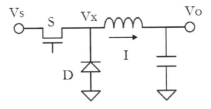

Fig.1148 Asynchronous buck converter

If the inductor current ever goes to zero then the output voltage will not be forced to either of these conditions. At this transition point the current just reaches zero as seen in figure 11.46. We have that

$$I_{Peak} = \frac{(V_s - V_o)}{L} kT \qquad (11.51)$$

During the ON time V_s-V_o across the inductor is thus at the continous-discontinous boundary the average current which must match the output current satisfies

$$I_{average} = \frac{I_{Peak}}{2} = \frac{(V_s - V_o)}{2L} t_{on} = \frac{V_s(1-k)k}{2L} T \qquad (11.52)$$

As for the continuous conduction analysis as shown in figure 11.49 and figure 11.50 we use the fact that the integral of voltage across the inductor is zero over a cycle of switching T. The transistor OFF time is now divided into segments of diode conduction $k_d T$ and zero conduction $k_o T$. The inductor average voltage thus gives

$$\frac{(V_s - V_o)}{L} kT = \frac{V_o}{L} k_d T \qquad (11.53)$$

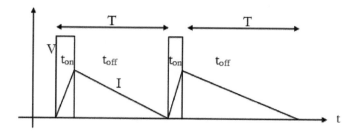

Figure 11.49 Continuous current conduction

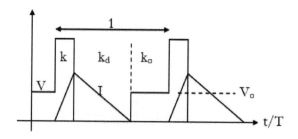

Figure 11.50 Discontinuous current conduction

$$V_o = V_s \frac{k}{k + k_d} \qquad (11.54)$$

For k +kd <1, we have that

$$I_{average} = \frac{I_{Peak}}{2}(k + k_d) \qquad (11.55)$$

The peak current is given as:

$$I_{peak} = \frac{V_o}{L} k_d T \qquad (11.56)$$

Since

$$I_o = \frac{V_o}{2L} k_d T(k + k_d) \quad (11.57)$$

$$I_o = \frac{V_i}{2L} k k_d T \quad (11.58)$$

We have that

$$k_d = \frac{2L I_o}{V_s k T} \quad (11.59)$$

And that

$$\frac{V_o}{V_s} = \frac{k^2}{k^2 + \frac{2L I_o}{V_s T}} \quad (11.60)$$

Or

$$k = \sqrt{\frac{2 L V_o I_o / T}{V_s (V_s - V_o)}} \quad (11.61)$$

As can be seen in above discussion, once the output current is high enough, the voltage ratio depends only on the duty-cycle k. At low currents the discontinuous operation tends to increase the output voltage of the converter towards V_s.

By detecting the reversal of inductor current in synchronous converter the switch S# also shuts both switches off during this time. This is known as the discontinuous mode (DCM) of operation. In non-synchronous regulators that use diodes instead of NMOS devices in the position of S#, the DCM occurs automatically.

For light loads in DCM mode, the duty-cycle is significantly lower than that in CCM mode. The analysis of currents and voltages earlier no longer apply. The MOS devices used as switches have lower limits on how fast they can be opened and closed. This minimum possible on-time (tp, min) of the PMOS device limits, in constant frequency PWM mode, the minimum load current (Io)min at which the output stays in regulation. If the output load current is reduced beyond (Io)min the output voltage will start to rise, limited only by over voltage protection or some other breakdown mechanism. This minimum load current is given by:

$$(I_o)_{min} = \frac{V_s - V_o}{2LTV_o} t_{p,min}^2 \tag{11.62}$$

11.4.2 SKIP MODE

The skip mode is a feature that allows the regulator to skip cycles to improve efficiency at light load. In the skip mode the circuit will not initiate a new charge cycle, simply allows the inductor current or inductor energy to discharge to zero. At this point the diode blocks any reverse-inductor current flow and the voltage across the inductor goes to zero such that it goes into the discontinuous operation. A new cycle is initiated only when the output voltage drops below the regulating threshold. While in skip mode and discontinuous operation, the switching frequency is proportional to the load current.

The skip mode can be easily realized in the diode rectifier switched-inductor dc/dc converter. However, the skip mode with a synchronous rectifier is, unfortunately, somewhat more complicated since inductor current can be reversed in the MOS switch if the gate is left ON. This will require a comparator to sense when the current through the inductor has been reversed and opens the switch, allowing the MOS device body diode to block the reverse current.

Skip mode may offer improved light-load efficiencies at the expense of noise as shown in figure 11.50 because the switching frequency is not fixed.

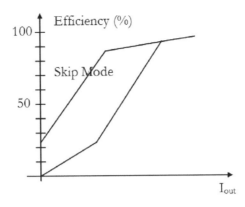

Fig. 11.50 Power efficiency improvement using skip mode

One solution is to use a forced-PWM control technique to maintain a constant switching frequency, and varies the ratio of charge cycle to discharge cycle as the operating parameters vary. Because the switching frequency is fixed, the noise spectrum is relatively narrow, thereby allowing using simple lowpass or notch filter techniques to minimize the peak-to-peak ripple voltage.

11.4.3 PFM MODE

An alternative approach to address the conversion efficiency issue is to use the pulse-frequency-modulation (PFM) technique. The PFM is based on two switching times (the maximum on-time and the minimum off-time) and two control loops (a voltage-regulation loop and a maximum peak-current, off-time loop). The PFM is characterized by the control pulses of variable frequency. The two one-shot circuits in the controller define the T_{on} (maximum on-time) and the T_{off} (minimum off-time). The Tin one-shot circuit activates the second one-shot, T_{off}. Whenever the comparator of the voltage loop detects that the output voltage is out of regulation, the Ton one-shot circuit is activated. The time of the pulse is fixed up to a maximum value. This pulse time can be reduced if the maximum peak-current loop detects that the current limit is surpassed.

The PFM mode is a type of discontinuous conduction mode. The quiescent current consumption of a PFM controller is limited only to the current needed to bias its reference and error comparator (usually tens of μA) that can be much lower than the PWM mode operation where the oscillator must be ON continuously, leading to a current consumption in the order of several milliamps. The PFM mode can be implemented with the single pulse per cycle or the multiple pulses per cycle method. By combining the PFM and the PWM modes in the same regulator and providing automatic switching, higher efficiency can be achieved over a wide output load current range.

11.4.4 LDO MODE

A LDO mode can also be included in the switched-inductor dc/dc converter for lower operation noise. In this operation mode, the dc/dc converter will serve as a linear low drop out (LDO) regulator

11.4.5 COMPARISON OF OPERATION MODES

Modern regulators may offer combined PFM and PWM and LDO operation modes. The regulator may switches the modes adaptively based on the relative priority of requirements. For example, in response to the load change, the regulator adjusts the duty-cycle or frequency depending on the PWM/PFM mode of operation. This translates to change in the currents in the various components and ultimately change in the input source current. At steady-state, the duty-cycle goes back to satisfy the $V_o = kV_i$ equation. Similar change in frequency or duty-cycle occurs in response to input supply voltage or current change, to keep the output voltage in regulation. The benefits and limitations of these operation modes are as follows:

- The PWM Mode: It offers the best efficiency at moderate to heavy load. PWM mode usually uses constant switching frequency and therefore has low switching noise.

- The PFM Mode: It offers high efficiency over a very wide range of load. PFM mode offers very low quiescent current with simple control. However it has increased output voltage ripple.

- The LDO Mode: It is a linear regulation and therefore has the lowest noise and very low quiescent current. Control loop of the LDO mode is very simple.

11.5 BEHAVIROAL DC/DC CONVERTER MODEL

The switched-inductor dc/dc converter circuit can be behaviorally modeled to speed-up the design, design analysis and design verification processes. In such modeling method, the operation of the circuit is "averaged" for each control clock period by ignoring the high frequency contents within each switching cycle. Only the envelope performance parameters are modeled in the behavioral model.

11.5.1 2-S SWITCHED-INDUCTOR MODEL

For a simple two-switch synchronous switched-inductor circuit shown in figure 11.51. By letting the steady-state inductor currents at the end of the S and S# phases are I_1 and I_2, we have that

$$V_A - V_B = L\frac{I_2 - I_1}{kT} \tag{11.63}$$

$$V_B = L\frac{I_2 - I_1}{(1-k)T} \tag{11.64}$$

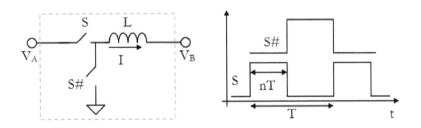

Fig. 11.51 Two-switch switched-inductor circuit model

We can see that such a two-switch switched-inductor network can be behaviorally modeled as a dc voltage transform as shown in figure 11.52.

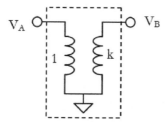

Fig.11.52 DC buck transformer model of two-switch inductor

$$V_A = kV_B \qquad (11.65)$$

Since the duty-cycle is equal to or less than 1, such a circuit realizes an A→B buck conversion.

It is interesting to see that such a synchronous circuit structure is reversible where terminal B can also be used as the input and the terminal A as the output as shown in figure 11.53 and figure 11.54.

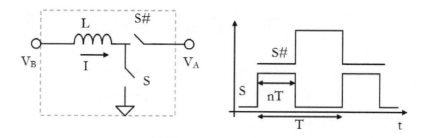

Fig.11.53 Boost switched inductor circuit

We can see that this circuit behaviorally serves as B→A boost converter.

$$V_B = V_A / k \tag{11.66}$$

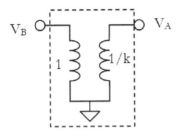

Fig.11.54 DC boost transformer model of two-switch inductor

Note that the above models can be extended into the diode rectified switched-inductor element with the S# replaced using diodes to keep the current flow in the inductors to be continuous as shown in figure 11.55 and figure 11.56. However such circuit modes are not reversible and they need to be modified when the current in the inductor is reversed (i.e. in the discontinuous current mode operation).

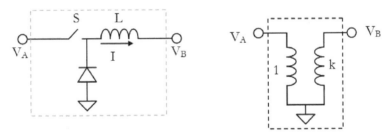

Fig.11.55 DC buck transformer model of diode rectified switched-inductor

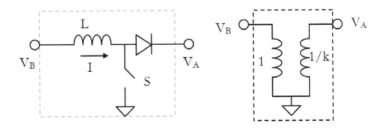

Fig.11.56 DC boost transformer model of diode rectified switched-inductor

11.5.2 4-S SWITCHED-INDUCTOR MODEL

A more general synchronous switched-inductor circuit element consisting of four switches controlled by a two phase no-overlap clocks S_1 S_2 with period T and duty-cycle k is shown in figure 11.57. In this circuit we use I to indicate the reference current direction. The actual current in the inductor can be positive or negative (reversible) based on the terminal voltage levels applied at the circuit terminals.

Under the steady-state condition the min and max currents of the inductor at the beginning and the end of the two clock phases can be expressed as:

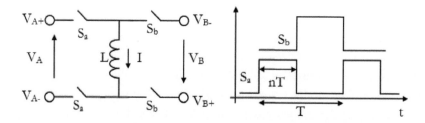

Fig. 11.57 Four-switch switched-inductor circuit structure

$$V_A = L\frac{(I_2 - I_1)}{kT} \quad (11.67)$$

$$V_B = L\frac{(I_2 - I_1)}{(1-k)T} \quad (11.68)$$

We can derive the terminal voltage relation under the steady-state condition as:

$$V_B = \frac{k}{1-k}V_A \quad (11.69)$$

It can be seen that such a switched-inductor circuit element realizes buck-boost voltage conversion that can be modeled using a floating dc voltage transform shown as figure 11.58.

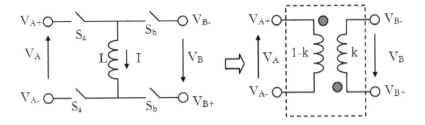

Fig.11.58 DC transformer model of four switch switched-inductor circuit

Note that since this is a synchronous switched-inductor circuit, the actual current can flow in either direction. As the result the voltage conversion in such circuit is reversible (i.e. terminal V_B can be used as the input and V_A as the output). This means that energy can transfer back and forth freely between the two terminals as soon as one of the terminal voltages falls below what specified in the steady-state terminal voltage relation equation. However when the diode elements are used to replace some switches in the above switched-inductor circuit, current may not be able to flow back and forth freely (e.g. discontinuous current mode) and the equivalent behavioral circuit mode may need to be modified to model the actual circuit operation.

11.5.3 CONVERSION BETWEEN 2-S AND 4-S MODELS

The four-switch switched-inductor circuit element model can be simplified to the two-switch switched-inductor element model when switches are only used in one terminal of the inductor. For example, if the switches are only placed on upper terminals of the inductor as shown in figure 11.59, the four-terminal transform models can be simplified as shown in figure 11.60.

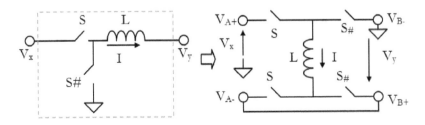

Fig.11.59 Simplified switched-inductor circuit structure

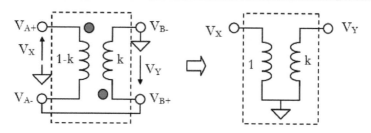

Fig.11.60 Simplified switched-inductor transformer model

For such circuit configuration, we have that

$$V_Y = kV_X \tag{11.70}$$

Since the two terminals are swappable, we can realize the step-down dc/dc conversion using terminal X as the input and Y as the output. Alternatively if Y is used as the input and X as the output, we may realize a step-up dc/dc conversion. The above model can be used to analyze the Buck converter as shown in figure 11.61.

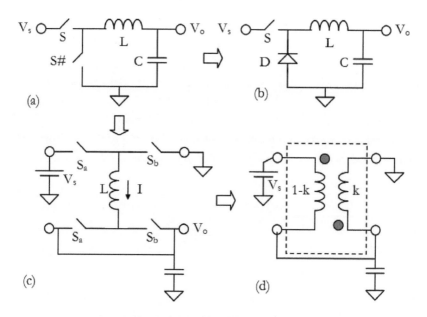

Fig.11.61 Dc transformer model of buck converter

For this circuit structure we have that

$$V_o = \frac{k}{1-k}(V_s - V_o) \tag{11.71}$$

Or

$$V_o = KV_s \tag{11.72}$$

Similarly a boost converter can also be behaviorally model using the basic switched-inductor element as shown in figure 11.62.

The output voltage of such circuit can be derived using the basically switched-inductor element as:

$$V_o = \frac{k}{1-k}V_s + V_s = \frac{1}{1-k}V_s \tag{11.73}$$

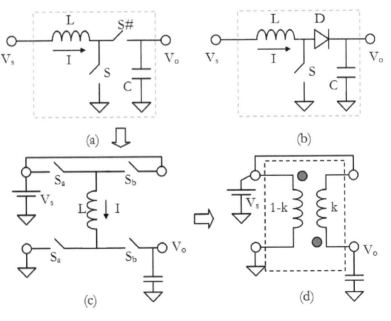

Fig.11.62 Dc transformer model of boost converter

11.5.4 VOLTAGE DOMAIN CONVERSION

By using the dc voltage transform model, the source and load resistances of a switched-inductor circuit network can be converted between the source and the load voltage domain as shown in figure 11.63.

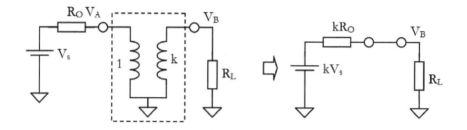

Fig.11.63 Impedance conversion based on transformer model (I)

It can be seen that when a source resistant at the input voltage domain is converted to the load voltage domain, the resistance value will be multiplied by a voltage conversion factor such that the power losses are the same in both models.

On the other hand, when the resistance is converted from the load voltage domain to the source voltage domain as shown in figure 11.64, the resistance is divided by the voltage conversion factor.

Using this dc transform model, we can analyze the buck and the boost dc/dc converter in the similar way.

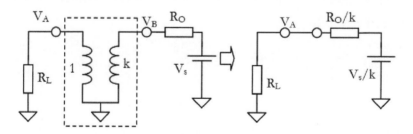

Fig.11.64 Impedance conversion based on transformer model (II)

11.5.5 STEADY-STATE DC/DC CONVERTER MODEL

Shown in figure 11.65 is an alternative steady-state model of the switched-inductor buck dc/dc converter.

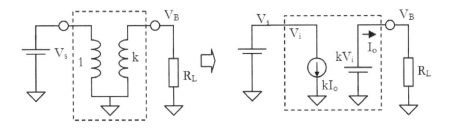

Fig.11.65 Transformer model of DC/DC converter

In this model, the input to the switched-inductor network is modeled using a dependent current source that is scaled from the output average current of the network with scaling factor set by the duty-cycle of the control clock of the dc/dc converter. On the other hand, the output of the switched-inductor network is modeled using a dependent scaled voltage source from the input voltage with scaling factor set by the duty-cycle of the control clock.

11.6 CLOSED-LOOP MODELING OF REGULATORS

A closed-loop DC/DC regulator as shown in figure 11.66 consists of a controller in the loop. The error voltage from the controller is then converted into the PWM to drive the switched-inductor network. This controller is used to compensate for regulator response and stability.

There are a few generally used regulator loop compensation types that can be used for stablizing the closed loop DC/DC converters.

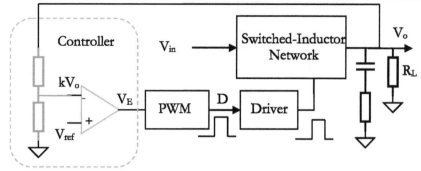

Fig.11.66 Switched-inductor DC/DC Regulator controller

- Type I compensation: In type I compensation as shown in figure 11.67 , only one pole is represented in the circuit that has maximum 90 degrees phase shift.

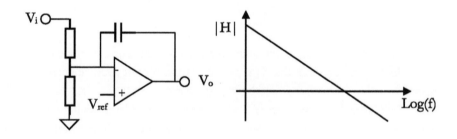

Fig. 11.67 MMM Type I loop compensation

- Type II compensation: In type II loop comensation as shown in figure 11.68 is based on single zero compensation.

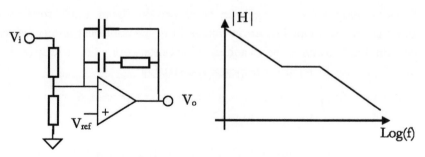

Fig. 11.68 Type II loop compensation

VLSI Modulation Circuits

- Type III compensation: In type III loop compensation as shown in figure 11.69 is based on double zero compensation

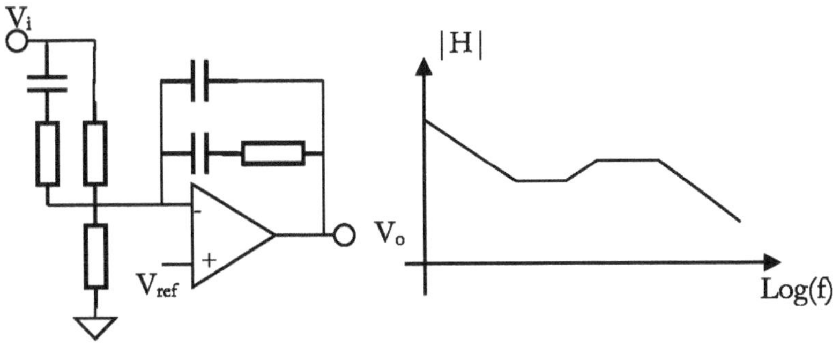

-

Fig. 11.69 Type III loop compensation

11.6.1 VOLTAGE MODE REGULATORS

Shown in figure 11.70 is a voltage-mode regulator control loop.

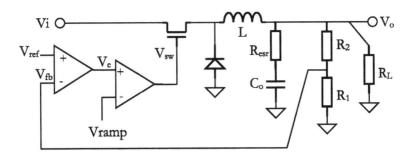

Fig 11.70 Voltage mode buck regulator circuit

The open loop model of the circuit is shown in figure 11.71

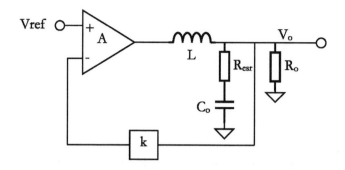

Fig 11.71 Equivalent voltage mode buck regulator circuit model

The open loop transfer function is given as:

$$H(s) = kA \frac{(R_{ESR}C_o s + 1)}{LC_o s^2 + (\frac{R}{R_o} + C_o R_{ESR})s + 1} \quad (11.74)$$

Shown in figure 11.72 is the typical Bode plot of the open-loop transfer function.

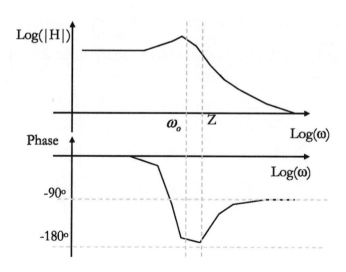

Fig.11.72 Bode plot of the voltage mode open-loop transfer function.

This transfer function has the following characteristics:

- The indutor and the load capacitor forms a two poles that might degrades the phase margin of the loop;
- The ESR will introduce a zero that can be used to compensate the loop stability if proper location of the zero is assigned related to the location of the two poles.

11.6.2 CURRENT MODE REGULATORS

Current mode regulators can be realized based on two control loops as shown in figure 11.73 Where the internal loop is a fast current mode feedback loop and the external slow loop is a voltage feedback loop.

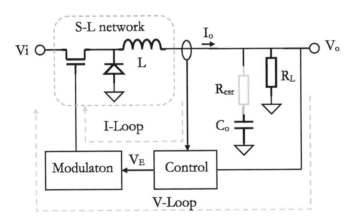

Fig 11.73 Current mode buck regulator circuit

By using the fast current loop the inductor can be eliminated from the regulator loop calculation.

Shown in figure 11.74 and figure 11.75 are a current mode DC/DC regulator and its fast current loop equivalent model. By using such model, the inductance effect is eliminated from the overall regulator loop model.

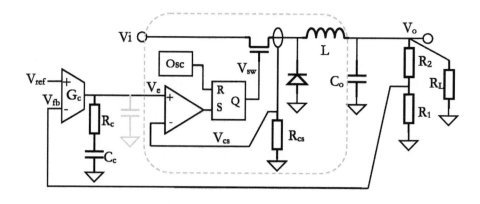

Fig 11.74 Current mode buck regulator circuit

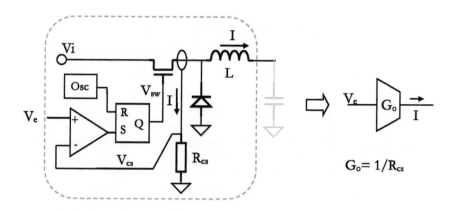

Fig 11.75 Fast current mode loop equivalent

Under this equiavlent model, the current mode loop can be eliminated from the voltage mode analysis as shown in figure 11.76.

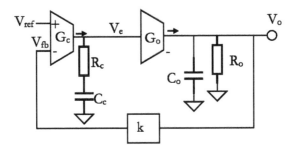

Fig 11.76 Equivalent Current mode buck regulator circuit model

The open-loop transfer function of the circuit can be expressed as:

$$H(s) = \frac{kG_c G_o R_o}{sC_c}\left(\frac{1+\frac{s}{z}}{1+\frac{s}{p}}\right) \tag{11.75}$$

Where

$$k \equiv \frac{R_1}{R_1 + R_2} \tag{11.76}$$

$$R_o \equiv R_L //(R_1 + R_2) \tag{11.77}$$

$$p \equiv \frac{1}{R_o C_o} \tag{11.78}$$

$$z \equiv \frac{1}{C_c R_c} \tag{11.79}$$

We can see that the compensation circuit provided by the RC and Cc offers the following property.

- Very high DC gain that improves the dc precision of the regulator.
- A zero in the open-loop transfer function for the phase margin the of feedback loop.

For stability of the closed-loop circuit is determined by the pole and zero location as follows:

- Case I: p= z, the pole and zero cancel each and the Bode Plots are given as figure 11.77 with 90 degree phase margin. The circuit is stable.

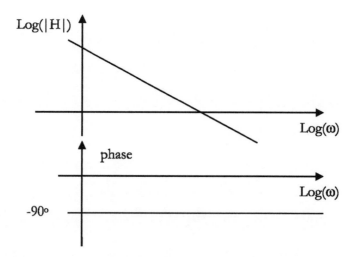

Fig.11.77 Bode Polts for p=z

- Case II (high impedance load condition): p> z, the zero comes before the pole. Bode Plots are given as figure 11.78 With 90 degree phase margin. The circuit is also stable.

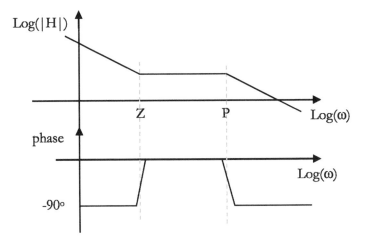

Fig.11.78 Bode Polts for p>z

- Case III (low impedance load condition): p< z, the pole comes before the zero and the Bode Plots are given as figure 11.79 With 90 degree phase margin. The circuit may not be stable. This condition suggests that the compensation circuit should compensate for the heavy load condition to ensure circuit stability.

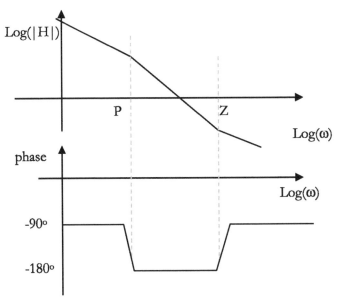

Fig.11.79 Bode Plots for p<z

11.7 CONVERTER PERFORMANCE MODELS

The performance of the dc/dc converter can be specified using several specific performance parameters.

11.7.1 ENERGY FACTOR

All power dc/dc converter circuits have a pumping circuit to transfer the energy from the source to some passive energy storage elements. The energy transfer from source to the load is not continuous that is controlled in discrete-time of clock period T. The energy transfer out of the source in per clock period can be calculated by a pumping energy (PE) factor as:

$$PE \equiv \int_0^T V_s i(t)dt = V_s \int_0^T i(t)dt = V_s I_s T \qquad (11.80)$$

Where Is is the average source current over a clock period.

The stored energy in the inductors and capacitors are given respectively as:

$$W_L = \frac{1}{2} L I_L^2 \qquad (11.81)$$

$$W_C = \frac{1}{2} C V_C^2 \qquad (11.82)$$

For the entire dc/dc converter circuit includes N_L inductors and N_C capacitor, a total energy storage parameter can be defined as:

$$SE = \sum_{j=1}^{N_L} W_{Lj} + \sum_{j=1}^{N_C} W_{Cj} \tag{11.83}$$

A capacitor/inductor stored energy ratio (CIR) can be defined as:

$$CIR \equiv \sum_{j=1}^{N_C} W_{Cj} / \sum_{j=1}^{N_L} W_{Lj} \tag{11.84}$$

The energy loss of the dc/dc convert can be defined using its energy loss per clock period as:

$$EL \equiv P_{Loss} T \tag{11.85}$$

A stored energy to pump energy ratio can be defined as:

$$EP \equiv \frac{SE}{PE} = \frac{\sum_{j=1}^{N_L} W_{Lj} + \sum_{j=1}^{N_C} W_{Cj}}{V_s I_s T} \tag{11.86}$$

11.7.2 POWER EFFICIENCY

For a pure resistive load and the voltage transfer gain of a switched-inductor dc/dc converter circuit can be defined as:

$$M \equiv \frac{V_o}{V_s} \tag{11.87}$$

The energy loss can be caused by connectors, inductor resistance, and capacitor wire and switching MOS devices. The total power loss (P_{Loss}) can be defined using the resistive loss (P_r), element loss (P_e) and the active device loss (P_d) as:

$$P_{Loss} = P_r + P_e + P_d \tag{11.88}$$

The power transfer efficiency η can be expressed as:

$$\eta \equiv \frac{P_o}{P_i} = \frac{P_i - P_{Loss}}{P_i} \tag{11.89}$$

11.7.3 TIME CONSTANT

The time constant of a dc/dc converter is defined as

$$\tau = \frac{2T \cdot EF}{1 + CIR}(1 + CIR\frac{1-\eta}{\eta}) \tag{11.90}$$

This time constant can be used to estimate the converter transient response, such as the settling time of the converter.

11.7.4 DAMPING TIME CONSTANT

The damping time constant of a switched-inductor dc/dc converter can be defined as

$$\tau_d = \frac{2T \cdot EF}{1+CIR} \cdot \frac{CIR}{\eta + CIR(1-\eta)} \qquad (11.91)$$

This time constant can be used to estimate the converter response with oscillation.

11.7.5 TIME CONSTANT RATIO

The time constant ratio of a dc/dc converter is defined as

$$\xi \equiv \frac{\tau_d}{\tau} = \frac{CIR}{\eta(1+CIR\frac{(1-\eta)}{\eta})^2} \qquad (11.92)$$

11.8 CONVERTER PARAMETER SELECTIONS

There are some generally used rules for selecting the circuit components in a VLSI switched-inductor dc/dc converter circuits based on the circuit parameters such as performance, cost and reliabilities.

Shown in figure 11.80 are current waveforms in circuit components of the given switched-inductor dc/dc converter circuit. It can be seen that when the switch S is ON, the current in the switch S is the same in the inductor. On the other hand, when the switch S# is ON, its current is the same as the inductor current. As the result, the peak current for the switches and the inductor are the same. The average (or DC) inductor current is then equal to the current going through the resistive load R. In addition, the ac inductor current is the same as current in the output capacitor C.

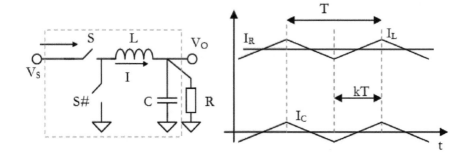

Fig.11.80 Current signal in DC/DC converter circuit elements

$$I_R = \frac{V_O}{R} = \overline{I_L} \qquad (11.93)$$

$$\Delta I_L = \frac{V_o(1-k)T}{L} \qquad (11.94)$$

11.8.1 SWITCHE SELECTION

The average and rms values of currents in the switches S and S# of the switched-inductor dc/dc converter circuit can be derived.

For the switch S, the current can be expressed as:

$$\overline{I_S} \approx k I_R \qquad (11.95)$$

$$\sqrt{\overline{I_S^2}} \approx \sqrt{k} I_R \qquad (11.96)$$

Similarly for switch S#, we have that

$$\overline{I_{S\#}} \approx (1-k)I_R \qquad (11.97)$$

$$\sqrt{\overline{I_{S\#}^2}} \approx \sqrt{1-k}\,I_R \qquad (11.98)$$

These equations can usually be used for the selection of the switches.

11.8.2 INDUCTOR SELECTION

There are a few commonly used inductor parameters for dc/dc converter inductor selection including the inductance (L), the dc resistance (DCR), the self-resonant frequency (SRF), the saturation current (Isat), and the RMS current (Irms).

- Inductance (L) is the primary functional parameter of an inductor. This is the value that is calculated by converter design to determine the inductor's ability to handle the desired output power and control ripple current.

- DC Resistance (DCR) is the resistance in a component due to the length and diameter of the winding wire used.

- Self-Resonant Frequency (SRF) is the frequency at which the inductance of an inductor winding resonates naturally with the distributed capacitance characteristic of that winding.

- Saturation Current (Isat) is the amount of current flowing through an inductor that causes the inductance to drop due to core saturation.

- RMS Current (Irms) is the amount of continuous current flowing through an inductor that causes the maximum allowable temperature rise.

A rule of thumb for selecting a value of inductance in above LC filter is to start with an inductor value that results in a peak-to-peak inductor current that is 10% of the full load current.

In the VLSI switched-inductor dc/dc converters, the inductor is usually selected with the minimum value that supports the maximum load current across the rated temperature. The temperature of an inductor in the dc/dc converter will usually increase due to effects such as self-heating and heat generated by other components in the system.

The inductor current rating is usually specified using the Isat which is the current at which the core saturates and Irms which is the current at a given temperature condition. Other conditions such as size are usually applied on top to further narrow down this selection.

A lower inductance usually results in higher inductor ripple current, so that the inductor current limit is engaged at a lower-than-expected load current. Higher ripple voltage at the output, which may be seen as excessive noise by the load. The lower crossover frequency pushing the regulator closer to instability. If an inductor is selected with lower Isat or Irms, the reduction in the inductor value at higher actual current level (e.g. Iout > Isat) could also have same issue as choosing a lower value inductor. Conversely, if an inductor with higher Isat or Irms is selected but it has higher DCR, it leads to higher dc losses in the inductor and consequently lowers overall efficiency. The higher ripple current, the first issue in the lower value inductor, also increases the ac loss in the inductor.

11.8.3 CAPACITOR SELECTION

The output capacitor provides storage capacity that reduces output voltages ripple. The input capacitor provides the storage capacity needed to reduce the input current ripple from propagating back to the source (e.g. battery) and the rest of the system. The rms value of the currents in the input and output capacitors can be derived as:

$$I_{C-RMS} = \frac{\Delta I}{\sqrt{3}} \tag{11.99}$$

We may use overshoot to get a value for C to control the output ripple voltage. This can usually be done by using a simple rule of thumb of making the characteristic impedance of the filter, Zo, equal the load resistor, which gives an overshoot of SQRT(2)= 1.41 and a slightly over-damped filter at full load. The following equation can be used to calculate the minimum value of the output capacitor as a function of the target output voltage and inductor current ripples.

$$C = \frac{\Delta I}{4f} \frac{1}{2\Delta V_o} \qquad (11.100)$$

11.8.4 LOSS CONTROL IN DC/DC CONVERTERS

There are two families of losses in a typical VLSI switched-inductor dc/dc converter including the conduction and switching losses.

- The conduction losses are driven by the equivalent resistance of switches and passive components.

- The switching losses are proportional to the switching frequency, including charging/discharging of capacitance at MOS gates and switch node, and inductive switching transitions, eddy-current and core losses.

- The inductor switching losses are due to the ripple current and equivalent series resistance (ESR) of the inductor at the switching frequency:

$$P_L = \frac{(\Delta I)^2}{3} R_{I-ESR} \qquad (11.101)$$

The power loss also includes oscillator and other miscellaneous losses in the regulator internal circuits. figure 11.81 shows models of the regulator including equivalent resistances of the switches and the inductor. These resistances contribute to the conductive losses.

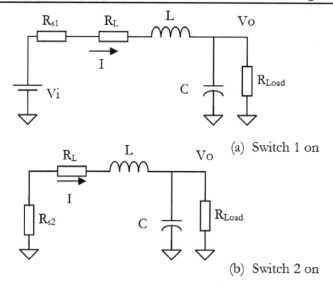

Fig.11.81 Losses in DC/DC converter circuit

The conduction loss in the dc/dc converter can be modeled using the steady-state dc/dc converter as shown in figure 11.82.

At low output current, majority of the losses are usually due to ripple current. At high output, the conduction losses dominate. The loss in the inductor is given as:

$$P_L = (I_R)^2 R_{I-DCR} \qquad (11.102)$$

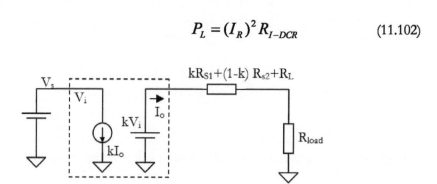

Fig.1182 Modeling of DC/DC converter resistive loss

11.8.5 ZERO VOLTAGE SWITCHING (ZVS)

Shown in figure 11.83 Is the switching capacitance loss model of the DC/DC converter.

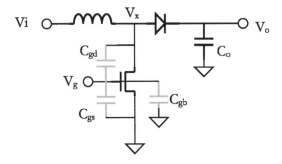

Fig 11.83 Equivalent power MOS gate capacitance model

The effective gate capacitance of the power gate can be modeled based on the gate capacitances and the Miller capacitance as:

$$(C_g)_{eq} = C_{gs} + C_{gd} + (1 + \frac{V_x}{V_{gs\,max}})C_{gd} \qquad (11.103)$$

It can be seen that voltage Vx contributes to the power NMOS gate loss due to the Miller capacitor effect. One method to reduce this is to employ the zero voltage switching (ZVS) technique where Vx is close to zero when the switch is turned on.

References:

[1] Jian Sun; Mitchell, D.M.; Greuel, M.F.; Krein, P.T.; Bass, R.M.; "Averaged modeling of PWM converters operating in discontinuous conduction mode," IEEE Transactions on Power Electronics, Volume: 16, Issue: 4, Page(s): 482 - 492, 2001.

[2] Bibian, S.; Hua Jin; "High performance predictive dead-beat digital controller for dc power supplies," IEEE Transactions on Power Electronics, Volume: 17, Issue: 3, Page(s): 420 - 427, 2002.

[3] Thottuvelil, V.J.; Verghese, G.C.; "Analysis and control design of paralleled dc/dc converters with current sharing," IEEE Transactions on Power Electronics, Volume: 13, Issue: 4 Page(s): 635 - 644, 1998.

[4] Siew-Chong Tan; Y. M Lai, C. K. Tse, M. K. H. Cheung, "Adaptive feedforward and feedback control schemes for sliding mode controlled power converters," IEEE Transactions on Power Electronics, Volume: 21, Issue: 1, Page(s): 182 - 192, 2006.

[5] On-Cheong Mak; Yue-Chung Wong; Ioinovici, A.; "Step-up dc power supply based on a switched-capacitor circuit," IEEE Transactions on Industrial Electronics, Volume: 42, Issue: 1, Page(s): 90 - 97, 1995.

[6] Mattavelli, P.; Rossetto, L.; Spiazzi, G.; Tenti, P.; "General-purpose sliding-mode controller for dc/dc converter applications," 24th Annual IEEE Power Electronics Specialists Conference, 1993. Page(s): 609 - 615

[7] K. M. Smith, Jr. K. M. Smedley, "Properties and synthesis of passive lossless soft-switching PWM converters," IEEE Transactions on Power Electronics, Volume: 14, Issue: 5, 1999, Page(s): 890 - 899.

[8] Jain, P.K.; St-Martin, A.; Edwards, G.; "Asymmetrical pulse-width-modulated resonant dc/dc converter topologies," IEEE Transactions on Power Electronics, Volume: 11, Issue: 3, 1996, Page(s): 413 - 422

[9] Jian Sun; Grotstollen, H.; "Symbolic analysis methods for averaged modeling of switching power converters," IEEE Transactions on Power Electronics, Volume: 12, Issue: 3, 1997, Page(s): 537 - 546

[10] Liu, K.-H.; Lee, F.C.Y.; "Zero-voltage switching technique in dc/dc converters," IEEE Transactions on Power Electronics, Volume: 5, Issue: 3, 1990, Page(s): 293-304.

[11] Changrong Liu; Johnson, A.; Jih-Sheng Lai; "A novel three-phase high-power soft-switched dc/dc converter for low-voltage fuel cell applications," Industry Applications, IEEE Transactions on, 2005, Page(s): 1691-1697.

[12] Canales, F.; Barbosa, P.; Lee, F.C.; "A zero-voltage and zero-current switching three-level dc/dc converter," IEEE Transactions on Power Electronics, Volume: 17, Issue: 6, 2002, Page(s): 898-904.

[13] Yaow-Ming Chen; Yuan-Chuan Liu; Sheng-Hsien Lin; "Double-Input PWM dc/dc Converter for High-/Low-Voltage Sources," Industrial Electronics, IEEE Transactions on Volume: 53, Issue: 5, 2006, Page(s): 1538-1545.

[14] Emadi, A.; "Modeling of power electronic loads in ac distribution systems using the generalized State-space averaging method," IEEE Transactions on Industrial Electronics, Volume: 51, Issue: 5, 2004, Page(s): 992-1000.

[15] Bech, M.M.; Pedersen, J.K.; Blaabjerg, F.; Trzynadlowski, A.M.; "A methodology for true comparison of analytical and measured frequency domain spectra in random PWM converters," IEEE Transactions on Power Electronics, Volume: 14, Issue: 3, 1999, Page(s): 578-586.

[16] Kutkut, N.H.; Divan, D.M.; Gascoigne, R.W.; "An improved full-bridge zero-voltage switching PWM converter using a two-inductor rectifier," IEEE Transactions on Industry Applications, Volume: 31, Issue: 1, 1995, Page(s): 119 - 126.

[17] J. Czogalla, Jieli Li; C. R.Sullivan, "Automotive application of multi-phase coupled-inductor DC-DC converter," Conference Record of the Industry Applications Conference, 2003. 38th IAS Annual Meeting, Volume: 3, 2003, Page(s): 1524 - 1529.

[18] X. Xu, A. M. Khambadkone, T.M. Leong, R. Oruganti, "A 1-MHz Zero-Voltage-Switching Asymmetrical Half-Bridge dc/dc Converter: Analysis and Design," IEEE Transactions on Power Electronics, Volume: 21, Issue: 1, 2006, Page(s): 105-113.

[19] Qun Zhao; Fengfeng Tao, F. C. Lee, Peng Xu, Jia Wei, "A simple and effective method to alleviate the rectifier reverse-recovery problem in continuous-current-mode boost converters," IEEE Transactions on Power Electronics, Volume: 16, Issue: 5, 2001, Page(s): 649-658.

[20] A. Emadi, M. Ehsani, "Negative impedance stabilizing controls for PWM DC-DC converters using feedback linearization techniques," 35th

Intersociety Energy Conversion Engineering Conference and Exhibit, 2000. (IECEC), Volume: 1, 2000, Page(s): 613 - 620.

CHAPTER 12
INTRODUCTION TO VLSI MIXER CIRCUITS

The VLSI mixer circuits that serve as the frequency translators are critical components for radio frequency (RF) and other electronic systems. In the RF applications, the baseband signals (such as audio band signals) are usually translated into dedicated radio frequency (RF) bands that are suitable for transmission in free air. At the receiver end, on the other hand, the received RF signals are frequency translated to basebands to recover original information. Such frequency up- and down-conversion operations are realized using the VLSI mixer circuits.

In general, a mixer converts signal power at one frequency into another frequency to ease signal processing or to enhance the quality of the signal processing and transmission. A fundamental reason for this frequency conversion is to allow handling the signal at a frequency that is suitable to actual VLSI circuit implementation.

12.1 VLSI MIXER FUNDAMENTAL

In the simplest form, a mixer is a multiplier. The mixer multiplies the input signals to produce output signal at new frequency bands. Such mixing operations are commonly used as the modulator and the demodulator in the transmitter and receiver paths of a communication system, where baseband signal is converted into RF signal and vice versa.

A basic single-end mixer is a 3-port active or passive device. It is designed to output both the sum and the difference of frequencies of two distinct input frequencies at the two input ports. The VLSI Mixer circuits are usually used as modulator, de-modulator or as phase detector. In wireless communication systems, the two signals inserted into the two input ports are usually the local oscillator (LO) signal, and the incoming (for the receiver path) or outgoing (for the transmitter path) signal. After the mixing process, the output signal is frequency shifted down by the LO frequency in the down-conversion mixer and shifted up by the LO frequency in an up-conversion mixer.

A mixer based on the analog multiplication is shown in figure 12.1. The output signal of the mixer contains two frequency terms including the sum and the difference of the input signal frequencies given as

$$V_o = A_1 \cos(\omega_1 t) A_2 \cos(\omega_2 t) = \frac{A_1 A_2}{2}[\cos(\omega_1 + \omega_2)t + \cos(\omega_1 - \omega_2)t] \quad (12.1)$$

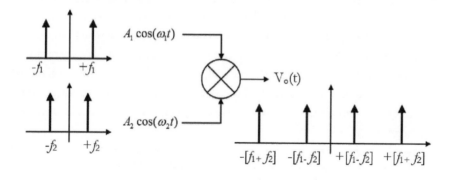

Fig.12.1 Analog mixer based on multiplication

In this mixer circuit, if one of the signal is a reference clock with a precisely controlled clock frequency that is usually provided by a local oscillator (LO), then the frequency band of the another mixer input terminal can be offset (shifted) by the frequency of the LO.

12.2 MIXER PREFORMANCE PARAMETERS

Although VLSI mixers are based on nonlinear operation of the VLSI circuit components, VLSI mixers are fundamentally linear devices in the signal path. The mixer shifts a signal from one frequency to another, keeping (faithfully) the properties of the initial signal (phase and amplitude), and therefore doing a linear transformation.

A VLSI mixer can be characterized by several commonly used performance parameters, such as the conversion gain, the linearity, the isolation, the dc offset and the noise level (spurs).

12.2.1 CONVERSION GAIN

The conversion gain is defined as the ratio of the output signal power to the input signal power. A conversion gain is usually specified in dB format. Another terminology that is commonly used for similar purpose is the conversion loss that is the inverse of the conversion gain.

$$G(dB) = 10\log(\frac{P_o}{P_i}) \qquad (12.2)$$

$$L(dB) = -10\log(\frac{P_o}{P_i}) \qquad (12.3)$$

Note that for a single sideband system, only one sideband is used. Therefore there is always a 3dB loss associated with such mixer structure. The additional loss can be caused by the circuit components.

12.2.2 NOISE FIGURE

Noise will usually be added to the signal during the frequency translation in VLSI mixer circuit. The noise factor is defined as the ratio of the input signal-to-noise ratio to the output signal-to-noise ratio at hot condition to describe the noise performance of the mixer circuit:

$$NF = 10\log(\frac{SNR_i}{SNR_o}) \qquad (12.4)$$

Note that the LO input power may also impact the noise factor of the mixer, where the mixer noise factor usually increase as the LO power decreases.

For a cascaded system the Friis noise figure equation for total noise figure of the cascaded devices is given as

$$F = F_1 + \frac{F_2 - 1}{G_1} + \frac{F_3 - 1}{G_1 G_2} + \frac{F_4 - 1}{G_1 G_2 G_3} + \ldots \qquad (12.5)$$

Note that all terms in the equation are numeric ratios and are not in dB. The overall noise figure for a cascade, expressed in dB is given as:

$$F_{dB} = 10\log(F). \qquad (12.6)$$

12.2.3 SIGNAL ISOLATION

The isolation is a parameter that is usually used to specify the isolation between the three ports (RF and IF, the LO and IF and the LO and RF ports) of the mixer. The isolation specifies the amount of power that leaks from one port to the other. Isolation is critical since the input signal at one port may impact the other and cause the performance degradation. For example, LO leakage may impact the incoming RF signal by interfering the LNA.

12.2.4 LINEARITY

Since mixer usually dealing with high power in the RF system, mixer linearity is important and it determines the system performance parameters of overall receiver or transmitter of a RF system. Due to mixer non-linearity, the mixer gain may degrade at high input signal power due to signal harmonic components. One commonly used way to specify the mixer linearity is to specify the signal power point where the gain drops by 1dB as shown in figure 12.2. An alternative method is to specify the power of the harmonic. The third-order intercept (IP3) is usually defined as the point the third order intermodulation distortion to intercept the input signal power.

The conversion compression is the RF input level above which the RF versus IF output curve deviates from linearity. Above this level, additional increases in the RF input level do not result in equal increases in the output level. The input power level at which the conversion loss increases by 1 dB is called the 1 dB compression point.

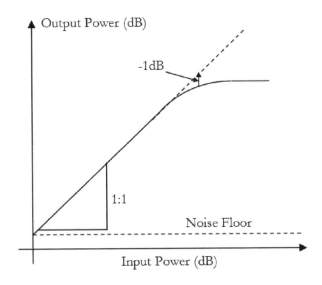

Fig.12.2 P1dB point of the mixer

The intercept point is a figure of merit for intermodulation product suppression of the mixer circuit. There are two commonly used specifications for such performance parameter: the input and the output intercept points (IIP and OIP, respectively). The input intercept point is the level of input RF power at which the output power levels of the undesired intermodulation products and IF products would be equal, that is, intercept each other if the mixer did not compress. As input RF power increases, the mixer compresses before the power level of the intermodulation products can increase to equal to the IF output power. So, the input and the output intercept points are theoretical and that are calculated by extrapolating the output power of the intermodulation and the IF products pass the 1-dB compression point until they equal each other. A high intercept point is desirable because it means the mixer can handle more input RF power before causing undesired products to rival the desired IF output product and essentially means that the mixer has a greater dynamic range. Dynamic range, 1-dB compression point, and intercept point are all interrelated, but it has been shown that, in general, no dB-for -dB rule of thumb exists to easily correlate 1-dB compression point with intercept point. The concept of intercept point can be applied to any intermodulation product, however, it normally refers to the two-tone, third-order intermodulation products. If the two input RF signals are incident at the mixer RF port, they cause the mixer to generate the following two-tone intermodulation products:

$$(\pm m_1 f_1 \pm m_2 f_2) \pm n f_{LO} \qquad (12.7)$$

where m_1, m_2, $n = 0, 1, 2, 3, \ldots$, m and n are integers. The two-tone, third-order intermodulation products have the following frequencies:

$$(\pm 2 f_1 \pm f_2) \pm f_{LO} \qquad (12.8)$$

$$(\pm f_1 \pm 2 f_2) \pm f_{LO} \qquad (12.9)$$

They are multiple third-order products because the sum of coefficients of f_1 and f_2 equals to 3. Notice that the order of intermodulation products refers only to coefficients of the RF inputs and does not include that of the LO. The order of the intermodulation product is important because a 1-dB change in the power level of each input RF signal causes the power level of each intermodulation product to change by an amount of dB equal to its order. A 1-dB change in power of each of the two input RF signals causes the power level of each two-tone third-order product to change by 3 dB. In general, the n-th order interception point is defined as

$$IIP = \frac{Inter\,modulation\,Suppresion(dBc)}{(order - 1)} + Input\,RF\,Power(dBm)$$

(12.10)

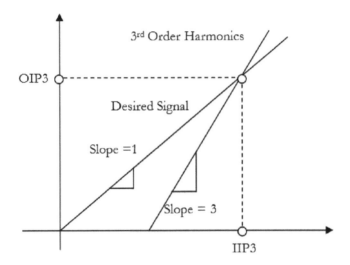

Fig.12.3 3rd order intercept point definition

The output and input intercept are related by the mixer conversion loss, or gain (for active mixers) as:

$$OIP(dBm) = IIP(dBm) + Mixer\,Conversion\,Gain(dB) \qquad (12.11)$$

12.2.5 SPURS

The mixing operation generates products of sum and difference of input as well as many unwanted spurious signal at the mixer outputs. These spurious signals may significantly impact the wideband system. For narrow band systems, only these in the passband will impact the system performance. With adequate bandpass filtering, most undesired spurs can be effectively attenuated. However, the IMD3 product is close to the desired signal so it is very difficult to filter out in real system.

12.2.6 DYNAMIC RANGE

The dynamic range is the amplitude range over which a mixer can operate without degradation of the performance. It is bounded by the conversion compression point for high input signals, and by the noise figure of the mixer for low level input signals. Since the thermal noise of each passive mixer is about the same, the conversion compression point normally determines the passive mixer's dynamic range. The 1-dB compression point is generally taken to be the top of the dynamic range of a mixer because the input RF power that is not converted into desired IF output power is instead converted into heat and higher order intermodulation products. The intermodulation products that begin to appear when RF power is increased beyond the 1-dB compression point can begin to obscure the desired IF output. Generally, the 1-dB compression point is 5 to 10 dB lower than the LO power, so a high level mixer has a higher 1-dB compression point than a low level mixer, and therefore a wider dynamic range.

12.2.7 IMAGE

The image is the signal that is located in the RF port with the frequency that is two times of the IF away from the signal itself. Such a signal will be overlapped with the desired signal in IF port after down conversion of mixer. The image signal can be minimized or eliminated using image rejection filter at RF or through a polyphase (I/Q) mixer or filtering in mixer or at the IF band.

12.2.8 DC OFFSET

The DC offset is a measure of the imbalance in mixer. The dc offset can be caused by the LO leakage and the device mismatch in the mixer circuit implementation. Isolation may impact the mixer dc offset by minimizing the self-mixing effect. This dc offset specification is important in the I/Q modulation and demodulation, especially in the zero-IF system using these modulation/demodulation schemes.

12.2.9 LO DRIVE LEVEL

Since the input power of the LO to mixer may impact the operation and performance of mixer. The insufficient or too high drive levels can degrade the performance of the mixer and entire system.

12.2.10 VOLTAGE STANDING WAVE RATIO

The VSWR is the measure of the impedance mismatch offered to the system by the mixer. It is usually specified over a given bandwidth as a function of LO power and temperature. The VSWR can be expressed as:

$$VSWR = \frac{1+|\rho|}{1-|\rho|} \tag{12.12}$$

where ρ, Z_L and Z_o are the reflection coefficient, the input impedance of the mixer, and the characteristic impedance of the system respectively. The reflection coefficient is given as:

$$\rho = \frac{Z_L - Z_o}{Z_L + Z_o} \tag{12.13}$$

Since VSWR does not include the phase of the reflection coefficient, the system designer may know if the input impedance is above or below the normal 50 Ω characteristic impedance. For example, if the LO port VSWR is 2:1, measured in a 50 Ω system, we do not know if the LO port input impedance is 25 Ω or 100 Ω since both these impedances give a VSWR of 2:1. Actually, the input impedance of a broadband mixer swept over a frequency range of an octave, or more, usually rotates through the low and high impedances, roughly producing a circle centered at 50 Ω in a Smith chart. Therefore, a given mixer having a LO VSWR of 2:1 over an octave bandwidth will have an input impedance varying from 25 Ω to 100 Ω, passing through an infinite number of complex impedance combinations as the LO frequency changes. The VSWR of the RF, LO, and IF ports are direct functions of the LO power. Change in the LO power alters the mixer operating point, resulting in a different impedance of all mixer ports, causing a corresponding change in VSWR. RF input power, which is usually much lower than LO input power, does not appreciably change the mixer bias point and consequently, has little effect on VSWR. When the mixer impedance changes, the input impedances of all three ports change. Hence, varying the LO power level will affect he VSWR of all three ports.

12.3 VLSI MIXER CIRCUIT EXAMPLES

Shown in figure 12.4 is an amplitude modulation (AM) super heterodyne radio RF front end example with a tunable input RF preselection. This tunable radio circuit allows both 1490kHz and 2400kH RF band to be down converted to the same IF frequency at 455kHz. This is realized by simultaneously tuning the LO at a constant difference of 455 kHz over the entire AM radio band of 550 - 1600 kHz to minimize the inference effects.

The second mixer circuit example shown in figure 12.5 is a FM radio employing a carrier frequency of 88 - 108 MHz. This radio does not use a 455kHz IF frequency since the image frequency would be only 910kHz from the desired FM station. It would be difficult to design a tuned RF amplifier in the 88 - 108 MHz range that rejects a station only 910kHz away from the desired signal since 910kHz is only about 1% different in frequency than the desired FM station. As the result, an IF of 11.7MHz is normally used in FM radios to allow adequate image rejection to be achieved by the tuned RF amplifier in the 88 - 108MHz band with reduced selectivity because of the higher IF bandwidth associated with a 11.7MHz IF filter.

VLSI Modulation Circuits

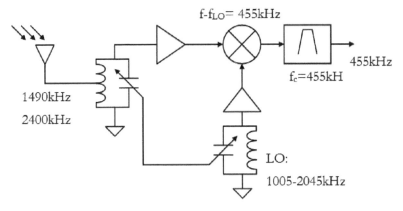

Fig12.4. AM Super heterodyne radio front end

The double conversion receiver as shown in figure is appropriate for VHF narrowband AM or FM operation and uses a 10.7MHz first IF for good image rejection, and a 455 kHz second IF for good selectivity.

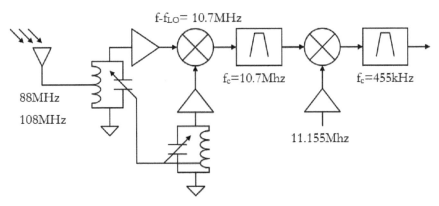

Fig12.5. FM Super heterodyne radio front end

12.4 VLSI MIXER CIRCUIT IMPLEMENTATIONS

VLSI mixers have been implemented in a few ways such as the single-device mixer, the single balanced mixer, and the double balanced mixer.

12.4.1 SINGLE DEVICE MIXER

It is interesting to observe that any general three terminal nonlinear VLSI components may provide certain mixing operation. This observation can be used to develop VLSI mixer circuit for various radio frequency and other applications. A VLSI single-device mixer uses one nonlinear component (such as diode or transistor) for the frequency translation operations.
Shown in figure 12.6 and 12.7 are two single device VLSI mixer circuit structures employing the diode devices.

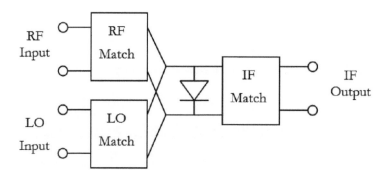

Fig12.6. VLSI diode based single device mixer (I)

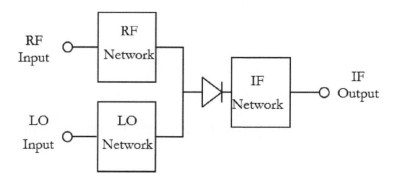

Fig12.7 VLSI diode based single device mixer (II)

Similarly, the nonlinear characteristics of the MOS device can also be used to construct the single device mixer circuit as shown in figure 12.8 and figure 12.9.

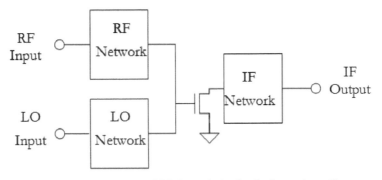

Fig12.8 VLSI MOS based single device mixer (I)

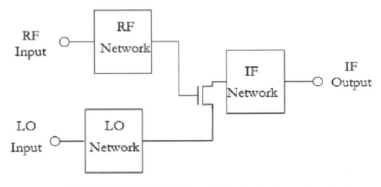

Fig12.9 VLSI MOS based single device mixer (II)

Note that the VLSI single device mixer has the disadvantage of not attenuating local oscillator AM noise and it always requires an injection filter. Single-device mixers therefore need to follow some general design rules for best performance. To get the maximum conversion gain, the LO node should be a short circuit at the RF and IF frequencies, while the RF node should be a short circuit at the LO frequency to prevent the LO leakage into the RF port.

12.4.2 BALANCED MIXER

The balanced VLSI mixers can be divided into two sub-families including the singly-balanced mixers (SBM) and the doubly-balanced mixers (DBM). The singly-balanced mixers use two devices, and are usually realized as two single device mixers connected via a 180-degree or 90-degree hybrid. The doubly-balanced mixers usually consist of four un-tuned devices interconnected by multiple hybrids, transformers or baluns. Shown in figure 12.10 and figure 12.11 are two VLSI balanced mixer structures.

The advantages of the balanced mixers over the single-device mixers include the capability to reject spurious responses and intermodulation products, better LO-to-RF, RF-to-IF and LO-to-IF isolation and better rejection of AM noise in the LO path. One major disadvantage of the balanced mixers is their higher LO power requirements. The balanced mixers are often used to separate the RF and LO ports when their frequencies overlap and filtering is impossible.

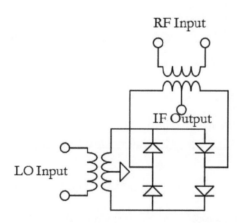

Fig12.10 VLSI diode based balanced mixer

VLSI Modulation Circuits

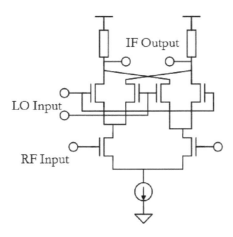

Fig12.11 VLSI MOS based balanced mixer

The doubly-balanced mixers usually have higher conversion loss (or lower gain) than the singly balanced mixers and lower limit in maximum frequencies. But they have broader bandwidth. The two most common types of doubly balanced mixers are the ring mixer and the star mixer.

Shown in figure 12.12 is a VLSI circuit implementation of the balanced mixer structure. This allows to directly integrating the Rx receiver to provide the audio frequency baseband signals.

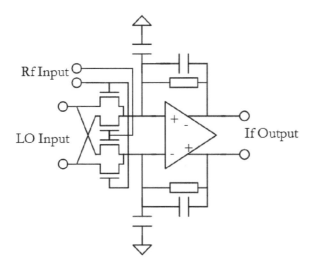

Fig12.12 VLSI integrated balanced mixer

12.4.3 IMAGE-REJECTION MIXER

The VLSI image-reject mixers employ the polyphase signal and circuit concepts that use a pair of balanced mixers. A VLSI image rejection mixer is especially useful for applications where the image and RF bands overlap, or the image is too close to the RF to be rejected by a RF filter. In the image rejection mixer as shown in figure 12.13, the LO ports of the balanced mixers are driven by a polyphase LO polyphase clocks that have 90 degrees phase difference.

In the image rejection mixer operation, I/Q (or polyphase) modulators are useful in discriminating and removing the lower sideband or the upper sideband images generated during frequency conversion, especially when sidebands are very close in frequency and the attenuation of one of the sidebands cannot be achieved with RF filtering. With an I/Q modulator, one of the sidebands is attenuated along with its carrier. The image rejection mixer can be used to enable the VLSI integration of the RF front-end employing the VLIF RF circuit architecture that avoids using the high-quality filter circuit at high frequency bands.

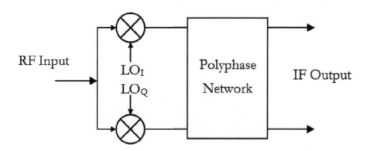

Fig12.13 VLSI image rejection mixer

12.4.4 SUB-HARMONIC MIXER

A sub-harmonic mixer uses the harmonics of the mixed RF and LO that allows the LO to operate at much lower frequency than the RF carrier. Such type of mixers are useful at higher frequency applications where it is difficult to produce a high quality LO signal at or near the carrier frequency with low phase noise, wide tuning range and high output power.

12.4.5 PHASE DECTION MIXER

One special application of the VLSI mixer is to detect the phase of a clock signal within electronic circuits, such as the VLSI phase locked-loop circuits. Since the output of a double-balanced mixer contains the sum and difference of the frequencies of the two input ports and if the two signals have identical frequencies, then their difference is a dc signal that is related to the phase difference of the two input clocks. This mixer property can be used for detecting the phase difference of two clock signals.

References:

[1] Loke Kun Tan; Samueli, H.; "A 200 MHz quadrature digital synthesizer/mixer in 0.8 μm CMOS," IEEE Journal of Solid-State Circuits, Volume: 30, Issue: 3, 1995, Page(s): 193-200.

[2] Safarian, A.Q.; Yazdi, A.; Heydari, P.; " Design and analysis of an ultra-wide band distributed CMOS mixer," IEEE Transactions on Very Large Scale Integration (VLSI) Systems, Volume: 13, Issue: 5, 2005, Page(s): 618-629.

[3] Ellinger, F.; Rodoni, L.C.; Sialm, G.; Kromer, C.; von Buren, G.; Schmatz, M.L.; Menolfi, C.; Toifl, T.; Morf, T.; Kossel, M.; Jackel, H.; "30-40-GHz drain-pumped passive-mixer MMIC fabricated on VLSI SOI CMOS technology," IEEE Transactions on Microwave Theory and Techniques, Volume: 52, Issue: 5, 2004, Page(s): 1382 - 1391.

[4] Kishore Rama Rao; Wilson, J.; Ismail, M.; "A CMOS RF front-end for a multistandard WLAN receiver," IEEE Microwave and Wireless Components Letters, Volume: 15, Issue: 5, 2005, Page(s): 321 - 323.

[5] Ellinger, F.; "26.5–30-GHz Resistive Mixer in 90-nm VLSI SOI CMOS Technology With High Linearity for WLAN," IEEE Transactions on Microwave Theory and Techniques, Volume: 53, Issue: 8, 2005, Page(s): 2559-2565.

[6] Stubbe, F.; Kishore, S.V.; Hull, C.; Delta Torre, V.; "A CMOS RF-receiver front-end for 1 GHz applications," Digest of Technical Papers, 1998 Symposium on VLSI Circuits, 1998. Page(s): 80 - 83.

[7] Svelto, F.; Deantoni, S.; Montagna, G.; Castello, R.; " Implementation of a CMOS LNA plus mixer for GPS applications with no external components," IEEE Transactions on Very Large Scale Integration (VLSI) Systems, Volume: 9, Issue: 1, 2001, Page(s): 100-104.

[8] Stanic, N.; Kinget, P.; Tsividis, Y.; "A 0.5 V 900 MHz CMOS Receiver Front End," Digest of Technical Papers. 2006 Symposium on VLSI Circuits, 2006, Page(s): 228-229.

[9] Behbahani, F.; Kishigami, Y.; Leete, J.; Abidi, A.A.; "CMOS 10 MHz-IF downconverter with on-chip broadband circuit for large image-suppression," Digest of Technical Papers. 1999 Symposium on VLSI Circuits, 1999, Page(s): 83-86.

[10] Karimi-Sanjaani, A.; Sjoland, H.; Abidi, A.A.; "A 2 GHz merged CMOS LNA and mixer for WCDMA," Digest of Technical Papers. 2001 Symposium on VLSI Circuits, 2001, Page(s): 19 - 22.

[11] Liscidini, A.; Brandolini, M.; Sanzogni, D.; Castello, R.; "A 0.13 µm CMOS front-end for DCS1800/UMTS/802.11b-g with multi-band positive feedback low noise amplifier," Digest of Technical Papers. 2005 Symposium on VLSI Circuits, 2005, Page(s): 406-409.

[12] Tan, L.K.; Samueli, H.; "A 200-MHz quadrature digital synthesizer/mixer in 0.8-µm CMOS," Proceedings of the IEEE 1994 Custom Integrated Circuits Conference, 1994, Page(s): 59-62.

[13] Samavati, H.; Rategh, H.R.; Lee, T.H.; "A 12.4 mW CMOS front-end for a 5 GHz wireless-LAN receiver, "1999. Digest of Technical Papers. 1999 Symposium on VLSI Circuits, 1999, Page(s): 87 - 90

[14] Montagna, G.; Castello, R.; Tonietto, R.; Valla, M.; Bietti, I.; "A 72mW CMOS 802.11a direct conversion receiver with 3.5dB NF and 200kHz 1/f noise corner," Digest of Technical Papers. 2004 Symposium on VLSI Circuits, 2004, Page(s): 16-19.

[15] Ugajin, M.; Yamagishi, A.; Kodate, J.; Harada, M.; Tsukahara, T.; "A 1-V CMOS/SOI bluetooth RF transceiver for compact mobile applications," Digest of Technical Papers. 2003 Symposium on VLSI Circuits, 2003, Page(s): 123 - 126.

[16] Komurasaki, H.; Sate, H.; Sasaki, N.; Miki, T.; "A 2V 1.9GHz Si Down-mixer With LC Phase Shifter," Digest of Technical Papers., 1997 Symposium on VLSI Circuits, 1997, Page(s): 91 - 92.

[17] Mitomo, T.; Fujimoto, R.; Ono, N.; Tachibana, R.; Hoshino, H.; Yoshihara, Y.; Tsutsumi, Y.; Seto, I.; "A 60-GHz CMOS Receiver with Frequency Synthesizer, "2007 IEEE Symposium on VLSI Circuits, 2007, Page(s): 172-173.

[18] Mirzaei, A.; Chen, X.; Yazdi, A.; Chiu, J.; Leete, J.; Darabi, H.; "A frequency translation technique for SAW-Less 3G receivers, "2009 Symposium on VLSI Circuits, 2009, Page(s): 280 - 281.

[19] Zhang, Ning; Kenneth, K. O; "CMOS frequency generation system for W-band radars, "2009 Symposium on VLSI Circuits, 2009, Page(s): 126-127.

[20] Sarkar, S.; Sen, P.; Raghavan, A.; Chakarborty, S.; Laskar, J.; " Development of 2.4 GHz RF transceiver front-end chipset in 0.25 μm CMOS," Proceedings of 16th International Conference on VLSI Design, 2003. Page(s): 42-47.

[21] Maxim, A.; Johns, R.; Dupue, S.; "0.13μ CMOS hybrid TV tuner using a calibrated image and harmonic rejection mixer, "2007 IEEE Symposium on VLSI Circuits, 2007, Page(s): 206-207.

[22] Levantino, S.; Wang, X.; Andreani, P.; Samori, C.; Lacaita, A.L.; "A circuit technique improving the image rejection of RF front-ends, "2004. Digest of Technical Papers. 2004 Symposium on VLSI Circuits, 2004, Page(s): 368-371.

[23] Klumperink, E.A.M.; Louwsma, S.M.; Wienk, G.J.M.; Nauta, B.; "A 1 Volt switched transconductor mixer in 0.18 μm CMOS," Digest of Technical Papers. 2003 Symposium on VLSI Circuits, 2003, Page(s): 227 - 230.

[24] Jie Long; Weber, R.J.; "A low voltage, low noise CMOS RF receiver front-end," Proceedings. 17th International Conference on VLSI Design,2004, Page(s): 393-397.

[25] Alam, S.K.; "A 2 GHz Low Power Down-conversion Quadrature Mixer in 0.18-μm CMOS, "20th International Conference on VLSI Design,2007, Page(s): 146 - 154

[26] Pui-In Mak; Martins, R.P.; "High-/Mixed-Voltage RF and Analog CMOS Circuits Come of Age," IEEE Circuits and Systems Magazine, Volume: 10, Issue: 4, 2010, Page(s): 27 - 39

[27] Full-CMOS 2.4 GHz wideband CDMA transmitter and receiver with direct conversion mixers and DC-offset cancellation

[28] Do, A.V.; Boon, C.C.; Manh Anh Do; Kiat Seng Yeo; Cabuk, A.; "A 1-V CMOS ultralow-power receiver front end for the IEEE 802.15.4 standard using tuned passive mixer output pole, "2010 18th IEEE/IFIP VLSI System on Chip Conference (VLSI-SoC), 2010, Page(s): 381 - 386.

[29] Bruckmann, Dieter; "Optimized digital signal processing for flexible receivers, "2002 IEEE International Conference on Acoustics, Speech, and Signal Processing (ICASSP), Volume: 4, 2002, Page(s): IV-3764 - IV-3767

[30] Gambini, S.; Crossley, J.; Alon, E.; Rabaey, J.; "A fully integrated, 300pJ/bit, dual mode wireless transceiver for cm-range interconnects,

"2010 IEEE Symposium on VLSI Circuits (VLSIC), 2010, Page(s): 31 - 32.

[31] Kienmayer, C.; Thuringer, R.; Tiebout, M.; Simburger, W.; Scholtz, A.L.; "An integrated 17 GHz front-end for ISM/WLAN applications in 0.13 µm CMOS," Digest of Technical Papers. 2004 Symposium on VLSI Circuits, 2004, Page(s): 12 - 15.

[32] Ikeuchi, O.; Saito, N.; Nauta, B.; "Quadrature sampling mixer topology for SAW-Less GPS receivers in 0.18µm CMOS, "2010 IEEE Symposium on VLSI Circuits (VLSIC), 2010, Page(s): 177-178.

[33] Harvey, J.; Harjani, R.; "An integrated quadrature mixer with improved image rejection at low voltage," Fourteenth International Conference on VLSI Design,2001, Page(s): 269-273.

[34] Tillman, F.; Troedsson, N.; Sjland, H.; "A 1.2 volt 1.8GHz CMOS quadrature front-end," Digest of Technical Papers. 2004 Symposium on VLSI Circuits, 2004, Page(s): 362 - 365.

[35] James, D.; Mendelsohn, N.; Fuchs, D.; "Digital audio mixer: A VLSI approach," IEEE International Conference on ICASSP '82. Acoustics, Speech, and Signal Processing, Volume: 7, 1982, Page(s): 77-80.

[36] Sandalci, C.K.; Kiaei, S.; "Analysis of adaptive CMOS down conversion mixers," Proceedings of the 8th Great Lakes Symposium on VLSI, 1998, Page(s): 118-121.

[37] Mahdavi, S.; Abidi, A.A.; "Fully integrated 2.2 mW CMOS front-end for a 900 MHz zero-IF wireless receiver," Digest of Technical Papers. 2001 Symposium on VLSI Circuits, 2001, Page(s): 251-252.

[38] Lie, D.Y.C.; Kennedy, J.; Livezey, D.; Yang, B.; Robinson, T.; Sornin, N.; Saint, C.; Larson, L.E.; "Circuit and System Design for a Homodyne W-CDMA Front-End Receiver RF IC, "2006 International Symposium on VLSI Design, Automation and Test, 2006, Page(s): 1-4.

[39] Heiberg, A.; Brown, T.; Mayaram, K.; Fiez, T.S.; "A 250 mV, 352 µW low-IF quadrature GPS receiver in 130 nm CMOS, "2010 IEEE Symposium on VLSI Circuits (VLSIC), 2010, Page(s): 135-136.

CHAPTER 13
VLSI SPREAD SPECTRUM CLOCKING CIRCUITS

The spread spectrum clocking (SSC) is a circuit technique that is used for a variety of applications, such as to establish secure communications, to increase resistance of natural interference and jamming, to prevent detection, and to limit power flux density and to reduce the electronic emission interference (EMI) or electronic emission compliance (EMC) effects of the signals and associated harmonics. The SSC circuit intentionally broadbands a normally narrowband signal through the frequency or the phase modulations.

The SSC circuit spreads the radiation energy in frequency domain such that the peak energy of the system is minimized. As the result of the frequency and phase modulation, the amplitudes of the harmonics of all the signals resulted from this clock can be reduced.

The VLSI SSC circuits are usually implemented employing the VLSI PLL circuits where the output clock can be modulated based on the VCO modulation, the feedback divider modulation, and the output clock phase modulation techniques.

13.1 SPREAD SPECTRUM CLOCKING PRINCIPLE

The spread spectrum clocking avoids EMI/EMC problems by shaping the system's electromagnetic emissions to comply with the electromagnetic compatibility (EMC) regulations. The SSC is a popular technique because it can be used to gain regulatory approval with only a simple modification to the equipment. The spread spectrum clocking has become more popular in portable electronics devices because of faster clock speeds, the increasing use of wireless communication and the increasing integration of high-resolution LCD displays in smaller and smaller devices. Because these devices are designed to be lightweight and inexpensive, passive EMI reduction measures such as capacitors or metal shielding is not a viable option. Active EMI reduction techniques such as spread spectrum clocking are necessary in these cases.

13.1.1 ELECTROMAGNETIC INTERFERENCE (EMI)

The electromagnetic interference (EMI) is defined as the undesired noise from a combination of conduction and electromagnetic radiation. A closely used term similar to the EMI is the electromagnetic compatibility (EMC) that is the ability of electronic equipment that is compatible to the low emission noise requirements.

The EMI is caused by the energy radiated from a periodic source in which most of the energy becomes a single fundamental frequency. The influence of these unwanted signals can manifest itself in the limited operation of other devices and systems. In some cases, the EMI-generated disturbance can make it impossible for these devices or systems to operate. Synchronous systems are common sources for generating excessive EMI. For example, within an electronic system, the traces in PCBs (printed-circuit boards) can generate EMI to affect other system components.

In the synchronous digital systems, such as the PCs and the TVs, a control clock signal is used to sync the entire system, such systems show an unavoidably narrow frequency spectrum because of its periodic nature and desired low jitter performance. A perfect clock signal would have all the energy concentrated at a single frequency and its harmonics, and would therefore radiate energy with extremely high spectral density, resulting in a frequency spectrum that interferes with the normal system operations (such as Rx in the Smartphone).

Many methods have been used to minimize the EMI effects, such as the shielding, the pulse shaping, the slew rate control, the low voltage differential signaling, the output staggering and the special layout techniques, etc. The shielding is the traditional method for minimizing EMI that uses grounded steel cabinet to contain and dissipate the energy radiated by the system where coated plastic cabinets with good ground were used to enclose electronic devices to achieve the same purpose. However the shielding technique became less effective as electronics proliferated and became increasingly smaller. Higher clock speeds of electronics included higher harmonic frequencies that forced the designer to use techniques such as the EMI filtering, and a careful circuit layout to reduce radiated emissions. This approach became costly and more difficult as the electronic system shrunk.

13.1.2 SPREAD SPECTRUM CLOCKING BASIC

The spread spectrum clocking (SSC) techniques have been used as the alternative to solve the EMI issue in many practical electronic systems where the SSC circuits shape the radiation energy in the frequency domain to pass the EMI regulations. The core principle of the spread spectrum is the use of noise-like carrier waves, and, as the name implies, bandwidths widening to minimize the peak radiations.

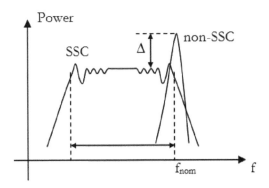

13.1 Power spectrum of SSC clock signal

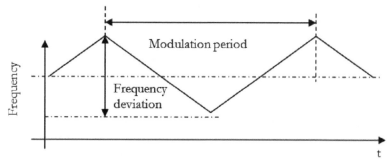

Fig.13.2 Time domain aspect of the SSC

Under this approach the EMI energy is spread across a range of frequencies using the frequency or phase modulation circuit techniques. By spreading or dithering the frequency of a system's clock, radiated emissions can be 'smeared' over a slightly wider spectrum with reduced peak-radiation that is usually regulated by the government institutes.

In the SSC circuits the frequency or phase of the clock signal is modulated by either a regular function such as a triangular wave, or by a pseudo-random function. Such technique reduces its peak spectral density to make them comply with the EMI/EMC regulation to gain regulatory approval with only a simple modification to the equipment.

It is important to note that the SSC method does not necessarily reduce the total energy in time domain. It works because the EMI receivers used by EMC testing laboratories divide the electromagnetic spectrum into frequency bands approximately 120kHz wide. A spread spectrum clocking distributes the energy into receiver's frequency bands therefore none of the band exceeds the EMI limits. The most common modulation techniques are center-spread and down-spread. The center-spread approach applies the modulated signal in such a way that the nominal frequency sits in the center of the modulated frequency range. That is, half of the modulated signals deviate above the nominal frequency, and the other half deviate below it. A down-spread approach also results in a range of deviated frequencies. However, in the down-spread approach, the modulated signals deviated below the nominal frequency.

For example, the PCI-express base specification provides guidelines for modulating the reference-clock input to PCIe devices. At a high level, the PCIe specification uses the down-spread approach when using a 30 to 33kHz-wave signal as the modulating frequency to the 100MHz clock, resulting in a frequency range of 99.5 to 100 MHz.

In the USB3.0 applications, a commonly used SSC method is to spread the 5Ghz clock frequency by 5000ppm (i.e. 25Mhz) with a triangular modulation profile of 33kHz (i.e. 30us modulation period).

13.2 SSC MODULATION MODELING

When the frequency of a clock is modulated, each individual harmonic is spread into a certain frequency band. The effect of frequency modulation on each harmonic can be modeled from the study of a sine waveform modulated with the desired modulation profile in time domain as:

$$V(t) = A\cos(\omega t + \phi(t)) \tag{13.1}$$

Where A is the amplitude of the original clock signal. ω is the angular frequency of the un-modulated clock and the ϕ is a time-dependent phase angle, representing the phase modulation.

$$\phi(t) = \int_0^t K_m V_m(\tau) d\tau \tag{13.2}$$

Where K_m is a factor controlling the peak of the frequency deviation. Vm(t) is a periodic function of angular frequency ω_m, giving the frequency modulation profile. There are several commonly used parameters to specify a SSC circuits such as the modulation frequency, the modulation index m_f, the rate of modulation δ, and the modulation profile.

13.2.1 MODULATION FREQUENCY

A periodic modulation waveform is deterministic and no impulsive in nature. In typical application, the divided version of the clock from a PLL is used to provide the clock for the modulation. A modulation frequency is selected to minimize any possible interference with existing applications. Historically, 31.25kHz was commonly used as the modulation frequency since the widest spread technology at that time the SSC was invented for analog TV and radio. This modulation frequency prevented interference with FM radio and minimized it for broadcast TV. The result of such modulation is a spread spectrum with a series of closely spaced harmonics spaced at about 32kHz.

In the mainstream computing industry, 30-50kHz modulation frequency range is typically used for the SSC circuit implementation. By dithering the clock, the frequency of the fundamental and its harmonics vary with time over a total spectrum consistent with the spread range as design specification. The amount of the time the clock stay at any one frequency will depend on the modulation profile, the modulation index, and the modulation frequency.

13.2.2 MODULATION INDEX

The modulation index is used to specify the related deviation of the frequency modulation with respect to the modulation frequency. The modulation index is defined as:

$$m_f \equiv \frac{\Delta\omega}{\omega_m} \qquad (13.3)$$

Where f_m is the modulation frequency. $\Delta\omega$ is the deviation of the modulated output clock frequency.

13.2.3 RATE OF MODULATION

The rate of modulation specifies how wide the energy of a single harmonic is spread related to the frequency of the clock. The rate of modulation is defined as:

$$m_f \equiv \frac{\Delta\omega}{\omega} \qquad (13.4)$$

13.2.4 MODULATION PROFILE

The modulation profile specifies the wave of the frequency change in time. The commonly used modulation profiles include the sinusoidal modulation, the triangular modulation, the exponential modulation, the two-point modulation, and the third-order modulation. Some of them are shown in figure 13.3.

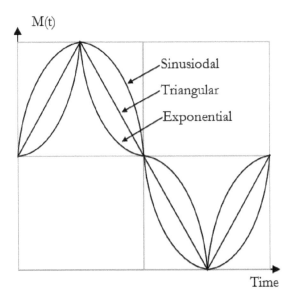

Fig.13.3 Typical SSC modulation profiles

The most effective EMI reduction is achieved when the frequency spectrum of the clock is flat that is primarily determined by the modulation profile of the SSC circuit.

Some studies show that the optimal reduction in EMI can be achieved by using a normalized periodic third-order modulation profile given as:

$$M(t) = \begin{cases} 0.45(t+1)^3 + 0.55(t+1) & -2 \leq t < 0 \\ -0.45(t-1)^3 - 0.55(t-1) & 0 \leq t \leq 2 \end{cases} \quad (13.5)$$

The above equation is normalized to a time period of [-2, 2].

13.2.5 SPREAD OF SPECTRUM UNDER SSC

Study shows that by using the SSC techniques, 98% of the energy of the fundamental component can be spread into a band B given by the equation as:

$$B = 2f_m(1+f_m) = 2(\Delta f + f_m) = \frac{1}{\pi}(\Delta\omega + \omega_m) \quad (13.6)$$

It can be seen that the bandwidth B_n, where a certain harmonic is spread up increase with harmonic order n. This is due to the fact that the high order harmonic may be overlapped and reduce the efficiency of modulation as:

$$B_n = 2f_m(1+nf_m) = 2(n\Delta f + f_m) = \frac{1}{\pi}(n\Delta\omega + \omega_m) \quad (13.7)$$

It can be seen that the spectrum amplitude reduction improves as the modulation index fm increases. But it is usually limited by the switch frequency variation.

It can also be seen that the triangular modulation provides the maximum flatness in spread spectrum and seems to give the maximum advantage for EMI reduction.

13.3 VLSI SSC CIRCUIT IMPLEMENTATIONS

The VLSI spread spectrum clocking techniques are commonly implemented using phase-locked loop circuits with proper modulation profile that provide control clock signal for the electronic systems. There are several commonly used SSC techniques, such as the direct VCO modulation, the two point modulation, the fractional-N $\Sigma\Delta$ modulation, and the delay-locked loop modulation. The SSC techniques can also be implemented in either the analog or digital domain.

13.3.1 DIRECT VCO MODULATION SSC CIRCUIT

Shown in figure 13.4 is a VLSI SSC circuit based on the direct VCO modulation architecture. This circuit employs a conventional VLSI PLL circuit where a second charge pump is included in the PLL. This charge pump is used to inject the frequency variation of the VCO output clock based on the pre-defined modulation profile, the modulation range, and the modulation frequency.

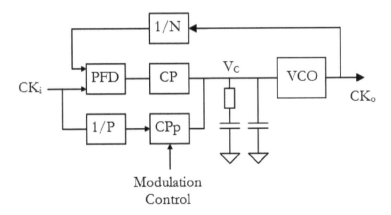

Fig.13.4 Direct VCO modulation SSC circuit

In this circuit implementation, the divided input by divider P is used as the periodic control clock for the modulation circuit. Such circuit allows adjusting the modulation index by programming the bias current in the charge pump.

Shown in figure 13.6 is a simplified s-domain linear model of this SSC circuit. The transfer function of the modulation input can be derived as:

$$\frac{\phi_o}{\phi_m} = \frac{\dfrac{K_{VCO}K_I}{s^2} + \dfrac{K_{VCO}K_P}{s}}{1 + \dfrac{K_{pd}}{N}[\dfrac{K_{VCO}K_I}{s^2} + \dfrac{K_{VCO}K_P}{s}]} \quad (13.8)$$

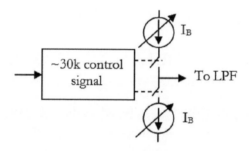

Fig.13.5 SSC injection charge pump

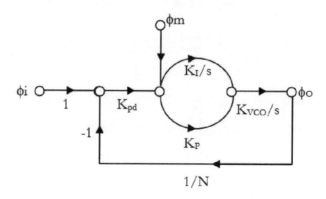

Fig.13.6 SSC injection charge pump

We can prove that such transfer function has a lowpass frequency response similar to the transfer function of the reference clock to the output path. For low frequency modulation within the bandwidth of the PLL, the modulation gain can be derived as:

$$\left|\frac{\phi_o}{\phi_m}\right|_{s\to 0} = \frac{N}{K_{pd}} \tag{13.9}$$

In a modified direct VCO modulation SSC circuit shown in figure 13.7, a resistor is added in the modulation path of the PLL loop filter. When the following constraint is applied in the design as:

$$R_1 C_1 = R_2 C_2 \tag{13.10}$$

We have that

$$\frac{V_C}{I_m} = \frac{1}{s(C_1 + C_2)} \tag{13.11}$$

The significance of above circuit structure is that a square wave type modulation current can be used to generate a triangular SSC modulation profile.

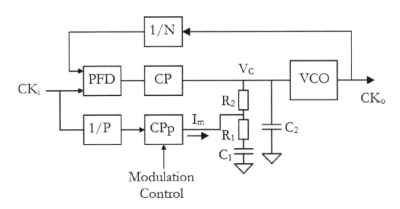

Fig.13.7 Triangular direct VCO modulation SSC circuit

Shown in figure 13.8 is an alternative VCO modulation SSC circuit structure.

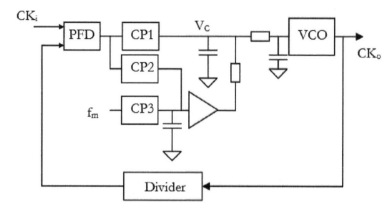

Fig.13.8 Alternative VCO modulation SSC circuit

13.3.2 VLSI FEEDBACK MODULATION SSC CIRCUITS

Shown in figure 13.9 is a VLSI SSC circuit implementation employing the modulation of the feedback dividing ratio. This circuit uses a multi-modular feedback divider controlled by a sigma-delta modulator. Such a circuit features a fractional-N frequency PLL. sigma-delta modulator is commonly used to provide randomized modulation of the feedback division ratio to minimize the noise tone generated by the modulation.

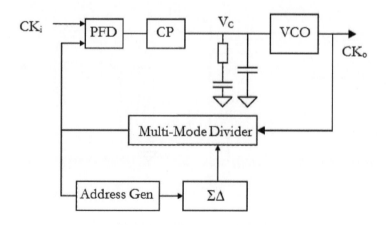

Fig.13.9 Frequency feedback modulation SSC circuit

Shown in figure 13.10 is a 1-1-1 MASH digital sigma-delta modulator for the feedback SSC modulation. Such circuit provides 3rd order sigma-delta modulation.

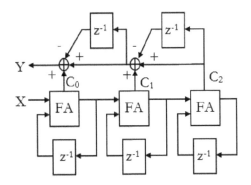

Fig.13.10 Digital 1-1-1 MASH Sigma-delta modulator

Shown in figure 13.11 is a multi-modular frequency divider to support the feedback SSC modulation. Such a circuit provides a programmable division ratio for the feedback modulation.

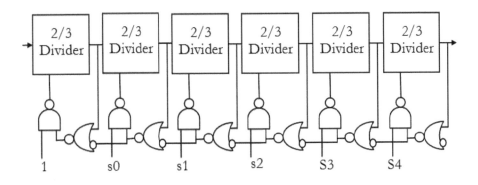

Fig.13.11 Multi-modular frequency divider

Feedback modulation SSC can also be implemented using feedback phase modulation method as shown in figure 13.12 where the phase, instead of the frequency of the PLL feedback is modulated.

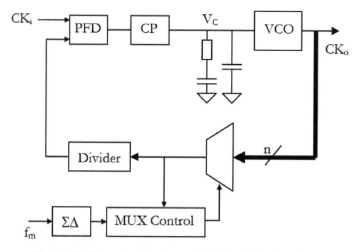

Fig.13.12 Phase feedback modulation SSC circuit

Such circuit can be realized using a programmable VCO phase selection or using delay line as shown in figure 13.13 such that the delay phase in the PLL feedback back path can be modulated based on certain modulation profile.

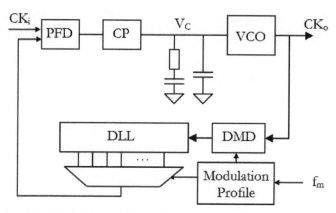

Fig.13.13 DLL based phase feedback modulation SSC circuit

Shown in figure 13.14 is a s-domain linear model of the feedback modulation SSC circuit.

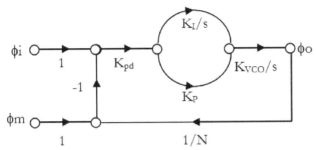

Fig.13.14 S-domain feedback SSC circuit model

The modulation transfer function under this modulation technique can be derived as:

$$\frac{\phi_o}{\phi_m} = \frac{-K_{PD}(\frac{K_{VCO}K_I}{s^2} + \frac{K_{VCO}K_P}{s})}{1 + \frac{K_{PD}}{N}[\frac{K_{VCO}K_I}{s^2} + \frac{K_{VCO}K_P}{s}]} \qquad (13.12)$$

Shown in figure 13.15 is a digital modulation profile generation circuit, where a digital integrator is used to convert the square input waveform into desired triangular modulation profile. The input data can also be programmable based on certain rule to generate other modulation profile needed.

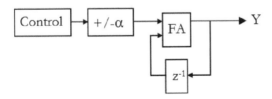

Fig.13.15 PWL Modulation profile generator

Shown in figure 13.16 is a programmable digital delay line circuit that can be used to implement the feedback SSC modulation.

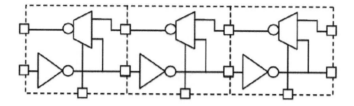

Fig.13.16 Digital programmable delay line

Reference:

[1] K. Hardin, et al, "Investigation Into the Interference Potential of Spread spectrum Clock Generation to Broadband Digital Communication," IEEE Transaction on Electromagnetic Compatibility, Vol. 45, No. 1, Feb. 2003.

[2] H. H. Chang, et al, "A Spread spectrum Clock Generator with Triangular Modulation," JSSC, Vol. 38, No. 4, April 2003.

[3] S. A. Belcella, et al, "EMI Reduction in Switched Power Converters by Means of Spread Spectrum Modulation Techniques," 2004 35th Annual IEEE Power Electronics Specialist Conference.

[4] Aoki, S., Amemiya, F., Kitani, A., Kuwabara, N., "Investigation of interferences between wireless LAN signal and disturbances from spread spectrum clocking," 2004 International Symposium on Electromagnetic Compatibility, Volume: 2, Page(s): 505 - 510.

[5] De Caro, D.; Romani, C.A.; Petra, N.; Strollo, A.G.M.; Parrella, C. "A 1.27 GHz, All-Digital Spread Spectrum Clock Generator/Synthesizer in 65 nm CMOS," IEEE Journal of Solid-State Circuits, Volume: 45, Issue: 5, Page(s): 1048 - 1060. 2010.

[6] Chi-Hsien Lin; Yen-Ying Huang; Shu-Rung Li; Yuan-Pu Cheng; Shyh-Jye Jou;"A spread spectrum clock generator with phase-rotation algorithm for 6Gbps clock and data recovery," IEEE 8th International Conference on ASIC, Page(s): 387 - 390, 2009.

[7] Komatsu, Y.; Ebuchi, T.; Hirata, T.; Yoshikawa, T.; "Bi-directional ac coupled interface with adaptive spread spectrum clock generator," 2007 IEEE Asian Solid-State Circuits Conference, Page(s): 71 - 74.

[8] Wei-Ta Chen, Jen-Chien Hsu, Hong-Wen Lune, Chau-chin Su, "A spread spectrum clock generator for SATA-II," ISCAS 2005. IEEE International Symposium on Circuits and Systems, Vol. 3. Page(s): 2643 - 2646.

[9] Won-Young Lee; Lee-Sup Kim; "A spread spectrum clock generator with spread ratio error reduction scheme for Display Port main link," 2009 IEEE International Symposium on Circuits and Systems, Page(s): 185 - 188.

[10] Hyung-Rok Lee; Ook Kim; Gijung Ahn; Deog-Kyoon Jeong; "A low-jitter 5000ppm spread spectrum clock generator for multi-channel SATA transceiver in 0.18μm CMOS," Digest of Technical Papers. 2005 IEEE International Solid-State Circuits Conference, Vol. 1. Page(s): 162 - 590.

[11] Coenen, M.; van Roermund, A.; "Discrete spread spectrum sampling (DSSS) to reduce RF emission and beat frequency issues," 2010 Asia-Pacific Symposium on Electromagnetic Compatibility (APEMC), Page(s): 342 - 345.

[12] Sadamura, H.; Daimon, T.; Shindo, T.; Kobayashi, H.; Myono, T.; Suzuki, T.; Kawai, S.; Iijima, T.; "Spread spectrum clocking in switching regulators to reduce EMI," 2002 IEEE Asia-Pacific Conference on ASIC, 2002. Proceedings. Page(s): 141 - 144.

[13] Won-Young Lee; Lee-Sup Kim; "A Spread Spectrum Clock Generator for Display Port Main Link," Express Briefs, IEEE Transactions on Circuits and Systems II: Volume: 58, Issue: 6, Page(s): 361 - 365, 2011.

[14] Kuo-Hsing Cheng; Cheng-Liang Hung; Chih-Hsien Chang; "A 0.77 ps RMS Jitter 6-GHz Spread spectrum Clock Generator Using a Compensated Phase-Rotating Technique," IEEE Journal of Solid-State Circuits, Volume: 46, Issue: 5, Page(s): 1198 - 1213, 2011.

[15] Kawamoto, T.; Takahashi, T.; Inada, H.; Noto, T.; "Low-jitter and Large-EMI-reduction Spread spectrum Clock Generator with Auto-calibration for Serial-ATA Applications," 2007 IEEE Custom Integrated Circuits Conference, Page(s): 345 - 348, 2007.

[16] Kawamoto, T.; Kokubo, M.; "A low-jitter 1.5-GHz and large-EMI reduction 10-dBm spread spectrum clock generator for serial-ATA," 2009 Asia and South Pacific Design Automation Conference, Page(s): 696 - 701, 2009.

[17] Hanumolu, P.K.; Gu-Yeon Wei; Un-Ku Moon; "A Wide-Tracking Range Clock and Data Recovery Circuit," IEEE Journal of Solid-State Circuits, Volume: 43, Issue: 2. Page(s): 425 - 439, 2008.

[18] Tang, A.T.K.; "Bandpass spread spectrum clocking for reduced clock spurs in autozeroed amplifiers," The 2001 IEEE International Symposium on Circuits and Systems, 2001.

[20] Badaroglu, M.; Wambacq, P.; Van der Plas, G.; Donnay, S.; Gielen, G.G.E.; De Man, H.J.; "Digital ground bounce reduction by supply current shaping and clock frequency Modulation," IEEE Transactions on

Computer-Aided Design of Integrated Circuits and Systems. Volume: 24, Issue: 1. Page(s): 65 - 76. 2005.

[21] Matsumoto, Y.; Fujii, K.; Sugiura, A.; "Estimating the Amplitude Reduction of Clock Harmonics Due to Frequency Modulatio," IEEE Transactions on Electromagnetic Compatibility, Volume: 48, Issue: 4, Page(s): 734 - 741, 2006.

[22] Hung-Sung Li; Yu-Chi Cheng; Puar, D.; "Dual-loop spread spectrum clock generator," Digest of Technical Papers. ISSCC. 1999 IEEE International Solid-State Circuits Conference, Page(s): 184 - 185.

[23] Hardin, K.; Oglesbee, R.A.; Fisher, F.; "Investigation into the interference potential of spread spectrum clock generation to broadband digital communications," IEEE Transactions on Electromagnetic Compatibility, Volume: 45, Issue: 1, Page(s): 10 - 21, 2003,

[24] Jenchien Hsu; Maohsuan Chou; Chauchin Su; "Built-in jitter measurement methodology for spread spectrum clock generators," 2008. IEEE International Symposium on VLSI Design, Automation and Test, Page(s): 67 - 72.

[25] Yi-Bin Hsieh; Yao-Huang Kao; "A Fully Integrated Spread spectrum Clock Generator by Using Direct VCO Modulation," Regular Papers, IEEE Transactions on Circuits and Systems I, Volume: 55, Issue: 7, Page(s): 1845 - 1853, 2008.

[26] Arnett, D.; "Suppressed carrier digital clocks," 1999 IEEE International Symposium on Electromagnetic Compatibility, Volume: 2, Page(s): 816 - 821, 1999.

[27] Moncunill-Geniz, F.X.; Pala-Schonwalder, P.; del Aguila-Lopez, F.; "New superregenerative architectures for direct-sequence spread spectrum communications," Express Briefs, IEEE Transactions on Circuits and Systems II: Volume: 52, Issue: 7, Page(s): 415 - 419, 2005.

[28] Chien, C.; Jain, R.; Cohen, E.G.; Samueli, H.; "A single-chip 12.7 Mchips/s digital IF BPSK direct sequence spread spectrum transceiver in 1.2 μm CMOS," IEEE Journal of Solid-State Circuits, Volume: 29, Issue: 12, Page(s): 1614 - 1623, 1994.

[30] Yao-Huang Kao; Yi-Bin Hsieh; "A Low-Power and High-Precision Spread Spectrum Clock Generator for Serial Advanced Technology Attachment Applications Using Two-Point Modulation," IEEE

Transactions on Electromagnetic Compatibility, Volume: 51, Issue: 2, Page(s): 245 - 254, 2009.

[31] Junfeng Zhou; Dehaene, W.; "A Synchronization-Free Spread Spectrum Clock Generation Technique for Automotive Applications," IEEE Transactions on Electromagnetic Compatibility, Volume: 53, Issue: 1, Page(s): 169 - 177, 2011.

[32] Hsiang-Hui Chang; I-Hui Hua; Shen-Iuan Liu; "A spread spectrum clock generator with triangular modulation," IEEE Journal of Solid-State Circuits, Volume: 38, Issue: 4, Page(s): 673 - 676, 2003.

[33] Win, M.Z.; "A unified spectral analysis of generalized time-hopping spread spectrum signals in the presence of timing jitter," IEEE Journal on Selected Areas in Communications, Volume: 20, Issue: 9, Page(s): 1664 - 1676, 2002.

[34] Yoonjae Lee; Mittra, R.; "Electromagnetic interference mitigation by using a spread spectrum approach," IEEE Transactions on Electromagnetic Compatibility, Volume: 44, Issue: 2, Page(s): 380 - 385, 2002.

[35] Sugawara, M.; Ishibashi, T.; Ogasawara, K.; Aoyama, M.; Zwerg, M.; Glowinski, S.; Kameyama, Y.; Yanagita, T.; Fukaishi, M.; Shimoyama, S.; Noma, T.; "1.5 Gbps, 5150 ppm spread spectrum SerDes PHY with a 0.3 mW, 1.5 Gbps level detector for serial ATA," Digest of Technical Papers, 2002 Symposium on VLSI Circuits, Page(s): 60 - 63, 2002.

[36] Ming-Luen Lieu; Tzi-Dar Chiueh; "A low-power digital matched filter for direct-sequence spread spectrum signal acquisition," IEEE Journal of Solid-State Circuits, Volume: 36, Issue: 6, Page(s): 933 - 943, 2001.

[37] Ebuchi, T.; Komatsu, Y.; Okamoto, T.; Arima, Y.; Yamada, Y.; Sogawa, K.; Okamoto, K.; Morie, T.; Hirata, T.; Dosho, S.; Yoshikawa, T.; "A 125–1250 MHz Process-Independent Adaptive Bandwidth Spread Spectrum Clock Generator With Digital Controlled Self-Calibration," IEEE Journal of Solid-State Circuits, Volume: 44, Issue: 3, Page(s): 763 - 774, 2009.

[38] Damphousse, S.; Ouici, K.; Rizki, A.; Mallinson, M.; "All Digital Spread Spectrum Clock Generator for EMI Reduction," IEEE Journal of Solid-State Circuits, Volume: 42, Issue: 1, Page(s): 145 - 150, 2007.

[39] Santolaria, A.; Balcells, J.; Gonzalez, D.; Gago, J.; Gil, S.D. "EMI reduction in switched power converters by means of spread spectrum

modulation techniques," 2004 IEEE 35th Annual Power Electronics Specialists Conference, Volume: 1, Page(s): 292 - 296.

[40] Hardin, K.B.; Fessler, J.T.; Bush, D.R.;"Spread spectrum clock generation for the reduction of radiated emissions," 1994 IEEE International Symposium on Electromagnetic Compatibility, Page(s): 227 - 231.

[41] Hardin, K.B.; Fessler, J.T.; Bush, D.R.; "A study of the interference potential of spread spectrum clock generation techniques," 1995 IEEE International Symposium on Electromagnetic Compatibility, Page(s): 624 - 629.

[42] Balcells, J.; Santolaria, A.; Orlandi, A.; Gonzalez, D.; Gago, J. ;"EMI reduction in switched power converters using frequency Modulation techniques," IEEE Transactions on Electromagnetic Compatibility, Volume: 47, Issue: 3, Page(s): 569 - 576, 2005.

CHAPTER 14
VLSI FRACTIONAL-N PHASE-LOCKED LOOPS

The fractional-N PLL is a special PLL family that offers wide bandwidth with narrow frequency spacing and alleviates PLL design constraints for phase noise and reference spur. Since their lower division ratio and higher phase detection frequencies than the integer-PLL counterparts, phase noise in fractional-N PLL can be suppressed to a higher degree. A fractional-N PLL provides a VLSI clock source with fast settling time, low frequency error, better spurious performance, and low phase noise that is suitable for various VLSI wireless applications.

The fractional PLLs provide many advantages over traditional integer PLLs such as generating almost any frequency from one crystal, the true fractional multiplication that can be determined by the word length of the fractional N divider, and maximize silicon performance.

14.1 VLSI PHASE-LOCKED LOOP BASIS

A PLL is a negative feedback loop in which the phase of a generated signal is forced to follow that of a reference signal. A basic VLSI charge pump base PLL as shown in figure 14.1 comprises a reference source, a phase frequency detector, a charge pump, a loop filter, a voltage controlled oscillator (VCO), a pre-divider, and a feedback divider.

The feedback divider (1/N) output of the VCO output clock is phase-compared with the pre-divided (1/M) reference at the phase frequency detector (PFD). The phase difference is used to control the pump-up or pump-down current sources in the charge pump. As a result, some charge is transferred to or taken away from the integrating capacitor in the loop filter. The amount of charge transferred is proportional to the magnitude of the phase difference. This, in turn, results in an adjustment in the tuning voltage of the VCO so that its phase is retarded or advanced. The loop is such designed that the phase error is corrected when the loop is in lock condition.

The PFD enables the right current source (i.e., pump-up current or pump-down current) to speed up or slow down the VCO in case of a frequency difference between the two incoming signals to the PFD. When the loop is locked, the frequency of the feedback clock from the VCO is equal to that of the input reference at the PFD inputs.

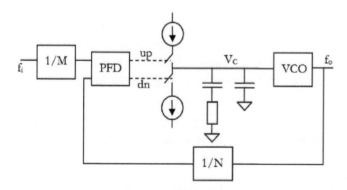

Fig.14.1 Basic charge-pump PLL structure

14.1.1 VLSI INTEGER PHASE-LOCKED LOOP

Most VLSI PLL circuits used today are based on the integer N and M to generate the frequencies of most on-chip clock, where the ratio of N/M can be used to set the frequency scaling factor as:

$$\frac{f_o}{f_i} = \frac{N}{M} \qquad (14.1)$$

In principle, any given frequency scaling factors can be realized by proper selection of integer N and M. However the M/N ratio may not always be able to generate the desired frequency with sufficient accuracy. Since the reference input at the PFD is a divided version of the reference clock, bandwidth of the PLL is dedicated by the input reference clock frequency (that is usually selected to be less than 1/10 of the PFD input reference clock frequency). Consequently, the integer-N PLL will have very low bandwidth if a large M must be used.

For example, if an output frequency of 13.576 MHz is needed to be generated from a reference input frequency of 27 MHz, the desired frequency scaling factor using integer division parameter is 3072/3375. This will require very large division factor M of 3375.

In order to achieve higher frequency resolution, multiple integer PLLs can sometimes be cascaded as shown in figure 14.2 to provide better frequency coverage.

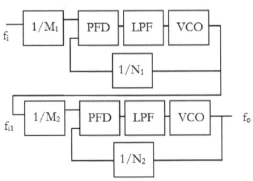

Fig.14.2 Cascaded PLL structures

For the cascaded PLL circuits, the frequency-scaling factor is equal to the product of the each PLL frequency-scaling factor as:

$$\frac{f_o}{f_i} = \frac{N_1}{M_1} \cdot \frac{N_2}{M_2} \tag{14.2}$$

With four integers in the frequency scaling equation, the selection of the scaling factor is more flexible and therefore it may offer better coverage of the frequency selection.

In the general case, if K PLLs are cascaded, the frequency-scaling factor is equal to the products of all PLLs as:

$$\frac{f_o}{f_i} = \frac{N_1}{M_1} \cdot \frac{N_2}{M_2} \cdot \ldots \frac{N_k}{M_k} \tag{14.3}$$

The cascade PLL offers more flexibility in frequency selection and better frequency accuracy. However such PLL structures suffer from several fundamental limitations:

- Because an integer PLL can only multiply by whole numbers, it is usually impossible to obtain all desired clock frequencies by using the integer PLL. This is very true for design applications that need extremely accurate frequency scaling factor. For example, if the 3072/3375 scaling factor is needed for output frequency of 13.576 MHz from a reference input frequency of 27 MHz, This will require very large division factor M (or M_1M_2) and N or (N_1N_2) that is also simply impossible to stay in range of each VCO operation.

- The cascade PLL adds significant design overheads in area, power and long term jitter than a single PLL.

- In addition, the cascaded integer PLLs make it more difficult to change settings on the fly without causing significant frequency swings.

14.1.2 VLSI FRACTIONAL-N PHASE-LOCKED LOOP

For the integer PLL, the frequency resolution of a PLL is limited by the phase detector rate. For example, if a 1 kHz phase detector rate is used, the frequency resolution of the PLL will also be 1 kHz since the out frequency of an integer PLL can only be integer multiplication of the phase detection frequency. For many applications that demand much finer frequency selection, fractional-N PLLs can be used to provide much more distinct output frequencies, compared to simpler integer-N PLLs using the same reference frequency.

The fractional-N PLLs are based on the variable division ratio dividers. A basic fractional-N PLL system is shown in figure 14.3.

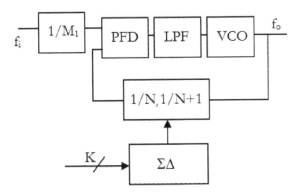

Fig.14.3 Fractional-N PLL structures

The division ratio in a fractional-N PLL is made to have a fractional component by changing the division ratio of the divider periodically between $1/N$ and $1/(N+1)$, so the average value contains a fractional element. For example, if a fractional value of 0.1 is required, then the division ratio is changed by 1 on every tenth cycle. If a fractional value of 0.01 is required, then the division ratio is changed by 1 on every one-hundredth cycle. Very low fractions can also be realized by digital way in the divider control. This offers much finer frequency resolution than the integer PLL.

The basic structure of a fractional-N PLL is the same as the integer PLL. However, a sigma-delta modulator is usually used to generate the fractional multiplier. The number of fractional bits may vary anywhere from 16 to 24 bits in practical applications, where the increasing in the number of bits increases the precision of the fractional value. For example, a 16-bit fraction will give five places decimal point and 24 bits will give eight decimal places of fractional precision. With a fractional-N PLL, sub hertz and sub PPM accuracy and step size can easily be achieved.

Unlike the integer PLLs, the sigma-delta fractional-N PLL inherently supports any input frequency, delivers any output frequency, and can do so without needing to cascade PLLs. This can reduce the cost for a product by being able to generate arbitrary frequency with one clock source. Using a sigma-delta fractional-N PLL in place of the more traditional integer PLL can reduce or eliminate the power and area concerns at the same time.

14.2 VLSI FRACTIONAL-N PLL IMPLEMENTATIONS

The heart of the VLSI fractional-N PLL is the programmable divider. This is the major difference between the integer PLL and fractional-N PLL circuit structures. However fractional PLLs tend to generate fractional spurs, because the deterministic alternating pattern used to switch the division ratios, where periodic phase shifts are inevitable and create large unwanted spurs in the output spectrum. It is often difficult to make sure that these spurs will be low enough to avoid degrading the performance to unacceptable levels. As the result, a sigma-delta modulator is usually needed to eliminate or minimize the spur problem of the fractional-N PLL.

The fractional-N PLL uses a sigma-delta modulator to modulate the division ratio, instead of a periodic N/(N+1) modulation pattern, the benefits of an arbitrarily fine frequency resolution can be retained while eliminating or minimizing these fractional spurs. In this case, the division ratio is continuously modulated by the output code of the sigma-delta modulator, where the average value of the division ratio provides the desired output frequency.

14.2.1 MULTI-MODULUS DIVIDER

The variability of the frequency divider in the fractional-N PLL is realized using the multi-modular divider circuit.

Shown in figure 14.4 is a modulus divider bitslice that can provide programmable 1/2, 1/3 division factors.

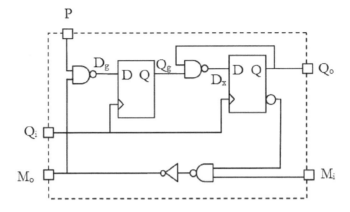

Fig.14.4 VLSI programmable divider

This circuit works as follows:

- If programming port P = "0", the gating DFF is disabled. The circuit serves as a 1/2 divider.

- If P = "1"and Mi = "0", the gating DFF is also disabled. The circuit will also serve as a 1/2 divider.

- If P = "1"and Mi = "1", the gating DFF is enabled. The circuit will stop the transition by one cycle when Qo = "0", resulting in a cycle-stretch (i.e. working as a 1/3 divider)

Shown in Table 14.1 is the logic table of the circuit with inputs P = "1"and Mi = "1".

Table 14.1 logic table of programmable bitslice for 1/3 operation

Qi cycle (rise)	Dg	Dx	Qg	Qo
0	1	0	1	1
1	0	1	1	0
2	1	1	0	1
3	1	0	1	1
4	0	1	1	0
5	1	1	0	1
6	1	0	1	1
7	0	1	1	0
8	1	1	0	1

Rows 0–2 and 3–5 and 6–8 are each grouped as "3".

It can be seen that there is one cycle stretch after Qo = "1".

Shown in figure 14.5 is the cascaded modulus divider using two programmable divider discussed above. The operation of this divider can be derived as follows:

- When both programming pins are set to "0" (i.e. [c1, c0] = [0, 0]), the gating paths are disabled and the divider has a division ratio of $2^2 = 4$.

- When the control vector are set to [c1, c0] = [0, 1], the second divider is set to divided-by-2 and the first divider has a cycle stretch that occurs one cycle after Ck1 AND Ck2 = "1". Therefore this circuit has a division ratio of $2^2+1=5$.

- When the control input vector is given as [c1, c0] = [1, 0], the second divider is set to divided-by-3 and the first divider is simply a divided-by-2 circuit. Therefore this circuit has a division ratio $2^2+2 = 6$.

- When the control input vector is given as [c1, c0] = [1, 1], the second divider has cycle stretch occurring after Cko =1. Similarly, the first divider has one cycle stretch at Ck1 AND Cko = "1". Therefore this circuit has a division ratio $2^2+3 = 7$.

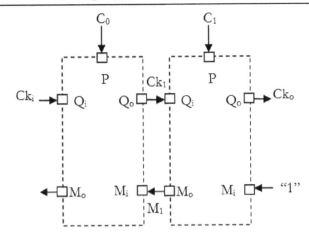

Fig.14.5 Cascaded VLSI programmable divider

Shown in Table 14.2 is the logic table of the cascaded modulus divider

Table 14.2 logic table of cascaded modulus divider

Qi cycle (rise)	Ck1	Cko (= Ck2)
0	0	0
1	1	1
2	1	1
3	0	1
4	1	1
5	1	1
6	0	1
7	1	0
8	0	0

(rows 1–7 bracketed as 7)

The division ratio of such multi-modulus divider can then be expressed as:

$$N = 2^2 + 2^1 C_1 + 2^0 C_0 \qquad (14.4)$$

Shown in figure 14.6 is a general-purpose 6-bit modulus divider circuit employing an array of the divider bitslice discussed above.

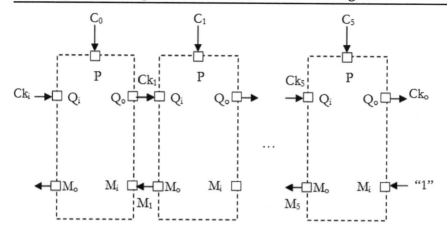

Fig.14.6 Cascaded VLSI programmable divider

The dividing ratio of such multi-modulus divider can be derived as:

$$N = 2^6 + 2^5 \cdot C_5 + 2^4 C_4 + 2^3 C_3 + 2^2 C_2 + 2^1 C_1 + 2^0 C_0 \qquad (14.5)$$

Shown in figure 14.7 is an alternative multi-modulus frequency divider circuit structure that can be used to realize the division factor given in above equation.

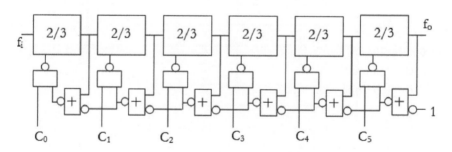

Fig.14.7 VLSI multi-modular divider

In an alternative circuit approach, the programmable frequency divider circuits can be constructed based on the binary divider and a cycle-stretch circuit technique. A VLSI programmable ratio between $1/2^N$ and $1/(2^N+M)$ is shown in figure 14.8. Such circuit consists of an N-bit asynchronous binary divider and a feedback cycle stretch path in the first stage employing a NAND gate, where the feedback path offers the so called cycle stretch operation since one additional clock cycle is inserted (cycle stretch) in the dividing process every time the hard-coded stretch location code is met during division counting.

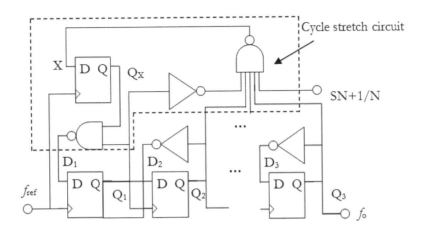

Fig.14.8 General VLSI programmable ratio divider circuit structure

Note that in the stretch location codes the Q_1 bit must be "0" to ensure negative feedback in the circuit loop for the proper circuit operation. Therefore the maximum number for M is limited to 2^{N-1}. For the above programmable dividers circuit, M = 1 is the commonly used in circuits in practical applications, such as the fractional-N PLL circuits.

Shown in figure 14.9 is a VLSI 1/8 (1/9) divider circuit where the stretch location code is given as "011". It can be seen that every time the counter reach the output state "011", the next state will be stretched to two clock cycles as shown in figure 14.10.

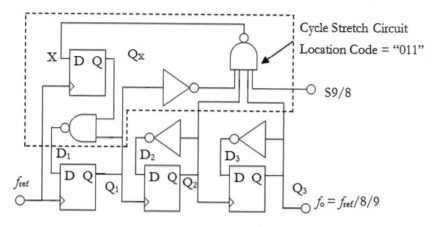

Fig.14.9 VLSI 1/8 1/9 programmable divider circuit structure

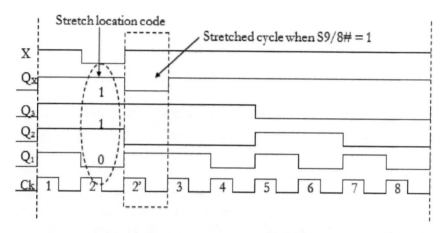

Fig.14.10 Timing diagram for 1/8 1/9 counter based on cycle stretch

The programmable frequency dividers for other dividing ratio can be realized based on this circuit technique. Show in figure 14.11 is a VLSI divide by 1/2 (1/3) divider where the cycle stretch location is given as "0". Normally such divider circuit serves as a divided-by-2 divider. However when the mode signal S2/3 is high, the divider counter inserts a cycle every time when Q_1 is "0". As the result, the divider circuit in this mode serves as a divide-by-3 circuit.

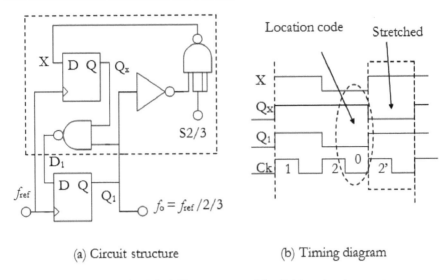

(a) Circuit structure　　　　　(b) Timing diagram

Fig.14.11 VLSI 1/2 1/3 programmable divider circuit structure

Similarly we can realize the 1/4 (1/5) divider using the circuit shown in figure 14.12, where the cycle stretch location code is set as "01" as shown in figure 14.13.

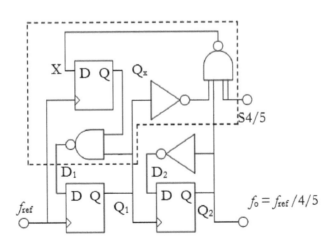

Fig.14.12 VLSI 1/4 1/5 programmable divider circuit structure

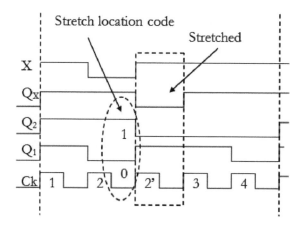

Fig.14.13 VLSI 1/4 1/5 programmable divider time diagram

14.2.2 MODULATOR

A conventional approach of controlling the division ratio is to use a digital accumulator as shown in figure 14.14. The accumulator uses the full-adder and the DFF arrays to add the contents of its input to its current output on each clock cycle of the clock. It works as the digital equivalent of an integrator and since the integral of the frequency is a phase, its output represents the relative phase of the fractional component.

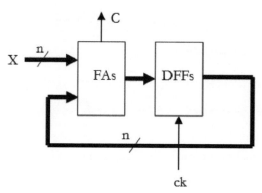

Fig.14.14 VLSI digital accumulator

The digital output of the accumulator is incremented for each cycle of the divider by the fractional frequency instruction. Every time the accumulator reaches its capacity it produces an overflow, which changes the divider division ratio.

Shown in figure 14.15 is the linear z-domain circuit model of the above simple modulator.

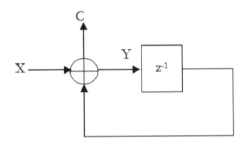

Fig.14.15 Z-Domain model of VLSI digital accumulator

The transfer function of the circuit can be derived as:

$$C = X + (z^{-1} - 1)Y \qquad (14.5)$$

The significances of the above equation are as follows:

- At low frequency range (i.e. $\omega T \ll 1$, where ω is the signal angular frequency and T is the control clock period), we have that

$$|(z^{-1} - 1)|_{z=e^{j\omega T}} = 2\sin(\omega T/2) \to 0 \qquad \omega T \to 0 \qquad (14.6)$$

- At the low frequency the carry C will represent the input value X

$$|C| \twoheadrightarrow |X| \quad \omega T \to 0 \tag{14.7}$$

- As the frequency increase, the Y term will contribute to the noise floor of the carry out put C. As the results, such modulator shows a first order sigma-delta modulation characteristics.

Reference:

[1] Riley, T.A.D.; Copeland, M.A.; Kwasniewski, T.A.; "Delta-sigma modulation in fractional-N frequency synthesis," IEEE Journal of Solid-State Circuits, Volume: 28, Issue: 5, 1993, Page(s): 553 - 559

[2] Miller, B.; Conley, R.J.; "A multiple modulator fractional divider," IEEE Transactions on Instrumentation and Measurement, Volume: 40, Issue: 3, 1991, Page(s): 578 - 583

[3] Pamarti, S.; Jansson, L.; Galton, I.; "A wideband 2.4-GHz delta-sigma fractional-NPLL with 1-Mb/s in-loop modulation," IEEE Journal of Solid-State Circuits, Volume: 39, Issue: 1, 2004, Page(s): 49 - 62

[4] Keliu Shu; Sanchez-Sinencio, E.; Silva-Martinez, J.; Embabi, S.H.K.; "A 2.4-GHz monolithic fractional-N frequency synthesizer with robust phase-switching prescaler and loop capacitance multiplier," IEEE Journal of Solid-State Circuits, Volume: 38, Issue: 6, 2003, Page(s): 866 - 874

[5] Chan-Hong Park; Ook Kim; Beomsup Kim; "A 1.8-GHz self-calibrated phase-locked loop with precise I/Q matching," IEEE Journal of Solid-State Circuits, Volume: 36, Issue: 5, 2001, Page(s): 777 - 783

[6] Perrott, M.H.; Trott, M.D.; Sodini, C.G.; "A modeling approach for Σ-Δ fractional-N frequency synthesizers allowing straightforward noise analysis," IEEE Journal of Solid-State Circuits, Volume: 37, Issue: 8, 2002, Page(s): 1028 - 1038

[7] Magoon, R.; Molnar, A.; Zachan, J.; Hatcher, G.; Rhee, W.; "A single-chip quad-band (850/900/1800/1900 MHz) direct conversion GSM/GPRS RF transceiver with integrated VCOs and fractional-n synthesizer," IEEE Journal of Solid-State Circuits, Volume: 37, Issue: 12, 2002, Page(s): 1710-1720.

[8] De Muer, B.; Steyaert, M.S.J.; "A CMOS monolithic $\Delta\Sigma$-controlled fractional-N frequency synthesizer for DCS-1800," IEEE Journal of Solid-State Circuits, Volume: 37, Issue: 7, 2002, Page(s): 835 - 844

[9] Yido Koo; Hyungki Huh; Yongsik Cho; Jeongwoo Lee; Joonbae Park; Kyeongho Lee; Deog-Kyoon Jeong; Wonchan Kim; "A fully integrated CMOS frequency synthesizer with charge-averaging charge pump and

dual-path loop filter for PCS- and cellular-CDMA wireless systems," IEEE Journal of Solid-State Circuits, Volume: 37, Issue: 5, 2002, Page(s): 536-542.

[10] Hegazi, E.; Abidi, A.A.; "A 17-mW transmitter and frequency synthesizer for 900-MHz GSM fully integrated in 0.35-μm CMOS," IEEE Journal of Solid-State Circuits, Volume: 38, Issue: 5, 2003, Page(s): 782 - 792

[11] Perrott, M.H.; "Fast and accurate behavioral simulation of fractional-N frequency synthesizers and other PLL/DLL circuits," Proceedings. 39th Design Automation Conference, 2002, Page(s): 498 - 503.

[12] Pengfei Zhang; Der, L.; Dawei Guo; Sever, I.; Bourdi, T.; Lam, C.; Zolfaghari, A.; Chen, J.; Gambetta, D.; Baohong Cheng; Gowder, S.; Hart, S.; Huynh, L.; Nguyen, T.; Razavi, B.; "A single-chip dual-band direct-conversion IEEE 802.11a/b/g WLAN transceiver in 0.18-μm CMOS," IEEE Journal of Solid-State Circuits, Volume: 40, Issue: 9, 2005, Page(s): 1932 - 1939.

[13] Riley, T.A.D.; Copeland, M.A.; "A simplified continuous phase modulator technique," IEEE Transactions on Circuits and Systems II: Analog and Digital Signal Processing, Volume: 41, Issue: 5, 1994, Page(s): 321 - 328

[14] Chun-Ming Hsu; Straayer, M.Z.; Perrott, M.H.; "A Low-Noise Wide-BW 3.6-GHz Digital Fractional-N Frequency Synthesizer With a Noise-Shaping Time-to-Digital Converter and Quantization Noise Cancellation," IEEE Journal of Solid-State Circuits, Volume: 43, Issue: 12, 2008, Page(s): 2776-2786.

[15] Kozak, M.; Kale, I.; "Rigorous analysis of delta-sigma modulators for fractional-N PLL frequency synthesis," Circuits and Systems I: Regular Papers, IEEE Transactions on Volume: 51, Issue: 6, 2004, Page(s): 1148-1162.

[16] Swaminathan, A.; Wang, K.J.; Galton, I.; "A Wide-Bandwidth 2.4 GHz ISM Band Fractional-N PLL With Adaptive Phase Noise Cancellation," IEEE Journal of Solid-State Circuits, Volume: 42, Issue: 12, 2007, Page(s): 2639-2650.

[17] Strange, J.; Atkinson, S.; "A direct conversion transceiver for multi-band GSM application," Digest of Papers. 2000 IEEE Radio Frequency Integrated Circuits (RFIC) Symposium, 2000, Page(s): 25 - 28

[18] van Zeijl, P.T.M.; Eikenbroek, J.-W.; Vervoort, P.-P.; Setty, S.; Tangenberg, J.; Shipton, G.; Kooistra, E.; Keekstra, I.; Belot, D.; "A Bluetooth radio in 0.18 µm CMOS," Digest of Technical Papers, 2002 IEEE International Solid-State Circuits Conference, Volume: 1, 2002, Page(s): 86 - 448 vol.1

[19] Meninger, S.E.; Perrott, M.H.; "A fractional- N frequency synthesizer architecture utilizing a mismatch compensated PFD/DAC structure for reduced quantization-induced phase noise," IEEE Transactions on Circuits and Systems II: Analog and Digital Signal Processing, Volume: 50, Issue: 11, 2003, Page(s): 839-849.

[20] Bram De Muer; Steyaert, M.S.J.; "On the analysis of ΔΣ fractional-N frequency synthesizers for high-spectral purity," IEEE Transactions on Circuits and Systems II: Analog and Digital Signal Processing, Volume: 50, Issue: 11, 2003, Page(s): 784-793.

[21] See Taur Lee; Sher Jiun Fang; Allstot, D.J.; Bellaouar, A.; Fridi, A.R.; Fontaine, P.A.; "A quad-band GSM-GPRS transmitter with digital auto-calibration," IEEE Journal of Solid-State Circuits, Volume: 39, Issue: 12, 2004, Page(s): 2200-2214.

[22] Wilingham, S.; Perrott, M.; Setterberg, B.; Grzegorek, A.; McFarland, B.; "An integrated 2.5 GHz ΣΔ frequency synthesizer with 5 µs settling and 2 Mb/s closed loop modulation," Digest of Technical Papers 2000 IEEE International Solid-State Circuits Conference, 2000, Page(s): 200 - 201, 457.

[24] Temporiti, E.; Albasini, G.; Bietti, I.; Castello, R.; Colombo, M.; "A 700-kHz bandwidth ΣΔ fractional synthesizer with spurs compensation and linearization techniques for WCDMA applications," IEEE Journal of Solid-State Circuits, Volume: 39, Issue: 9, 2004, Page(s): 1446-1454.

[25] Kozak, M.; Kale, I.; "A pipelined noise shaping coder for fractional-N frequency synthesis," IEEE Transactions on Instrumentation and Measurement, Volume: 50, Issue: 5, 2001, Page(s): 1154-1161.

[26] Weltin-Wu, C.; Temporiti, E.; Baldi, D.; Svelto, F.; "A 3GHz Fractional-N All-Digital PLL with Precise Time-to-Digital Converter Calibration and Mismatch Correction," Digest of Technical Papers. IEEE International Solid-State Circuits Conference, 2008, Page(s): 344-618.

[27] Rhee, W.; Ali, A.; "An on-chip phase compensation technique in fractional-N frequency synthesis," Proceedings of the 1999 IEEE International Symposium on Circuits and Systems, 1999. ISCAS '99. Volume: 3, 1999, Page(s): 363-366.

[28] Swaminathan, A.; Wang, K.J.; Galton, I.; "A Wide-Bandwidth 2.4GHz ISM-Band Fractional-N PLL with Adaptive Phase-Noise Cancellation," Digest of Technical Papers. IEEE International Solid-State Circuits Conference, 2007. ISSCC, 2007, Page(s): 302-604.

[29] Zarei, H.; Shoaei, O.; Fakraie, S.M.; Zakeri, M.M.; "A low-power multi-modulus divider in 0.6 μm digital CMOS technology," Proceedings of the 12th International Conference on Microelectronics, 2000, Page(s): 359-362.

[30] Sulaiman, M.S.; Khan, N.; "A novel low-power high-speed programmable dual modulus divider for PLL-based frequency synthesizer," Proceedings, IEEE International Conference on Semiconductor Electronics, 2002. Page(s): 77-81.

[31] Chabloz, J.; Ruffieux, D.; Enz, C.; "A low-power programmable dynamic frequency divider," 34th European Solid-State Circuits Conference, 2008, Page(s): 370-373.

APPENDIX I: VLSI AMPLIFIER FAMILIES

The VLSI amplifiers are commonly used as basic VLSI signal processing elements for scaling operations. A VLSI amplifier can be used to enlarge a relatively small (either in magnitude or power) input signal into a larger one. There are various VLSI amplifier circuit types, such as the operational amplifiers (Opamp), the small signal amplifiers, the large signal and the power amplifiers. These VLSI amplifier circuits provide various circuit characteristics (e.g. linearity, bandwidth, gain, and efficiency).

The VLSI amplifier circuits can be classified based on their operation conditions such as conduction angles. Under such classification method, VLSI amplifiers can be classified into well-defined circuit types, including the class-A, the class-B, the class-AB, the class-D, the class-E, and the class-G amplifiers. Each VLSI amplifier type offers specific performance characteristics that can be used selectively for practical applications.

A.1 VLSI AMPLIFIER BASIS

Shown in figure A.1 is an ideal VLSI continuous time amplifier model, where the input signal is enlarged to provide a higher voltage, current or power signal output.

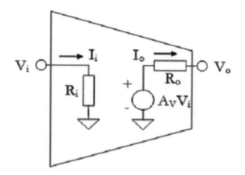

Fig. A.1 Continuous time amplifier model

A.1.1 AMPLIFIER GAIN

The amplification gain is a key performance parameter of a VLSI amplifier that represents the relationship between the signal magnitudes (voltage, current or power) of the output with respect to its input. There are three commonly used amplifier gain definitions, including the voltage gain A_V, the current gain A_i and the power gain A_P as defined below.

The voltage gain of the amplifier is defined as:

$$A_V \equiv \frac{V_o}{V_i} \qquad (A.1)$$

The current gain is defined as

$$A_i \equiv \frac{I_o}{I_i} \qquad (A.2)$$

The power gain is defined as

$$A_p \equiv \frac{P_o}{P_i} \qquad (A.3)$$

The amplification gains are also commonly expressed in Decibels or dB forms as:

The voltage gain in dB:

$$A_V(dB) \equiv 20\log(\frac{V_o}{V_i}) \qquad (A.4)$$

The current gain in dB:

$$A_i(dB) \equiv 20\log(\frac{I_o}{I_i}) \qquad (A.5)$$

The power gain in dB:

$$A_p(dB) \equiv 10\log(\frac{P_o}{P_i}) \qquad (A.6)$$

Note that based on input and output signal type, VLSI amplifiers can be classified as the voltage amplifiers for converting small input voltages into a much larger output voltage, the power amplifiers for driving loads such as the motor or high power loudspeakers, and the pre-amplifiers or instrumentation amplifiers for amplifying very low signal voltage levels of only a few micro-volts (µV) from sensors, audio, and RF signals.

A.1.2 POWER EFFICIENCY

The VLSI amplifier works on the basic principle that converting the dc power drawn from the power supply into an ac voltage signal delivered to the load. The power efficiency of a typical VLSI amplifier can never be 100% as some of its power is lost in the form of heat since the amplifier itself consumes power during the amplification process, where the power efficiency of an amplifier is defined as:

$$\eta \equiv \frac{P_o}{P_{DC}} \tag{A.7}$$

Where P_{DC} is the power the amplifier taken from the dc power supply and P_o is the output ac power of the amplifier.

A.1.3 AMPLIFIER LINEARITY

For an ideal amplifier, there is a linear relationship between the steady-state input and the output of the signal as given by the transfer function of the amplifier as:

$$\begin{cases} V_{in}(t) = V\cos(\omega t) = \operatorname{Re}(Ve^{j\omega t}) \\ V_{out}(t) = \operatorname{Re}(H(s)|_{s=j\omega} \cdot Ve^{j\omega t}) \end{cases} \tag{A.8}$$

For a practical VLSI circuit, the output not only contains the linear term of the input, but also the higher order terms:

$$\begin{cases} V_{in}(t) = V_i \cos(\omega t) = \text{Re}(V_i e^{j\omega t}) \\ V_{out}(t) = a_o + a_1 V_{in} + a_2 V_{in}^2 + a_3 V_{in}^3 + ... \end{cases} \quad (A.9)$$

These high-order terms will generate various harmonics of the input signal. The outcomes of such a phenomenon are twofold: (1) power loss in the desired signal frequency band and (2) the unwanted power injection to the signal band from adjacent signal/noise bands. Both of them result in degradations of the signal-to-noise ratio (SNR) of the amplifier. Mathematically, the coefficient a_i in the above equation provides the measure of the nonlinearity of a given amplifier. In practical cases, other equivalent linearity measures of the amplifier are also commonly used such as the total harmonic distortion (THD), the second- (HD2) and third-order (HD3) parameters commonly used in baseband systems, the second- (IP2) and third-order intercept point (IP3), and -1dB compression point (CP$_{-1dB}$) in the RF systems.

For a given input signal of single frequency ω, the amplifier output will usually contain the following dc value and the harmonics:

$$\begin{cases} V_{in}(t) = V_i \cos(\omega t) \\ V_{out}(t) = V_o + V_1 \cos(\omega t + \theta_1) \\ + V_2 \cos(2\omega t + \theta_2) + V_3 \cos(3\omega t + \theta_3) + ... \end{cases} \quad (A.10)$$

The second (HD2) and third (HD3) order harmonic distortions and the total harmonic distortion (THD) of a signal are defined using the following ratios:

$$\begin{cases} HD2 = \sqrt{\dfrac{V_2^2}{V_1^2}} \times 100\% \\ HD3 = \sqrt{\dfrac{V_3^2}{V_1^2}} \times 100\% \\ THD = \left(\dfrac{\sqrt{V_2^2 + V_3^2 + V_4^2 + ...}}{\sqrt{V_1^2}}\right) \times 100\% \end{cases} \quad (A.11)$$

The THD in dB format is given as:

$$THD = 10 \cdot \log(\frac{V_2^2 + V_3^2 + V_4^2 + \ldots}{V_1^2}) \quad \text{(dB)} \quad \text{(A.12)}$$

For nonlinear amplifier circuits, we have that

$$\begin{aligned}V_{out}(t) &= a_o + a_1 V_{in}(t) + a_2 \cdot V_{in}(t)^2 + a_3 \cdot V_{in}(t)^3 + \ldots \\ &= \left(a_o + \frac{a_2}{2}V_i^2 + \ldots\right) + \left(1 + \frac{3}{4}\frac{a_3 V_i^2}{a_1}\right) a_1 V_i \cos(\omega t) \\ &+ \left(\frac{a_2}{2}\right)V_i^2 \cos(2\omega t) + (\frac{a_3}{4})V_i^3 \cos(3\omega t) + \ldots\end{aligned} \quad \text{(A.13)}$$

We have the relationship between a_i and the harmonic distortions as follows:

$$HD2 = \left(\frac{a_2}{2a_1}\right)V_i \quad \text{(A.14)}$$

$$HD3 = (\frac{a_3}{4a_1})V_i^2 \quad \text{(A.15)}$$

We can also see that due to nonlinear effects, the amplifier gain at the fundamental frequency will be compressed versus the ideal amplifier. The -1dB compress point, when the output signal is 1dB different from the ideal circuit, is given as

$$20\log\left(1 + \frac{3}{4}\frac{a_3 V_i^2}{a_1}\right) = -1 \text{ (dB)} \quad \text{(A.16)}$$

or

$$V_i = \sqrt{\frac{1-10^{-\frac{1}{20}}}{\frac{3}{4}\left|\frac{a_3}{a_1}\right|}} \cdot \quad \text{(A.17)}$$

A.1.4 Intermodulation

For ideal amplifier, signals of different frequencies are isolated from each other in frequency domain. When nonlinear effects are presented in the amplifier circuit, signals of different frequencies will be mixed to create all possible intermodulation products. As a result, noise from the adjacent band will be injected into the signal frequency band of interest.

Such a phenomenon can be modeled by applying the following two-tone input signal:

$$V_{in}(t) = V_i(\cos(\omega_1 t) + \cos(\omega_2 t)) \quad \text{(A.18)}$$

The output of the nonlinear circuit with the two-tone input is given as follows:

$$\begin{aligned}
V_{out}(t) &= a_o + a_1 V_{in}(t) + a_2 \cdot V_{in}(t)^2 + a_3 \cdot V_{in}(t)^3 + \ldots \\
&= (a_o + a_2 V_i^2 + \ldots) \\
&\quad + (a_1 V_i + \frac{3}{2} a_3 V_i^3)(\cos(\omega_1 t) + \cos(\omega_2 t)) \\
&\quad + \frac{3}{4} a_3 V_i^3 (\cos((2\omega_1 - \omega_2)t) + \cos((2\omega_2 - \omega_1)t)) \\
&\quad + \frac{1}{2} a_2 V_i^2 (\cos(2\omega_1 t) + \cos(2\omega_2 t)) + (\cos((\omega_1 + \omega_2)t) + \cos((\omega_1 - \omega_2)t)) \\
&\quad + \frac{1}{4} a_3 V_i^3 (\cos(3\omega_1 t) + \cos(3\omega_2 t)) + \frac{3}{4} a_3 V_i^3 (\cos((2\omega_1 + \omega_2)t) + \cos((2\omega_2 + \omega_2)t))
\end{aligned}$$

(A.19)

It can be seen that for signal with frequency ω, any two signals or noises, which have frequencies meeting the following equation, will inject noise into the signal band.

$$\pm(N\omega_1 \pm M\omega_2) = \omega \tag{A.20}$$

N and M are two integers. In most narrow-band applications, the intermodulation terms created by the following combinations are commonly used to characterize the linearity of the system as follows:

$$\begin{cases} 2\omega_1 - \omega_2 = \omega \\ 2\omega_2 - \omega_1 = \omega \end{cases} \tag{A.21}$$

$$\frac{3}{4} a_3 V_i^3 \left(\cos((2\omega_1 - \omega_2)t) + \cos((2\omega_2 - \omega_1)t) \right) \tag{A.22}$$

The magnitude of this term to the fundamental is usually defined as the IM3 of the system:

$$IM_3 = \frac{\frac{3}{4} a_3 V_i^3}{a_1 V_i} = \frac{3}{4} \frac{a_3 V_i^2}{a_1} = 3HD_3 \tag{A.23}$$

For most RF amplifier circuit applications, a commonly used performance metric of system linearity is the third-order intercept point, which is defined as the point where the third order intermodulation term equals to the fundamental as

$$\frac{\frac{3}{4} a_3 V_{IIP3}^3}{a_1 V_{IIP3}} = 1 \tag{A.24}$$

or

$$V_{IIP3} = \sqrt{\frac{4}{3}\left|\frac{a_1}{a_3}\right|} \qquad (A.25)$$

or in power as

$$IIP3 = 10\log\left(\frac{(V_{IIP3})^2}{2R}/(mW)\right) = 10\log\left(\frac{2}{3R}\left|\frac{a_1}{a_3}\right|/(mW)\right)(dBm). \qquad (A.26)$$

Similarly, we can also define IIP2 as the second harmonic equal to the fundamental as

$$a_2 V_{IIP2}^2 = a_1 V_{IIP2} \qquad (A.27)$$

or

$$V_{IIP2} = \left|\frac{a_1}{a_2}\right|. \qquad (A.28)$$

A.2 VLSI AMPLIFIER CLASSES

The VLSI amplifier circuits can be classified into families known as classes based on their conductance angle within a cycle and that is directly related to their power efficiencies. These VLSI amplifier families are noted by the names of the class-A, the class-AB, the class-B, the class-C, the class-D, the class-E, the class-F, the class-G, and the class-H, etc. Each VLSI amplifier circuit class is featured by its specific device bias and input signal level for proper operation. Each amplifier class offers specific performance characteristics such as power efficiency and linearity. Shown in Table A.1 is a list of some commonly used VLSI amplifier classes that are defined by their driver conducting angles of current flow. It can be seen that the class-A amplifier has full cycle of signal conduction and that the class-B amplifier has half cycle of signal conduction. On the other hand, the class-AB and the class-C amplifiers have larger and smaller than half cycle signal conduction respectively, where the conduction angles directly impact the amplifier efficiency and signal quality (e.g. distortion).

Table A.1 Amplifier classes

Class	A	B	C	AB	D
Conduction Angle	360°	180°	<90°	180° <360°	0°
Bias Point	center of load line	at X axis	below X-axis	between load line and x-axis	Full on/off
Efficiency	25-30%	70-80%	>80%	50-70%	80-100%
Linearity	good	poor	poor	good	poor*

* Note that linearity of the class-D amplifier can be very good employing the post lowpass filters after the class-D amplification.

A.2.1 CLASS-A AMPLIFIER

In a class-A amplifier as shown in figure A.1.2, the active element remains conducting (i.e. in linear range) all of the time that offers 100% (or conduction angle = 360°) of signal conduction. Most VLSI small signal linear amplifiers are designed using class-A family to offer more linear and less complex circuit implementations than other types at the penalty of efficiency. This type of amplifier is most commonly used in small-signal stages or for low-power applications (such as driving headphones).

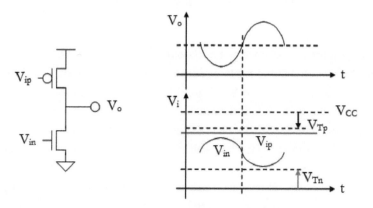

Fig. A1.2 VLSI class-A amplifier

In class-A amplifier the entire input signal waveform is faithfully reproduced at the amplifier's output as the transistor is perfectly biased within active (linear) region of the amplifier circuit. This then results in the ac input signal being perfectly centered between the amplifiers upper and lower signal limits.

The amplifying devices in the class-A amplifier conduct over the whole of the input cycle such that the output signal is an exact scaled replica of the input with no clipping.

The VLSI class-A amplifiers offer several advantages including

- Higher design simplicity than other classes. For example class-AB and B amplifier circuits require two devices (push-pull output) to handle both halves of the waveform, and the circuits to keep the quiescent bias optimal during temperature changes. A class-A amplifier can use either single-ended or push-pull and bias is usually less critical.

- The amplifying element is biased for that the device is always conducting to some extent, normally implying the quiescent (small-signal) drain current for MOS device is close to the most linear portion of its characteristic curve giving the least signal distortion.

- Because the device is never shut off completely there is no turn on time, little problem with charge storage, and generally better high frequency performance and feedback loop stability.

- The point at which the device comes closest to being cut off (and so significant change in gain, hence non-linearity) is not close to zero signal, so the problem of crossover distortion associated with class-AB and B amplifiers is avoided in class-A amplifier.

However, VLSI class-A amplifiers also have limitations such as

- Lower power efficiency. A theoretical maximum of 50% is obtainable with inductive output coupling and only 25% with capacitive coupling, unless a square law output stages are used. In a power amplifier this not only wastes power and limits battery operation, it may place restrictions on the output devices that can be used with increased costs.

- The power inefficiency comes not just from the fact that the device is always conducting, it also comes from that the standing current is roughly half the maximum output current together with the problem that a large part of the power supply voltage is developed across the output device at low signal levels. If high output powers are needed from a class-A circuit, the power waste (and the accompanying heat) will become significant. For every watt delivered to the load, the amplifier itself will, at best, dissipate another watt. For large powers this means very large and expensive power supplies and heat sinking.

A.2.2 CLASS-B AMPLIFIER

In VLSI class-B amplifiers, 50% of the input signal is used (i.e. conduction angle = 180°). The active element is on and works in its linear range half of the time. In the rest of the cycle, it is turned off.

The class-B amplifiers only amplify half of the input wave cycle, thus creating a large amount of distortion. The high distortion comes at the benefits of greatly improved efficiency than class-A amplifier. VLSI class-B amplifier has a maximum theoretical efficiency of 78.5% (i.e., $\pi/4$). This is because the amplifying element is switched off altogether half of the time.

A single class-B element is rarely found in practice, though it has been used for driving the loudspeaker in some early applications, and it can be used in RF power amplifier where the distortion levels are less important. Unlike the class-A amplifier that uses a single transistor for its output stage, the class-B amplifier usually uses two complimentary transistors (PMOS and NMOS) for each half of the output waveform. One transistor conducts for the positive half of the waveform and another conducts for the negative half of the waveform. This means that each transistor spends half of its time in the active region and half its time in the cut-off region thereby amplifying only 50% of the input signal. A class-B operation has no dc bias voltage, instead only the transistor conducts when the input signal is greater than the threshold voltage of the MOS device. Therefore, at zero input there is zero output. This then results in only half the input signal being presented at the amplifier's output with a greater efficiency.

Since in a class-B amplifier, no dc current is used to bias the transistors, so for the output transistors to start to conduct each half of the waveform, both positive and negative, they need the input voltage Vi to be greater than the device threshold required for a MOS transistor to start conducting. Then the lower part of the output waveform which is below this threshold window will not be reproduced accurately resulting in a distorted area of the output waveform as one transistor turns "OFF" waiting for the other to turn back "ON". The result is that there is a small part of the output waveform at the zero voltage cross over point, which will be distorted. A practical circuit using class-B elements is the push-pull stage, such as the very simplified complementary pair arrangement shown in figure A1.3. Here, complementary or quasi-complementary devices are each used for amplifying the opposite halves of the input signal, which are then recombined at the output. This arrangement gives excellent efficiency, but can suffer from the drawback of high distortion in the cross-over region as one output device has to take over supplying power exactly as the other finishes. This is called crossover distortion.

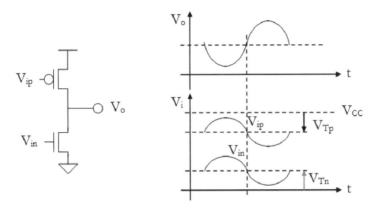

Fig. A1.3 VLSI class-B amplifier

A.2.3 CLASS-AB AMPLIFIER

The VLSI class-AB amplifier is a compromise between the class-A and the class-B configurations where the two active elements conduct more than half of the time as a means to reduce the cross-over distortions of class-B amplifiers. While class-AB operation still uses two complementary transistors in its output stage, a very small biasing voltage is applied to the gate of the transistor to bias it close to the cut-off region when no input signal is present. An input signal will cause the transistor to operate as normal in its active region therefore eliminating any crossover distortion, which is present in class-B configurations. A small bias current will flow in the class-AB amplifier when there is no input signal but it is much less than that for the class-A amplifier circuit. This means that the transistor will be "ON" for more than half a cycle of the waveform. This type of amplifier configuration improves both the efficiency and linearity of the amplifier circuit compared to a pure class-A configuration.

In the VLSI class-AB amplifier operation as shown in figure A.1.4, each device operates the same way as in the class-B amplifier over half the waveform, but also conducts a small amount on the other half. As a result, the region where both devices simultaneously are nearly off (i.e. the "dead zone") is reduced. The result is that when the waveforms from the two devices are combined, the crossover is greatly minimized or eliminated altogether.

The class-AB amplifier trade-off some power efficiency over the class-B amplifier in favor of linearity, thus is less power efficient (below 78.5%). However, it is typically much more efficient than the class-A amplifier.

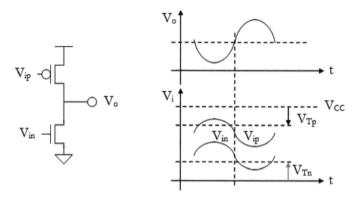

Fig. A1.4 VLSI class-AB amplifier

A.2.4 CLASS-C AMPLIFIER

In the VLSI class-C amplifier, less than 50% of the input signal is used (i.e. conduction angle < 180°). The advantage is the potentially high efficiency, but at the disadvantage of higher distortion. In practical applications, a tuned filter can usually be used to filter out the distortion. The class-C amplifier is mostly used in RF applications such as in FM transmitters where a linear amplification is not necessary. However the voltage can be AM-modulated, the RF output too, with fairly low distortion.

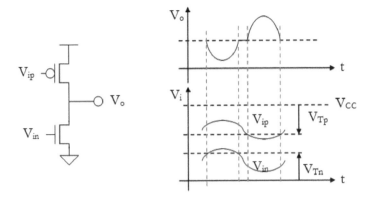

Fig. A1.5 VLSI class-C amplifier

A.2.5 CLASS-D AMPLIFIER

The class-D amplifiers offer the benefits of smaller size and also lower weight. A class-D amplifier is a switching amplifier using the pulse-width modulation (PWM) or the pulse-code modulation (PCM) circuits. The operation of a class-D amplifier is mainly in the digital domain that can be realized by the VLSI digital-like circuits. The modulation frequency in a class-D amplifier is much higher than audio, usually ranging from 200khz to 1MHz. The benefits class-D amplification are less power consumption, smaller and less weight. The size and weight get even lower for class-D amplifiers up to 5 - 20 W because heat sinks aren't needed.

In the class-D amplifier the input signal is converted to a sequence of higher voltage output pulses. The averaged-over-time power values of these pulses are directly proportional to the instantaneous amplitude of the input signal. The frequency of the output pulses is typically ten or more times than the highest frequency in the input signal to be amplified. The output pulses contain inaccurate spectral components (that is, the pulse frequency and its harmonics) which can be removed by a low-pass passive filter. The resulting filtered signal is then an amplified replica of the input.

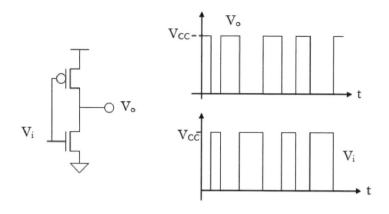

Fig. A1.6 VLSI Class-D amplifier

The class-D amplifiers use pulse width modulation, pulse density modulation (sometimes referred to pulse frequency modulation) or more advanced form of modulation such as sigma-delta modulation. The output stages such as those used in pulse generators are examples of class-D amplifiers. The term class-D is usually applied to devices intended to reproduce signals with a bandwidth well below the switching frequency.

The main advantages of a class-D amplifier are high power efficiency and fully digital compatible circuit implementation. Because the output pulses have fixed amplitude, the switching elements (usually MOS devices) are switched either completely on or completely off, rather than operated in linear mode. A MOS device operates with the lowest resistance when fully-on and thus has the lowest power dissipation when in that condition, except when fully off. When operated in a linear mode, the MOS has variable amounts of resistance that vary linearly with the input voltage and the resistance is something other than the minimum possible, therefore more electrical energy is dissipated as heat. Compared to class-A/B amplifier operations, class-D's lower losses allow the use of a smaller heat sink for the MOS while also reducing the amount of ac power supply power required. Thus, class-D amplifiers do not need large or heavy power supply transformers or heat sinks, so they are smaller and more compact in size than an equivalent class-AB amplifier.

A.2.6 CLASS-E AND CLASS-F AMPLIFIERS

The class-E and class-F amplifiers are highly efficient switching power amplifiers, typically used at such high frequencies that the switching time become comparable to the duty time. In a simple RF class-E amplifier circuit shown in figure A1.7, the transistor is connected via a serial LC circuit to the load, and connected via a large inductor L_1 to the supply voltage. A capacitor C_1 is connected between the transistor drain and the ground.

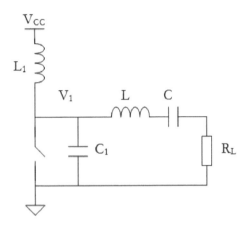

Fig.A1.7 VLSI Class-E amplifier

When the transistor in the class-E amplifier is on, it pushes through the serial LC circuit into the load and some current begins to flow to the parallel LC circuit to ground. Then the serial LC circuit swings back and compensates the current into the parallel LC circuit. At the point the current through the transistor is zero the switch is switched off. Both LC circuits are now filled with energy in C and L. The whole circuit performs a damped oscillation. The damping by the load has been adjusted so that sometime later the energy from the L is gone into the load, but the energy in both C peaks at the original value. The switch can be switched on again when the voltage across the transistor is zero.

When the switch is open, the voltage V_1 across capacitor C_1 is charged with expression given as:

$$C_1 \frac{dV_1}{dt} = I_{d1}(1 - a\sin(\omega_s t + \phi)) \qquad (A1.29)$$

Where ω_s is the signal frequency, I_{d1} is the dc drain current, a and ϕ are constants. V_1 can be expressed as:

$$V_1(t) = I_{d1}(\omega_s t) + a[\cos(\omega_s t + \phi) - \cos(\phi)] \qquad (A1.30)$$

Optimum operation of class-E amplifier requires two conditions:

$$\frac{dV_1}{dt}\bigg|_{t=T_s/2} = 0 \qquad (A1.31)$$

$$V_1|_{t=T_s/2} = 0 \qquad (A1.32)$$

These conditions avoid power dissipation due to either shorting the capacitor C_1 while has value or nonzero switching time at transition. By using these conditions, constant a and ϕ can be calculated as:

$$a = 1.86 \qquad (A1.33)$$

$$\phi = -32.5° \qquad (A1.34)$$

The voltage V_1 and the capacitor current I_1 can be derived for the entire cycle

$$V_1(t) = \begin{cases} \frac{I_{d1}}{\omega_s C_1}[\omega_s t + a[\cos(\omega_s t + \phi) - \cos(\phi)] & 0 \le \omega_s t \le \pi \\ 0 & 0 \le \omega_s t \le 2\pi \end{cases} \qquad (A1.35)$$

$$I_1(t) = \begin{cases} 0 & 0 \leq \omega_s t \leq \pi \\ I_{d1}(1 - a\sin(\omega_s t + \phi)) & 0 \leq \omega_s t \leq 2\pi \end{cases} \quad \text{(A1.36)}$$

The load is given as

$$Z_{netL} = \frac{V_1}{I_1} = \frac{0.28}{\omega_s C_1} e^{j49°} \quad \text{(A1.37)}$$

The input impedance of the load network in the class-E amplifier is given by

$$Z_{netL} = j\omega_s L + \frac{1}{j\omega_s C} + R \quad \text{(A1.38)}$$

The load component values in VLSI class-E amplifier can be obtained by equating the real and imaginary parts of above equations.

$$C_1 = \frac{1}{\omega_s R(\frac{\pi^2}{4} + 1)(\frac{\pi}{2})} \quad \text{(A1.39)}$$

$$C \approx C_1(\frac{5.447}{Q})(1 + \frac{1.153}{Q - 1.153}) \quad \text{(A1.40)}$$

$$Q = \frac{\omega_s L}{R} \quad \text{(A1.41)}$$

With load, frequency, and duty-cycle (0.5) as given parameters and the constraint that the voltage is not only restored, but peaks at the original voltage, the four parameters (L_1, L, C_1 and C) are determined. The class-E amplifier takes the finite on resistance into account and tries to make the current touch the bottom at zero. This means that the voltage and the current at the transistor are symmetric with respect to time. The Fourier transform allows an elegant formulation to generate the complicated LC networks and says that the first harmonic is passed into the load, all even harmonics are shorted and all higher odd harmonics are open.

VLSI class-E amplifier uses a significant amount of second-harmonic voltage. The second harmonic can be used to reduce the overlap of edges with finite sharpness. For this to work, energy on the second harmonic has to flow from the load into the transistor, and no source for this is visible in the circuit diagram. In reality, the impedance is mostly reactive and the only reason for it is that Class E is a Class F amplifier with a much simplified load network and thus has to deal with imperfections.

The function of the shunt capacitance in class-E amplifier is to reduce the peak voltage appearing across the MOS device when the device is in the off state, and to spread the width of the "off" pulse. The shunt capacitor is also part of the output-matching network.

A.2.7 CLASS-G AND CLASS-H AMPLIFIERS

The class-G and the class-H amplifiers offer enhanced power efficiency compared with class-AB amplifier that are commonly used in audio power amplifications where the heat sinks and power transformers would be prohibitively large and costly without the efficiency enhancement. The class-G and class-H amplifiers are sometimes used interchangeably to refer to different designs, based on the manufacturer.

These amplifier families employ several power rails at different voltages and adaptively switch between them as the signal output approaches each level. Thus, the amplifier increases efficiency by reducing the wasted power at the output transistors. These amplifiers offer higher power efficient than class-AB amplifiers but lower power efficiency than the class-D amplifiers. However these amplifiers have minimized EMI effects compared with class-D amplifiers.

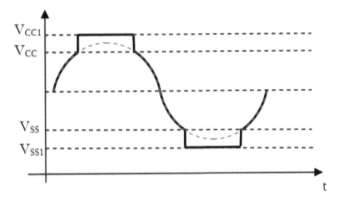

Fig.A.8 Output waveform of class-G amplifier

The class-H amplifiers extend idea of the class-G to create an infinitely variable supply rail. This is done by modulating the supply rails so that the rails are only a few volts larger than the output signal at any given time. The output stage operates at its maximum efficiency all the time. Significant efficiency gains can be achieved but with the drawback of more complicated supply design and reduced THD performance.

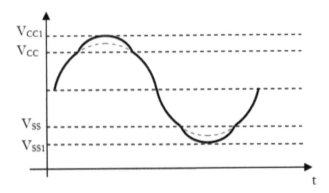

Fig.A.9 Output wavefrom of class-H amplifier

Reference:

[1] Hanington, G.; Pin-Fan Chen; Asbeck, P.M.; Larson, L.E.; "High-efficiency power amplifier using dynamic power-supply voltage for CDMA applications," IEEE Transactions on Microwave Theory and Techniques, Volume: 47, Issue: 8, 1999, Page(s): 1471 - 1476.

[2] Sokal, N.O.; Sokal, A.D.; "Class E-A new class of high-efficiency tuned single-ended switching power amplifiers," IEEE Journal of Solid-State Circuits, Volume: 10, Issue: 3. 1975, Page(s): 168 - 176.

[3] Su, D.K.; McFarland, W.J.; "An IC for linearizing RF power amplifiers using envelope elimination and restoration," IEEE Journal of Solid-State Circuits, Volume: 33, Issue: 12, 1998, Page(s): 2252 - 2258.

[4] Athas, W.C.; Svensson, L.J.; Koller, J.G.; Tzartzanis, N.; Ying-Chin Chou, E.; "Low-power digital systems based on adiabatic-switching principles," IEEE Transactions on Very Large Scale Integration (VLSI) Systems, Volume: 2, Issue: 4, 1994, Page(s): 398-407.

[5] Diez, S.; Schmidt, C.; Ludwig, R.; Weber, H.G.; Obermann, K.; Kindt, S.; Koltchanov, I.; Petermann, K.; "Four-wave mixing in semiconductor optical amplifiers for frequency conversion and fast optical switching," IEEE Journal of Selected Topics in Quantum Electronics, Volume: 3, Issue: 5, 1997, Page(s): 1131-1145.

[6] Crols, J.; Steyaert, M.; "Switched-opamp: an approach to realize full CMOS switched-capacitor circuits at very low power supply voltages," IEEE Journal of Solid-State Circuits, Volume: 29, Issue: 8, 1994, Page(s): 936-942.

[7] Changsik Yoo; Qiuting Huang; "A common-gate switched 0.9-W class-E power amplifier with 41% PAE in 0.25-μm CMOS," IEEE Journal of Solid-State Circuits, Volume: 36, Issue: 5, 2001, Page(s): 823 - 830

[8] Sahu, B.; Rincon-Mora, G.A.; "A high-efficiency linear RF power amplifier with a power-tracking dynamically adaptive buck-boost supply," IEEE Transactions on Microwave Theory and Techniques, Volume: 52, Issue: 1, Part: 1, 2004, Page(s): 112-120.

[9] Berkhout, M.; "An integrated 200-W class-D audio amplifier," IEEE Journal of Solid-State Circuits, Volume: 38, Issue: 7, 2003, Page(s): 1198-1206.

[10] El-Hamamsy, S.-A.; "Design of high-efficiency RF class-D power amplifier," Power Electronics, IEEE Transactions on Volume: 9, Issue: 3, 1994, Page(s): 297-308.

[11] Jayaraman, A.; Chen, P.F.; Hanington, G.; Larson, L.; Asbeck, P.; "Linear high-efficiency microwave power amplifiers using bandpass delta-sigma modulators," IEEE Microwave and Guided Wave Letters, Volume: 8, Issue: 3, 1998, Page(s): 121-123.

[12] Mazzanti, A.; Larcher, L.; Brama, R.; Svelto, F.; "Analysis of reliability and power efficiency in cascode class-E PAs," IEEE Journal of Solid-State Circuits, Volume: 41, Issue: 5, 2006, Page(s): 1222-1229.

[13] Mader, T.B.; Popovic, Z.B.; "The transmission-line high-efficiency class-E amplifier," Microwave and Guided Wave Letters, IEEE Volume: 5, Issue: 9, 1995, Page(s): 290-292.

[14] Marco, L.; Poveda, A.; Alarcon, E.; Maksimovic, D.; "Bandwidth limits in PWM switching amplifiers," 2006 IEEE International Symposium on Circuits and Systems, 2006, Page(s): 4 pp.-5326.

[15] Asbeck, P.M.; Larson, L.E.; Galton, I.G.; "Synergistic design of DSP and power amplifiers for wireless communications," IEEE Transactions on Microwave Theory and Techniques, Volume: 49, Issue: 11, 2001, Page(s): 2163-2169.

[16] Rodriguez, E.; Guinjoan, F.; Poveda, A.; Alarcon, E.; El-Aroudi, A.; "Characterizing fast-scale instability in a buck-based switching amplifier for wideband tracking," IEEE International Symposium on Circuits and Systems, 2008, Page(s): 3262-3265.

[17] Kee, S.D.; Aoki, I.; Hajimiri, A.; Rutledge, D.; "The class-E/F family of ZVS switching amplifiers," IEEE Transactions on Microwave Theory and Techniques, Volume: 51, Issue: 6, 2003, Page(s): 1677-1690.

[18] Farrell, R.; "An efficient parallel structure for $\Sigma\Delta$ modulators for use in high speed switching power amplifiers," 18th European Conference on Circuit Theory and Design, 2007, Page(s): 711-714.

[19] Kitchen, J. N.; Deligoz, I.; Kiaei, S.; Bakkaloglu, B.; "Polar SiGe Class E and F Amplifiers Using Switch-Mode Supply Modulation," IEEE Transactions on Microwave Theory and Techniques, Volume: 55, Issue: 5, 2007, Page(s): 845-856.

[20] Ertl, H.; Kolar, J.W.; Zach, F.C.; "Basic considerations and topologies of switched-mode assisted linear power amplifiers, Eleventh Annua Applied Power Electronics Conference and Exposition, Volume: 1, 1996, Page(s): 207 - 213 vol.1.

[21] Krabbenborg, B.; Berkhout, M.; "Closed-Loop class-D Amplifier With Nonlinear Loop Integrators," IEEE Journal of Solid-State Circuits, Volume: 45, Issue: 7, 2010, Page(s): 1389-1398.

[22] Franco, M.; Katz, A.; "class-E Silicon Carbide VHF Power Amplifier," IEEE/MTT-S International Microwave Symposium, 2007, Page(s): 19-22.

[23] Acar, M.; Annema, A.J.; Nauta, B.; "Variable-Voltage class-E Power Amplifiers," IEEE/MTT-S International Microwave Symposium, 2007, Page(s): 1095-1098.

[24] Blakey, P.A.; "$\pi/2$-mode operation of class-E power amplifiers," ISSSE '07. International Symposium on Signals, Systems and Electronics, 2007, Page(s): 205-208.

[25] Albulet, M.; Zulinski, R.E.; "Effect of switch duty ratio on the performance of class-E amplifiers and frequency multipliers," IEEE Transactions on Circuits and Systems I: Fundamental Theory and Applications, Volume: 45, Issue: 4, 1998, Page(s): 325-335.

[26] Adduci, P.; Botti, E.; Dallago, E.; Venchi, G.; "PWM Power Audio Amplifier With Voltage/Current Mixed Feedback for High-Efficiency Speakers," IEEE Transactions on Industrial Electronics, Volume: 54, Issue: 2, 2007, Page(s): 1141 - 1149

Biography

Dr. Hongjiang Song currently serves as an adjunct professor at Arizona State University where he has been offering VLSI analog signal processing and high-speed I/O circuit courses since 2002. He also served several other academic positions, such as faculty of Yunnan University and analog/RF circuits track chair for the IEEE International SOCC Conference. He has published a number of books and technical papers in VLSI analog circuits. He holds 30+ US patents in VLSI circuits. Dr. Song has been with Intel Corporation since 1994, working on various leading-edge VLSI circuit developments in CPU, chipset, high-speed I/O, RF, and analog circuits. Dr. Song's key research areas and interests include the high-speed/low-power circuits, RF, A/D, D/A, and sigma-delta converter circuits; PLL, DLL, and data-recovery circuits; transceiver circuits; digital and analog signal processing circuits; power management and reference circuits; and power amplifier (linear and class-D) circuits.

Book Summary

VLSI Modulation Circuits

Signal Processing, Data Conversion, and Power Management

This text covers various modeling and design topics related to the VLSI modulation circuits and their applications in VLSI signal processing, data conversion, and power management, including the modulation techniques, such as the PWM modulation, the sigma-delta modulation, and the VLSI circuit techniques, such as class-D amplifier circuits, lock-in amplifier circuits, switched-capacitor charge pump circuits, DC/DC converter circuits, chopper circuits, mixer circuit, and fractional-N PLL circuits.

VLSI CIRCUIT DESIGN SERIES

VLSI Analog Signal Processing Circuits – Algorithm, Architecture, Modeling, and Circuit Implementation. by Hongjiang Song, ISBN #978-1-4363-7740-9. (2009).

This is the textbook for the **VLSI Switched-Capacitor Filter and Analog Signal Processing Circuit Design** class (EEE598) the author offered at Arizona State Uinversity covering VLSI passive, active-RC, MOS-C, Gm-C, CTI, SC, SI analog filters and signal processing circuit techniques.

VLSI High-Speed I/O Circuits – Theoretical Basis, Architecture, Modeling, and Circuit Implementation. by Hongjiang Song, ISBN #978-1-4415-5987-6. (2010).

This is the textbook for the **VLSI high-speed I/O circuits** class (EEE598) the author offered in Arizona State University covering the analysis, modeling, and implementation of VLSI high-speed I/O circuits, such as timing models, jitter analysis, transmitter, receiver, equalizer, phase-locked loop (PLL), and data recovery circuit designs.

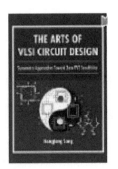

The Arts of VLSI Circuit Design – Symmetry Approaches Toward Zero PVT Sensitivities. by Hongjiang Song, ISBN #978-1-4568-7468-7. (2011).

This is the textbook for the **Structural VLSI Analog Circuit Design** class (EEE598) the author offered in Arizona State Uinversity covering various the state-of-the-arts symmetry based low PVT sensitivity circuit design techniques for basic VLSI circuit elements, circuit blocks and systems.

VLSI Analog Signal Processing Circuits – Problems and Solution Keys. by Hongjiang Song, ISBN #978-1-304-74949-9. (2013).

This book includes exam sheets and solution keys for the **Switched-Capacitor and Analog Signal Processing Circuit Design** class (EEE598) offered in Arizona State University in the past ten years covering VLSI passive, active-RC, MOS-C, Gm-C, CTI, SC and SI circuit techniques.

VLSI High-Speed I/O Circuits – Problems, Projects and Questions. by Hongjiang Song, ISBN #978-1-4415-5987-6. (2014).

This book includes a collection of commonly seen class homework problems, design projects, and interview questions for VLSI high-speed I/O circuit design, modeling and implementations.

The Arts of VLSI Opamp Circuit Design by Hongjiang Song, ISBN # 978-1-312-05130-0.(2014).

This book provides an introduction to structural VLSI opamp circuit design concepts, techniques and practices employing symmetry principles.

VLSI Modulation Circuits – Signal Processing, Data Conversion, and Power Management by Hongjiang Song, ISBN#978-1-312-21861-1. (2014).

This is a textbook developed for a VLSI circuit course at Arizona State University covering a special signal processing circuit family employing modulation circuit techniques with applications in signal processing, data conversion, and power management.